Table of Contents

Foreword: Pre-Clovis in the Americas by J. M. Adovasio and L...

Pre-Clovis in the Americas: Characterizing Early Sites, Material Culture, and Origins by Alison T. Stenger 5

Inundated Landscapes and the Colonization of the Northeastern Gulf of Mexico by C. Andrew Hemmings and J. M. Adovasio 16

Loess, Landscape Evolution, and Pre-Clovis on the Delmarva Peninsula by John S. Wah, Darrin L. Lowery, and Daniel P. Wagner 32

Regional Variability in Latest Pleistocene and Holocene Sea-Level Rise Across the California-Oregon-Washington and Bering Sea Continental Shelves by Jorie Clark, Jerry X. Mitrovica, and Jay Alder 49

Meadowcroft Rockshelter: Retrospect by J. M. Adovasio and David R. Pedler 63

Modeling Cactus Hill (44SX202) by Michael F. Johnson 77

Pre-Fishtail Settlement in the Southern Cone ca. 15,000 – 13,100 yr. cal. BP: synthesis, evaluation, and discussion of the evidence by Rafael Suárez 153

Plant Fiber Technologies and the Initial Colonization of the New World by J. M. Adovasio ..192

Origin and Antiquity of a Western North American Stemmed Point Tradition: A Pre-Clovis Perspective by David G. Rice 208

Paleoenvironments and Paleoindians in Eastern South America by Astolfo Gomez de Mello Araujo 221

Submerged Lithic Tools Indicate Alternative Procurement Strategies by Alison T. Stenger 262

Cover designed by Alison Stenger & Dennis Honse
Compilation and Production by Dennis Honse
Cover images courtesy of The Smithsonian Institution

Foreword

J. M. Adovasio and David R. Pedler

If any of the contributions in this volume had appeared 40 years ago, none of them would have been accorded serious consideration, let alone been published, because the Clovis-first model for the colonization of the New World was then at its zenith and any disagreement with that model's chronological and behavioral tenets met with nearly universal opprobrium in the North American scholarly press. In other words, with the exception of a handful of obviously demented heretics and marginalized apostates, no serious archaeologists doubted that sometime around 13,000 cal yr BP, a small group of intrepid hunters crossed the interior of the Bering Platform and arrived in the fauna-rich paradise of the unglaciated Bering Refugium. Nor would most archaeologists have doubted that shortly after their arrival, these highly mobile and highly skilled hunters moved swiftly south through the corridor between the Cordilleran and Laurentide ice sheets, proceeding all the way to the hemisphere's Southern Cone. This incredibly rapid passage through the Americas—thought to have occurred within the mere span of only 500 radiocarbon years, filling both continents with monolithic genetic and technological progeny—was presumed to have been marked by a trail of carcasses of over 30 genera of late Pleistocene fauna, all of which were swiftly driven to extinction via stone-tipped spears. For indeed, if these dead animals marked the hunters' wake, the "signature" Clovis fluted projectile pointed the way.

Not surprisingly, especially given the overwhelmingly androcentric composition of the scholarly community at that time, North American archaeologists fell head over heels for this exciting, lithic tool-dominated scenario—a fanciful rendering that allowed no room for women, the young or old of either gender, non-durable artifacts, or the open discussion of alternative scenarios. This milieu in turn permitted the Clovis-first model to evolve rather rapidly from hypothesis and theory to received wisdom and ultimately to a quasi-religious dogma that over time became progressively entrenched by the debunking of hundreds of Pre-Clovis claimant sites, each enjoying a brief renown before being re-consigned to obscurity by the Clovis-first cognoscente.

Interestingly, virtually from the beginning, the Clovis-first model was challenged by a few European and many more South American scholars who pointed out that no such rapid, long distance, terrestrial colonization event (either with, or without, attendant massive macro-faunal extinction) had ever occurred in human, or even hominid, history. It was further noted that certain lithic industries in South America were at least as old as Clovis, and therefore not possibly derived from it. These observations were dismissed by the North American archaeological community as uninformed at best, and intellectually deficient at worst.

This situation would change dramatically in the mid- to late 1970s with the excavation, analysis, and preliminary publication of the carefully collected data sets from Meadowcroft Rockshelter in Pennsylvania and Monte Verde in Chile. Unlike any of the several hundred post-1933 and pre-1973 pre-Clovis claimants, these sites—and, most particularly, Monte Verde with its two-volume final report— would shift the pendulum slowly but inexorably away from Clovis-first. Both sites and their excavators were and still are subjected to furious attacks by acolytes of Clovis-first who may have seen, in one or both of them, the end of a venerable but fatally flawed paradigm. This position has nonetheless continued to be vigorously maintained in those quarters, despite the fact that many other sites in the Americas have since produced materials as old, if not older, than those recovered from the two seminal localities. Not surprisingly, like Meadowcroft and Monte Verde, each of the new sites and/or re-dated "older" sites has been the subject of virulent criticism from an ever dwindling but ever more strident coterie of Clovis-first spear carriers. Yet many sites have survived the critical onslaught, and the inquiry into the point(s) of origin and the timing of the arrival of the first Americans has been reopened and revitalized.

This newly open inquiry has led to a majority consensus that rather than a single (if not singular and homogenous) peopling event about 13,000 cal yr BP as described above, there have been multiple peopling pulses, some of which may have occurred before the Last Glacial Maximum, about 26,500–19,000 cal yr BP. Moreover, it appears that these sequent and perhaps overlapping pulses may well have involved different populations with different genetic and linguistic profiles entering the New World via different routes. In addition to the venerable interior route through the Mackenzie River valley, a western coastal entry (as first proposed by Fladmark in 1979) seems ever more certain. These coastal migrants may well have skirted the southern margins of Beringia by a combination of boat and pedestrian travel and thence moved all the way down the coast of the Americas without ever venturing inland. Alternatively, some of migrants could have traversed the Isthmus of Panama, moved up the Gulf Coast, and thereafter followed the Florida coasts and proceeded north along the Eastern seaboard. Of course, all of these routes may have been employed, either simultaneously or sequentially. It should also be stressed that whatever the route(s) taken, it is becoming progressively clearer that the populations involved in each entrada employed different durable technologies and pursued lifeways, in some cases, very different from Clovis and which may well have failed, leaving no technological or genetic progeny whatsoever.

The symposium that occasioned volume was held on 9–10 November 2012 at the National Museum of National History in Washington, DC, and involved 15 presentations concerning pre-Clovis archaeology as it relates to submerged contexts on the continental shelves, long-established and newly discovered sites across North America and the Southern Cone of South America, bioanthropological inquiry, and the roles played by various artifact classes, both durable and non-durable. Though some of the contributors to the Smithsonian symposium unfortunately were not able to participate in this publication, this volume nonetheless serves as a snapshot of the current, reinvigorated state of affairs in pre-Clovis research. This is not to say, of course, that an all-encompassing replacement hypothesis has succeeded Clovis-first. As the papers in this volume show, no such single hypothesis exists, and given the complexity of the peopling process it is probable that no one hypothesis will ever satisfactorily and definitively explain the late Pleistocene colonization of the New World.

Pre-Clovis in the Americas: Characterizing Early Sites, Material Culture, and Origins

Alison T. Stenger, Institute for Archaeological Studies, Portland, Or. USA

ABSTRACT
The number of pre-Clovis sites and materials that have been documented provide far more than the mere validation that sites older than Clovis exist. Some seemingly similar pre-Clovis features, tools, materials, and technologies have been reported from many different regions. The task now is to determine what similarities or differences are reflected in these early materials, and what this can tell us about the people who made them. Additionally, a vast array of occupation environments has been described. The differing economies, and cultural preferences, that this indicates may be suggestive of distinct pre-Clovis entries.

This paper will focus upon these topics, but it will also include several additional issues. These include human groups and tool types that may have descended from a common but distant ancestor, as well as those specimens that appear to be unique. Questions about selected analysis methods, and index species thus far ignored, will also be introduced.

The purpose of the conference and this publication is to expand our familiarity with pre-Clovis. Simply put, to determine what defines pre-Clovis, other than age. To accomplish this, we will consider multiple sites with dates older than 11,050 RCYBP (radiocarbon years before present).[1] Sites, features, and cultural material predating that age threshold will be described in detail by other authors in this volume. Further, the vast array of environments that supported early populations will also be addressed, as the early human arrivals in the Americas are tied to these landscapes.

The sites reported at this conference span an extremely lengthy period of time. We now know that the Americas were occupied 20,000 years before Clovis aged cultures emerged. Well dated sites now extend beyond 31,000 years of age in both North and South America, and multiple sites dating from 30,000-17,000 are not uncommon.[2] While some areas are now submerged, others are well above sea level. Importantly, terrestrial coastline and submerged sites are consistently older than inland sites. This speaks directly to older theories that suggested inland migration routes for the initial peopling of the Americas, as the dating of early sites does not validate such inland models. Further, as

---

[1] Clovis dates are currently described as 11,500-10,900 ^{14}C yr B.P., adjusted to 11,050-10,800 ^{14}C yr B.P. (Waters and Stafford, 2007).
[2] Florida has a date of 31,550 +/- 240 BP (Stanford, 2012), nine sites have now been reported as dated from 27,900 +/- 230 BP to 17,820 +/-170 along the East Coast, as far north as Chesapeake Bay (Lowery, 2012), and sites such as Meadowcroft in Pennsylvania are exceedingly well documented to 16,175 +/- 975 RCYBP, SI-2354 (Adovasio and Donahue 1990; Adovasio 2012).

demonstrated by geoarchaeological investigators, and geologists, ice free corridors are not part of the northern Pleistocene landscape (Clark, 2012; Bryson and DeWall, 2007).

Early habitation and use areas are tremendously varied. Numerous geophysical regions on both continents support early sites, and multiple ecosystems are utilized. Pre-Clovis cultural material occurs in rock shelters, on valley floors, and upon coastal plains in North America (Fig. 1).[3] The variety of early site environments identified in South America is also becoming increasingly diverse (Araujo, 2012; Schneider, 2009). So many sites have now been identified, in nearly every environment, that the widespread occurrence of early populations is clearly indicated. A geographic preference, however, is not. This is an important fact, as it tells us a great deal about early entrants. The broad distribution also suggests multiple origins, rather than a single entry.

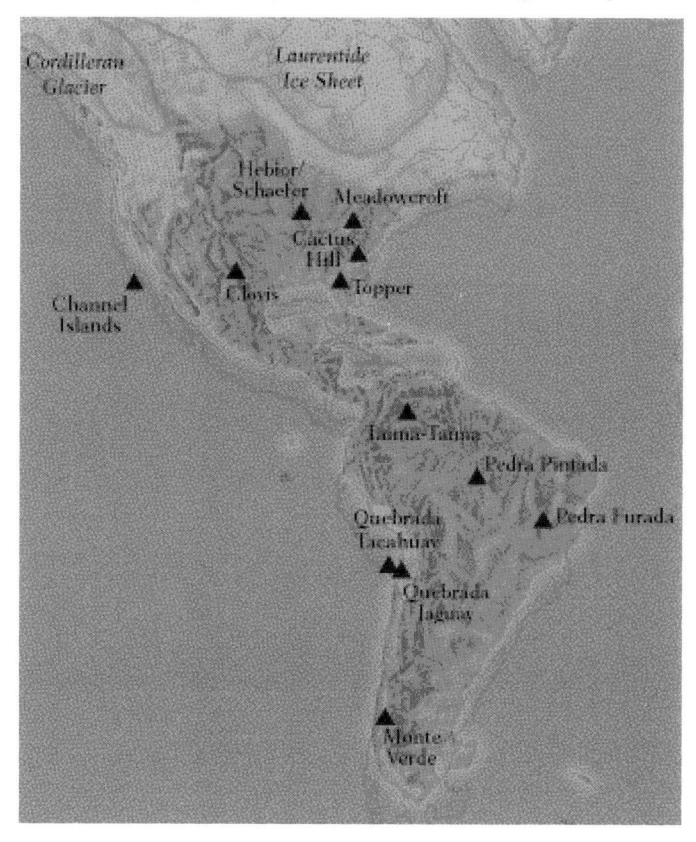

Figure 1. Map of some Pre-Clovis sites in the Americas, showing their broad distribution. Many different regions and different environments were utilized. Paisley Cave and the Stafek site in Oregon, and On Your Knees Cave in Alaska, are not shown, but the location of the original Clovis type site is depicted. Illustration courtesy of the Smithsonian Institution.

---

[3] Documentation of these early sites is provided by the authors in this publication, plus others such as Dillehay (1997) and Stanford (2012).

While dry land sites such as Meadowcroft and Monte Verde have long been documented and tightly dated, more underwater resources are being recognized. Cultural areas now include numerous saturated environments, including oceans, estuaries, rivers, and even high elevation lakes.[4]

Site features include some interesting challenges. This is due in part to their diversity, but also because of the many different types of environments in which they are observed. Features can occur as isolates, or with a distribution that covers a limited area or group of sites. Alternatively, some features may occur only within 300 miles of the current shoreline, but on several coasts.[5] The interpretation, too, can be difficult, as some features do not occur at more recent sites. This lack of a more current proxy suggests changing life ways, and eliminates the possibility of looking to the historic or proto-historic record for explanation.

It is potentially significant that, like some site features, early cultural materials often evidence stylistic or technological attributes not observed in more recent materials. This change in technologies and tool types encourages the consideration of changing populations over time. As many early forms do not occur at more recent sites, despite the continued presence of many of the same food resources, a change in human populations is indicated.[6]

One of the exceptions to changing types over time is the Western Stemmed point. This type demonstrates that styles and technologies can extend over time, and in a large area. Thus, when other types from other areas do not persist, a replacement of populations may well be indicated.

Despite the challenge of identifying parent populations, the discussions need to happen if founding groups are to be recognized. One important issue is whether habitation and subsistence patterns, as well as artifacts, can reflect population origins.

---

[4] Submerged sites are documented from 6,400' in elevation, down to sea level (Hemmings, 2012; Vastag, 2012; Stenger, 1997).

[5] An early feature that is without explanation is a paper thin circular stain that is reported only in South Carolina and in Oregon (Lyttle, 2005; Goodyear, 2005; Stenger, 2012, 2005). Both site areas are terrestrial, and appear to be pre-Clovis.

[6] While extinction of some megafaunal species occurred, many large terrestrial mammals continued forward into modern times. Further, at least some early human groups were not reliant upon megafauna, but instead focused upon many alternative food sources. This is well defined at the Gault Site (Speth et al, 2010).

The results of continued efforts to compare and contrast material culture would suggest that this is true. The styles and technologies of some woven and twined materials, and some lithics, compare well between Old and New World traditions (Adovasio, 2012; Stanford, 2012; Bradley, 2012).

The large, bipointed lithic biface styles of Europe have now been documented along segments of the east coast of North America (Fig. 2). Based upon the dating of organics associated with some of these lithics, the large temporal separation between the paleoamerican and Solutrean sites has been eliminated (Stanford and Bradley, 2012). The analysis of woven and twined cultural materials is similarly linking populations and technologies between the Old World and the New World (Adovasio, 2012).

Figure 2. The large, bipointed lithic biface styles of Europe have now been documented along segments of the east coast of North America. The Solutrean style is no longer recognized exclusively in Europe. This figure courtesy of Chip Clark, and Dennis Stanford, Smithsonian Institution.

Underlying this line of inquiry regarding founding populations is the question of whether most culture bound behaviors translate well into new environments. Thus, do any of the early sites represent a new population that quickly failed? Can that account for the scarcity of some artifact types, and the seemingly unique features that are documented? Would this explain the older skeletal material, and specifically cranial elements, that do not compare well with more recent population attributes?

The study of founding populations includes the analysis of mtDNA, when it is viable, as well as morphological studies. Yet the validity of the data is problematic. Importantly, morphological phylogeny is often at odds with molecular phylogeny. In other words, when studying ancient individuals, the results of cranial studies often contradict mtDNA findings (Chatters, 2012). Conflicting analytical results occur in other populations, as well. The anatomical and DNA studies of widely disparate species, from bears to lizards, yield the very same types of contradictory findings (Hailer et all, 2012; Losos et al, 2012).[7]

Added to the complexity of classifying early people in the New World is the lack of biological relationships with past populations. Some early New World human mtDNA reflects populations that have not been genetically mapped elsewhere (Baker 2,000). These ancient individuals appear to be an indirect ad-mixture of several different populations, but without a direct relationship to any known groups (Chatters, 2012).

A human hair from a pre-Clovis site in Oregon is an example of previously unmapped human DNA. The hair is morphologically human (Fig. 3). Genetically, it is *Homo sapiens sapiens*. Yet, while this specimen is from a fully modern human, it is unlike the mtDNA from any modern population group whose DNA has been mapped (Fig. 4). Just as we observe in the cranial studies of early Americans, the ancient people are distinct from contemporary populations.

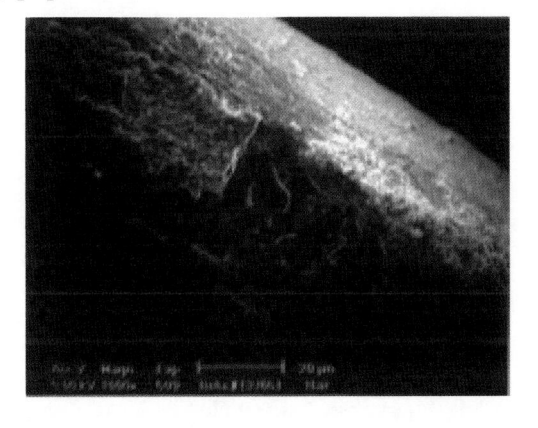

---

[7] As stated in the Losos article, "State-of-the-art molecular and morphological phylogenies…differ fundamentally."

Figure 3. Photomicrogaph of the first Pleistocene human hair documented within Mill Creek drainage in Woodburn, Oregon. Authentication provided by Intermountain Forensics Laboratory, and image courtesy of Beta Analytic Radiocarbon Laboratory.

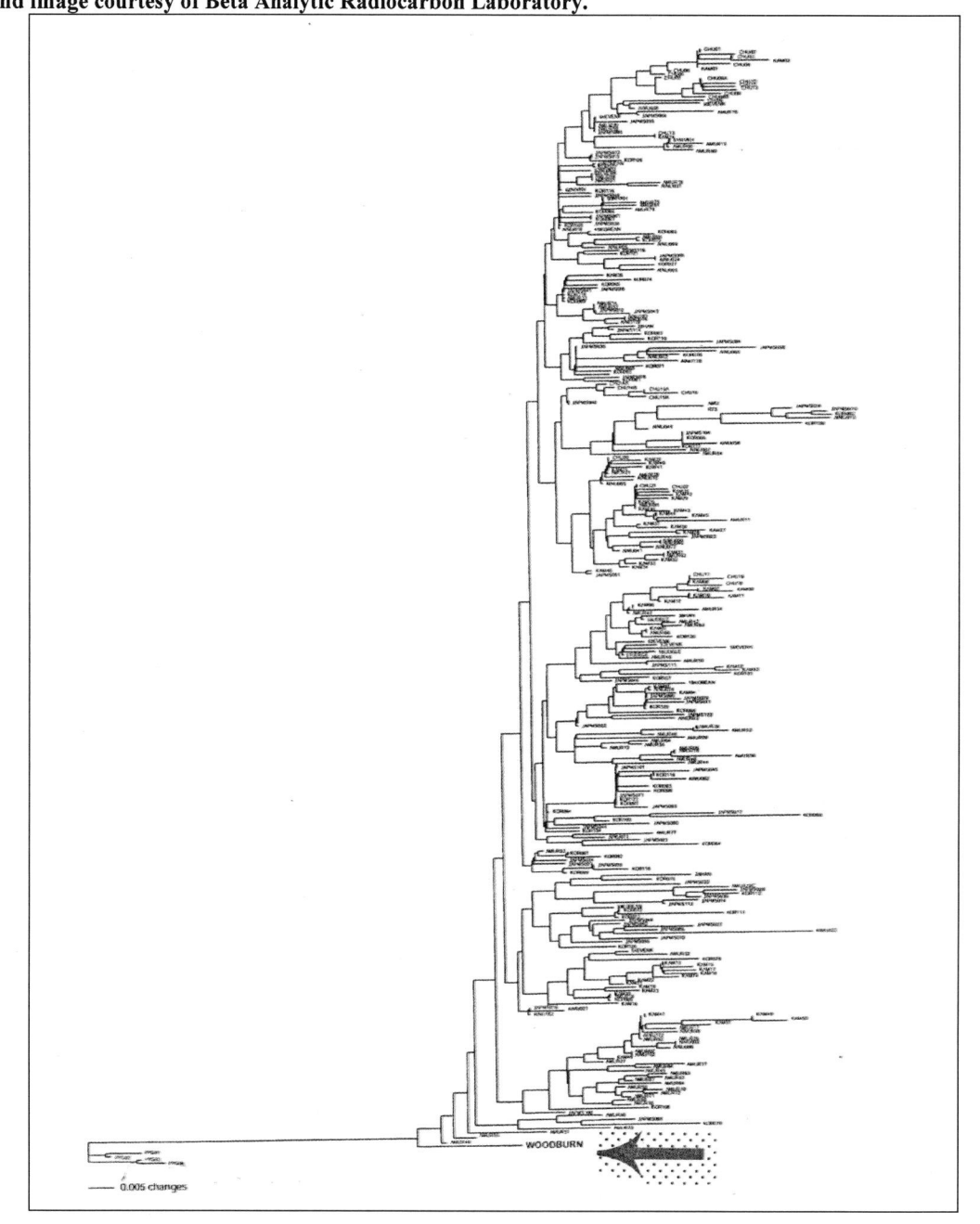

Figure 4. When the mtDNA from an ancient hair in western Oregon was analyzed, it proved to be unrelated to any previously identified population (Baker, 2000). The mtDNA from that hair is identified by an arrow at the base of this graphic, and labeled as Woodburn.

When researching the origin of the first Americans, the question of biological affinity is crucial. The artifacts and how they were used allows for an association with specific cultures, but to positively identify a population, biological attributes need to be evident (Jantz and Owsley, 2001; Jantz and Owsley, 2005). Thus, the current lack of corroboration between cranial studies and mtDNA suggest that a new approach to analyzing the DNA from early people may be useful. The most informative method of study, to date, is the examination of nuclear DNA.[8] Due to cost, however, Y-chromosomal studies in combination with mtDNA are expected to lend plausible answers (Oppenheimer, 2012). This approach, of looking at more than mtDNA, may resolve the contradictory issues that sometimes exist between DNA and cranial studies.

Biological relationships between past and present population groups also need to be reconsidered. Reports of the findings from mtDNA studies often state direct relationships between groups, when descent from a common ancestor may be far more accurate. Two populations may actually have no direct relationship to each other.

The analysis of North American specimens is particularly troublesome, as the interpretation that is most often used is, at best, misleading. For example, early American remains are now often associated through mtDNA to areas such as Siberia and Beringia. Two profound questions accompany such an association. The first is whether the Siberian or Beringian populations are directly related to the American remains from below Canada. Or, do the groups from these very different geographic regions have an ancestor in common, but no direct ties to each other?

The second issue is even more problematic. Some scientists seem able to identify > 8,000 year old mtDNA from Siberian that associates with early American mtDNA (Gilbert et al, 2008). However, other notable scientists find no such association of early remains. These researchers reveal a total absence of similar material, between North America and Siberia, older than a few thousand years (Oppenheimer, 2012; Stanford and Bradley, 2012). This issue, in itself, suggests that mtDNA analysis is actually still in its early developmental stages, and very prone to interpretive error.

---

[8] Nuclear DNA demonstrated very different results from mtDNA in bear populations, suggesting that mtDNA results may be extremely misleading, or even fully incorrect (Hailer et al, 2012).

The often incorrect description of mtDNA findings is shared with the public, and with special interest groups. As some public policies are formed based upon these results, it is important that the results are accurately conveyed. The lack of consensus among molecular biologists makes this impossible.

The many types of sites described in this publication, and the data obtained from them, should allow for many new aspects of site investigation. This may include the recognition of unanticipated DNA, and the detection of ancient bacteria and viruses. Additionally, in future work, pathogenic DNA sequences need to be considered. What pathogens have we missed, that could be pivotal in the survival of a Pleistocene species? Importantly for field personnel and site interpretation is the topic of continued viability of pathogens. Which pathogens may have survived millennia? The Mimi and Mega viruses are two examples of quite ancient pathogens that have only been identified in the past few years (Van Etten, 2011; Randerson, 2003). The previous lack of recognition was due only to the size of these viruses. As they far exceeded known viral sizes, no one expected them to exist. While this is reminiscent of "Clovis First" thinking, and not looking for evidence of older cultures, it is also a warning that we may be overlooking significant data. Other indicators of ancient origins do exist.

Due to the sensitivity of many pathogens, they may prove useful as index species, and hint at New World population origins. Studies of the hookworm parasite have shown its intolerance for environments that are too cold. Thus, the discovery of dead hookworms in pre-Columbian remains from colder environments strongly suggests population movement (Meggers, 2006; Jett, 2004, 2007).

All of these issues, from DNA and parasites to the study of site distribution, need to be addressed if we are to successfully synthesize the existing body of knowledge on pre-Clovis in the Americas. These data, now substantial in volume, will finally allow researchers to define paleoamerican populations.

The accurate histories of both North and South America are dependent upon conferences and publications such as this one. It is through the efforts of the Smithsonian Institution in hosting the Pre-Clovis in the Americas conference that the communication of these very important, new ideas can be communicated between scientists and the interested public.

References

Adovasio, James, 2012. Plant Fiber Technologies and the Initial Colonization of the New world. Pre-Clovis in the Americas conference, Smithsonian Institution, November 9-10.

Bradley, Bruce, 2012. Older Than Clovis Bifacial Technologies of Eastern North America. Pre-Clovis in the Americas conference, Smithsonian Institution, November 9-10.

Baker, Lori, 2000. Email to the author, 7/3. Results of sequencing of mtDNA of Woodburn hair, with emphasis on amplification of C, and failure of D to amplify.

Bryson, Reid A., Katherine McEnaney DeWall, 2007. Environments of the Northwest Entrada: An Examination of Climate, Environment, and Timing in the Peopling of the Americas from the Northwest. Proceedings of the International Science Conference, Houston, 35:59.

Chatters, James, 2012. Email to the author. 1/24/12, 5:26 p.m.

Chatters, James. 2010. Peopling of the Americas via Multiple Migration Routes from Beringia. Syposio Internacional El hombre temprano en America, J.C. Jimenez, A. Gonzalez Gonzalez, and F.J. Agullar Arellano, eds.:52-55.

Clark, Jorie, Jerry Mitrovica, Jay Alder, 2012. Regional Variability in Deglacial Sea-Level Rise Across the Oregon-Washington and Bering Sea Continental Shelves. . Pre-Clovis in the Americas conference, Smithsonian Institution, November 9-10.

Collins, Michael B., 2002. The Gault site, Texas, and Clovis Research. Athena Review 3(2):31-41, 100-101.

Collins, Michael B., 2007. Thinking Small and Falling Short—A Critique of Myopia in Archaeological Inquiry. Proceedings of the International Science Conference, Science in Archaeology, Houston:201-213.

Donahue, Jack, and J. M. Adovasio, 1990. Evolution of Sandstone Rockshelters in Eastern North America. Archaeological Geology of North America, edited by Norman P. Lasca and Jack Donahue. Geological Society of America, Boulder: 231–251.

Gilbert, M.Thomas, Dennis L.Jenkins, AndersGötherstrom, NuriaNaveran, Juan J. Sanchez, MichaelHofreiter, Philip FrancisThomsen, JonasBinladen, Thomas F. G.Higham, Robert M.Yohe II, Robert Parr, Linda ScottCummings, EskeWillerslev, 2008. DNA from Pre-Clovis Human Coprolites in Oregon, North America. April:1 www.sciencexpress.org.

Goodyear, Albert, 2005. Personal communication to author.

Hailer, Frank, Verena E. Kutschera, Bjorn M. Hallstrom, Denise Klassert, Steven R. Fain, Jennifer A. Leonard, Ulfur Arnason, Axel Janke, 2102. Nuclear Genomic Sequences Reveal that Polar Bears Are an Old and Distinct Bear Lineage. SCIENCE 20 April, V. 336[6079]:344-347

Jantz, Richard L. and Douglas W. Owsley, 2001. Variation among Early North American Crania. American Journal of Physical anthropology, 114:146-155.

Jantz, Richard L. and Douglas W. Owsley, 2005. Circumpacific Populations and the Peopling of the New World: Evidence from Cranial Morphometrics. Paleoamericdan origins: clovis and Beyond, Robson Bonnichsen, ed., College Station:267-275.

Jett, Stephen C., 2004. No Plague in the Land? Infectious Diseases and Their Implications for the Pre-columbian-Transoceanic-Contacts Controversy. Migration and Diffusion—An International Journal 5,19:6-31.

Jett, Stephen C., 2007. Genesis, Genes, Germs, and Geography: the Implications of Genetic and Human-Disease Distributions for Founding and Later Old World Entries into the Americas. Proceedings of the International Science Conference, Houston: 100-133.

Losos, Jonathan B., David M. Hillis, Harry W. Greene, 2012. Who Speaks with a Forked Tongue? State-of-the-art molecular and morphological phylogenies for lizards differ fundamentally. SCIENCE 338, 1428-1429, 14 Dec.

Lowery, Darin, 2012. Pedologic and Geological Protocols for Understanding the Archaeology of Exploration: A Middle Atlantic Pre-Clovis Case Study. Pre-Clovis in the Americas conference, Smithsonian Institution, November 9-10.

Lyttle, Patricia, 2005. Personal communication to author.

Meggers, Betty, 2006. TransPacific Voyages from Japan to America. Paths Across the Pacific Conference, Sitka, Alaska; Video recording of proceedings, North Star Television Network. Paths Across the Pacific, Sitka, Alaska.

Oppenheimer, Stephen, 2012. Nuclear DNA and First American Origins. Pre-Clovis in the Americas conference, Smithsonian Institution, November 9-10.

Randerson, James, 2003. Massive virus discovered in water tower. New Scientist, March 27:1900

Schneider, Alan L., 2007. Personal correspondence after discussions with Tom Dillehay, regarding older component of Monte Verde.

Speth, John D., Khori Newlander, Andrew A. White, Ashley K. Lemke, Lars E. Anderson, 2010. Early Paleoindian big-game hunting in North America: Provisioning or Politics? Elsevier Ltd and INQUA, Quaternary International: 1-29

Stanford, Dennis J., 2012. Welcoming remarks. Pre-Clovis in the Americas conference, Smithsonian Institution, November 9-10.

Stanford, Dennis J. and Bruce Bradley, 2012. Across Atlantic Ice. University of California Press:246-248.

Stenger, Alison T., 1997. Pauline and East Lakes: Underwater Survey for Submerged Cultural Resources, U.S. Forest Service. Institute for Archaeological Studies, Portland.

Stenger, Alison T., 2005. Lake on the Trail: Archaeological Investigations of a Potential PaleoIndian site. Institute for Archaeological Studies, Portland.

Van Etten, James L., 2011. Giant Viruses. American Scientist 99(4):304.

Waters, Michael R. and Thomas W. Stafford, Jr., 2007. Redefining the Age of Clovis: Implications for the Peopling of the Americas. Science 315(5815), 1122-1126.

Inundated Landscapes and the Colonization of the Northeastern Gulf of Mexico

C. Andrew Hemmings and J. M. Adovasio

Mercyhurst Archaeological Institute
Mercyhurst University
Erie, Pennsylvania 16546

Abstract

The inundated terrestrial Pleistocene landscape of Florida has been examined with increasingly more sophisticated techniques and equipment for over 100 years. Recent systematic remote sensing has identified nearly 175km of the buried Paleo-Suwannee River Channel, now traced to the SE corner of the Florida Middle Grounds. Scuba diver sampling of bedrock exposures of knappable stone and initial attempts to excavate adjacent areas are yielding important, incremental, data demonstrating that the buried channel is intact and has not been exposed to the ravages of modern seafloor activity during the Holocene. Future research plans and expectations are discussed. In addition, broader implications for the peopling of the New World after the LGM are derived from our nuanced understanding of this reconstructed landscape and the potential for this environment to have supported a rich Pleistocene faunule, and early human arrivals, as sea level rise drove them shoreward until about 5,000 years ago.

Presented at the Conference "Pre-Clovis in the America"
Smithsonian Institution, Washington DC
9-10 November 2012

Introduction

The year 2007 saw the bicentennial of the National Oceanic and Atmospheric Administration (NOAA) and its scientific predecessor, the Coastal Survey, while also notably marking another and not unrelated anniversary. Early in the fall of 1807, President Thomas Jefferson dispatched William Clark to the fossil quarry at Big Bone Lick in Kentucky. His assignment was to collect, if possible, a complete mastodon skeleton. This unusual mission proved to be eminently successful and resulted in the retrieval of numerous well-preserved paleontological specimens which are curated today in New York, Philadelphia, Monticello, and Paris, France.

In the course of excavating the faunal assemblage, five fluted points were recovered. Three of these have been housed at the Cincinnati Museum of Natural History since 1817. We note this artifactual recovery for two reasons. First, though then unrecognized as such, the Big Bone Lick points are among the very earliest legitimate Paleoindian materials found in the Americas. Second, and more to the point of our talk today, the NOAA funding of our 2008–2012 project marked a unique alliance between two scientific disciplines near and dear to the interests of our third president, just as each began its third century of inquiry.

By the middle of the twentieth century, points like those recovered at Big Bone Lick were viewed by the great majority of the North American archaeological community as the signature artifacts of the first occupants of the New World. Moreover, the makers of these points (now named "Clovis" after their initial occurrence in a stratified context near Clovis, New Mexico) became central players in a highly imaginative peopling scenario called, in recent years, "Clovis First."

According to this paradigm, a small group of migrants crossed the interior of the now-submerged Bering Platform about 12,000 14C yr BP. After a brief sojourn in the unglaciated Bering Refugium, these pioneers were thought to have passed down the ice-free corridor between the Cordilleran and Laurentide ice sheets, thence across virtually the length and breadth of the entire unglaciated New World, arriving at the tip of South America within a scant 400–500 radiocarbon years or less. In addition to its chronological implications, which explicitly presumed no earlier migrants to the Americas, the Clovis-first model also posited an essentially uniform technology and lifestyle for these putatively First Americans. In its later renderings, Clovis was thought to represent or constitute a unitary archaeological "culture" or complex whose members were rapidly moving, specialized, spear-wielding hunters who preyed upon and ultimately caused the extinction of some 40 species of late Ice Age beasts.

The temporal or chronological underpinnings of the Clovis-first hypothesis began to seriously unravel with a series of pivotal discoveries in both North and South America. Intensive multi-year, multi-disciplinary excavations at Meadowcroft Rockshelter in southwestern Pennsylvania (Adovasio et al. 1975, 1977a, 1977b, 1978, 1989, 1990, 1992; Carlisle and Adovasio 1982) and Monte Verde in coastal Chile (Dillehay 1989, 1997) demonstrated to the satisfaction of most scholars that humans had entered the New World at least 2,000–3,000 radiocarbon years before the Clovis era. Moreover, the recovered artifacts and ecological data from theses sites suggested different technologies and very different adaptations and subsistence strategies than those posited for Clovis (Adovasio and Pedler 1996, 2000, 2003). More recently, excavations at sites like Cactus Hill in Virginia (McAvoy and McAvoy 1997; McAvoy et al. 2000), Gault in Texas (Collins 2002), and a series of early coastal sites in Peru (Dillehay 2000) have confirmed pre-Clovis migration pulses to the New World and lifeways not entirely consistent with those typified by the Clovis First scenario.

As a direct consequence of the excavations and attendant analyses at the sites just noted, it has become increasingly clear that the peopling of the western hemisphere was a much more lengthy and complicated process than was previously imagined. Rather than a single entry route across the interior of the Bering landform, multiple colonization episodes employing multiple routes are now considered to have been highly likely for the human entry into the New World. Moreover, rather than embodying a unitary cultural adaptation, the first migrants to the New World are presently thought to have included a series of quite different lifestyles and subsistence strategies, many of which are broad-spectrum and generalized, and not focused or specialized as had previously been thought.

Within this fluid paradigm occasioned by recent research in New World migration, coastal environments, water transportation, and aquatic and marine resource utilization are receiving much more attention than has previously been the case. Significantly, this attention has extended into presently submerged coastal environments.

Research Summary

In order to pursue this growing interest in submerged coastal environments, in 2007 we submitted to NOAA a successful proposal which postulated that now inundated environments in littoral zones such as the eastern Gulf of Mexico should contain evidence of earlier coastlines with intact beach features, such as dune ridges, as well as indications of the initial anthropogenic use of such coastal environments. We also postulated that it should be possible to identify the channels and related geomorphological features of Florida's west coast karst river systems which intersected these submerged coastlines during the late Pleistocene. As such, and by extrapolation from terrestrial analogues, these rivers would have been magnets for human occupation/utilization and, by extension, early human habitations should be located in close juxtaposition to them. We suggested that the eastern Gulf of Mexico—specifically within an area paralleling the arc of the Ocala Uplift zone—was an ideal locus for the location, identification, and exploration of inundated coastal environments for five compelling reasons.

First, the onshore concentrations of sites in Florida older than 10,500 14C yr BP cluster along the Ocala Uplift zone, which contains high-quality, crypto-crystalline chert outcrops and numerous perennial water sources (Dunbar 1991:193, Figure 7; Thulman 2009). These notably include the underwater Page-Ladson, Sloth Hole, and Butler sites as well as the terrestrial locus at Wakulla Springs, all of which are older than 11,000 14C yr BP (Dunbar 1991, 2006; Hemmings 2005).

Second, similar clusters of sites have been documented offshore along channel and sink features in nearshore settings across the northern Gulf of Mexico (Faught 2004). By extrapolation, earlier sites should be located further offshore on earlier, inundated portions of this landscape.

Third, inundated archaeological sites in this portion of the Gulf of Mexico should be well preserved and accessible for a variety of reasons. This portion of the gulf is considered to be sand starved and the rivers of Florida that discharge into this area do not carry a heavy sediment load, so any sites near or on the inundated channels or sinkholes should not be deeply buried.

Fourth, the bathymetric gradient in this portion of the Gulf of Mexico is low, suggesting that during the meltwater phase of the terminal Pleistocene, comparatively quiet water inundated the flat continental shelf without destroying surface features or riverine filled sinkholes and channels. The low bathymetric gradient in turn suggests that the potential for site disturbance or destruction by storms or hurricanes should be at least partially mitigated. This is particularly important for the area east of the Florida Middle Grounds, which would have been shielded by barrier islands until ca 9,300 14C yr BP and lies outside of the normal gulf gyre.

Fifth, changes in the rate of sea level rise may also have affected the potential for archaeological site preservation, specifically in the case of very rapid, short-duration rates of sea level rise, which would have had the potential to reduce or limit the erosive effects of wave action on terrestrial soils. As discussed below, this is particularly true during Clovis times when sea levels rose dramatically.

Our investigations in the eastern Gulf of Mexico have generated more than two terabytes of remote sensing data in nineteen loci within three subareas within the greater study area (Figure 1). This field research has confirmed many of our earlier hypotheses, in that we have successfully: (1) located the intact, buried, Paleo-Suwannee River channel in nine locations adding 175 km to its length, placing it near the southeastern edge of the Florida Middle Grounds (Figure 2); (2) traced the Paleo-Aucilla-St Marks River system almost 50 km; (3) identified numerous additional (possibly tributary) buried stream and river channels, some with clearly visible overbank deposits; (4) located and identified hundreds of infilled, stratified sinkhole-like features with obvious, but as yet untapped, potential for organic and artifactual preservation; and (5) mapped nearly 10 km2 of intact shallow-water, nearshore, sand

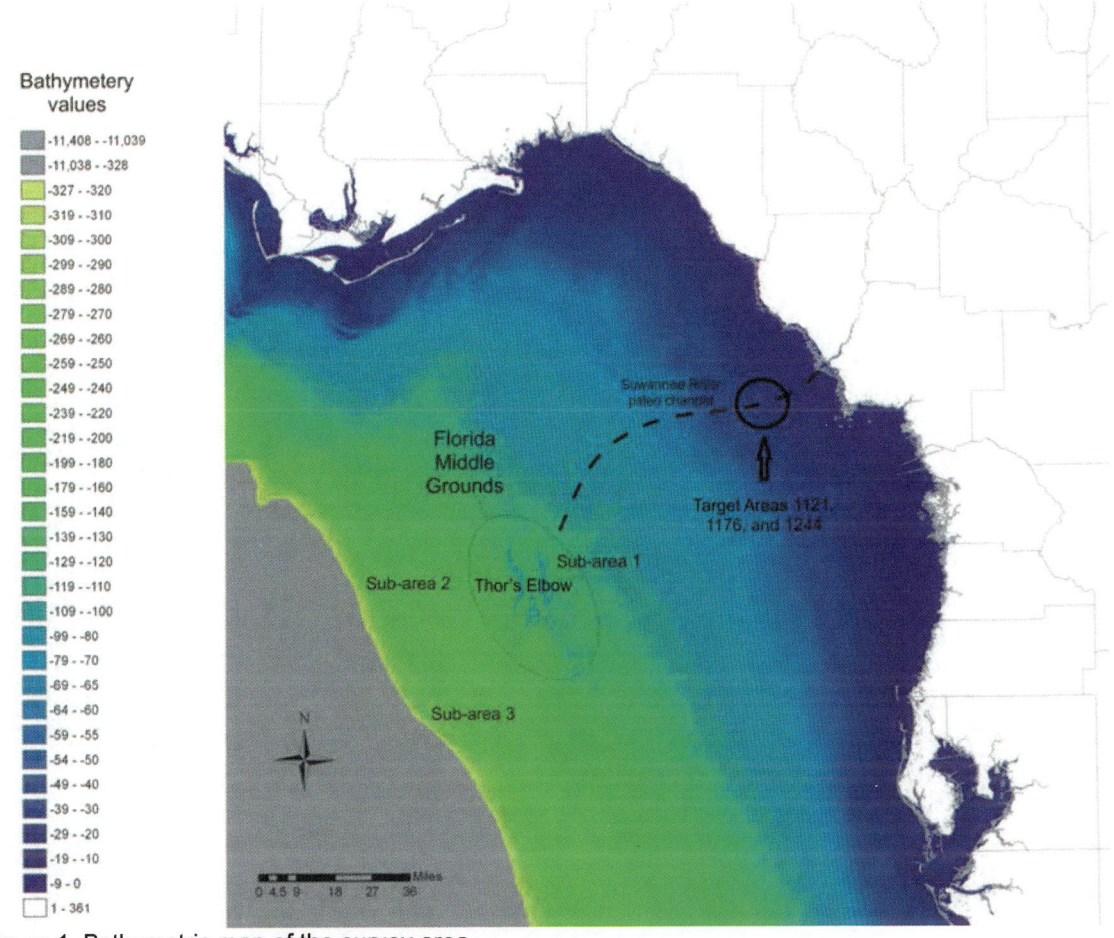

Figure 1. Bathymetric map of the survey area.

Figure 2. Side scan sonar image of possible west bank of the paleo-Suwanee river channel.

ripples/ridges just inside the Last Glacial Maximum (LGM) shoreline at a depth of over 90 m of water, (6) located and collected knappable bedrock stone exposures in multiple locations at three sites (7) excavated portions of the Holocene seafloor sand adjacent to the Paleo-Suwannee River channel at the Brownstone Site with the highest quality stone found to date (8) thereby uncovering bedrock that has not been exposed to the deteriorating effects of seawater and the colonization of marine organisms.

Significantly, and as we predicted in our 2007 NOAA proposal, the LGM shoreline is extraordinarily well preserved (Figure 3), as are the observed portions of the later shorelines closer to the modern west coast of Florida. We greatly increased our diving component of the project in 2009 and directly examined 16 locations ranging 12–40 m deep. Bedrock exposures of limestone and chert were taken in six different places where the sand-shell hash or modern biomantling could be penetrated. The three sites adjacent to the Paleo-Suwannee River channel with surficial exposures of de-silificied chert hold some of the greatest potential for our future work as we continue to probe these locations with a dredge (Figure 4). The infilled karst features with multiple layers of distinct strata are also of great interest archaeologically and methodologically. Such features were located in each of the study subareas, and analogous onshore phenomena of this kind suggest that their submerged counterparts offer similarly great potential as archaeological and/or paleontological loci.

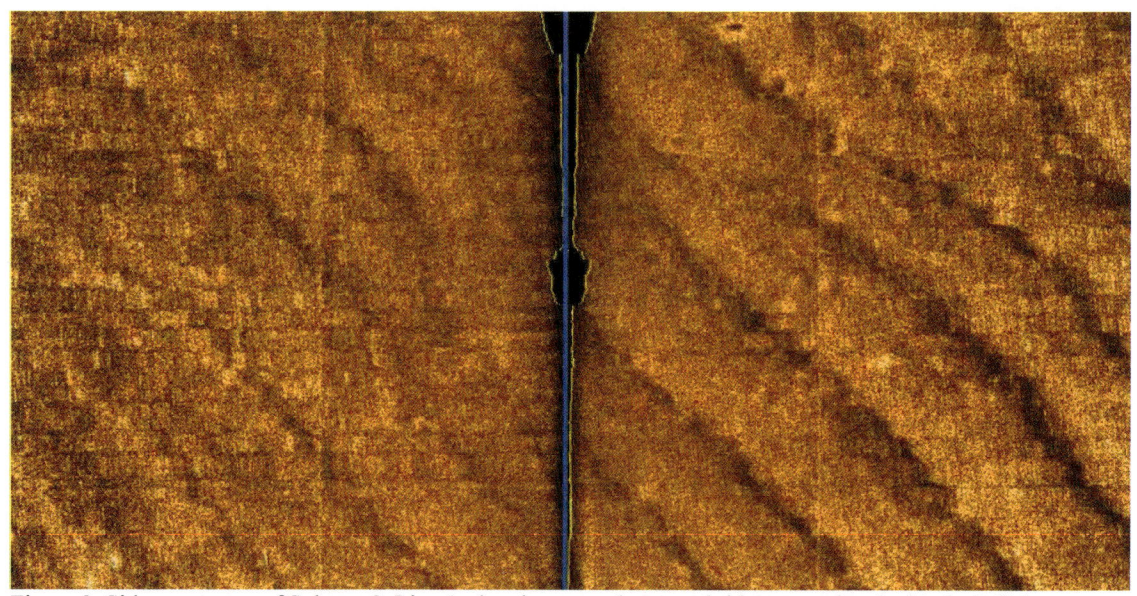

Figure 3. Side scan sonar of Subarea 3, Line 1, showing near-shore sand ridges associated with the LGM coast.

An important methodological epiphany was the practical demonstration during our 2008 fieldwork that, via multiple forms of data collection, we were able to locate and identify high-interest targets that were not visible to one or another remote sensing device alone. Now using a combined side scan sonar and sub bottom profiler, we can demonstrate exactly how this works. However, it is important to note that even with the combined system our patch test experiment around Ray Hole Spring demonstrates that the orientation to buried features may be even more important. In other areas, we overcame this factor by intersecting the Paleo-Suwannee at a right angle rather than trying to follow the meander as had been done previously from the mouth by other researchers (Wright et al. 2005).

Our continued explorations in 2011 and 2012 focused on attempting to excavate into the Paleo-Suwannee River channel at the Brownstone site, roughly 21 miles west of Cedar Key or 25 miles from the mouth of the Suwannee River. Distressingly slow, but very important, progress has been made in the last two years of fieldwork. Consistently rough seas have hindered our ability to excavate due to diver safety concerns, ship movement, and equipment breakdowns. The last two abbreviated field seasons lasted under nine days with far less excavation occurring than had been anticipated.

Figure 4. Divers examining a chert specimen from a now inundated outcrop.

In 2011 we relocated the bedrock exposure of dolostone or very grainy chert along the western edge of the Paleo-Suwannee River, now called the Brownstone Site (see figure 4). After collecting several samples of better quality stone than first found in 2009 we started to excavate into the buried channel using a 6" induction dredge. Our strategy was to follow the bedrock steps through the marine sands and hopefully expose the differentiated sediments visible in the sub-bottom profile data. We were able to excavate nearly a cubic meter to a depth of 110cm before increasingly large waves made it impossible for us to remain on site. One critical observation is very encouraging despite the limited time we had to excavate. The freshly exposed bedrock limestone/chert was white and smooth. The bedrock just below the modern surface has not been exposed to salt water or marine organisms, at any time in the past, to deteriorate or be colonized by burrowing creatures. This observation was confirmed in two more locations along the western Paleo-channel boundary by excavations in 2012 as well.

Three additional dives were made at the Brownstone Site in December, 2011 to collect additional stone samples and explore the extent of the knappable stone exposure. Better quality stone is clearly limited to a patch roughly 75 meters long on the west bank though the northern site boundary is not yet as clearly defined as is the southern.

Exposed limestone/chert, the rock sticking out of the thin layer of sand outside of the Paleo-channel, even up to the edge of the Paleo-Suwannee River channel, may in fact not be scoured. Rather, intact exposures of similar elevated post forms can be seen along the Aucilla River near Page/Ladson today.

The 2012 excavations reinforced previous observations regarding the channel edge, quality of knappable stone being geographically restricted, and immediately adjacent to terraced bedrock "steps" as first observed in our sub-bottom profile data. Posts used to mark our work areas with bouys hit bedrock at depths of 5 and 8 feet just 5 and 9 feet east of the exposed bedrock chert outcrop we excavated against.

When we are able to penetrate the marine sands to 4 to 6 feet we should then be able to directly sample the next stratigraphic unit observed in the sub bottom imagery.

After multiple reconfigurations of our dredges and hoses we have a working system that can overcome the depth and distance resistance we face from leaving the screen on deck. Temporary moorings will be deployed to securely hold the ship in place much closer to our work area than has been possible thus far. With each incremental advancement in our knowledge we are closer to sampling non-marine sediments of increasing age and proximity to Pleistocene terrestrial layers.

Two additional beneficial results of this fieldwork should also be noted. First, our research team has successfully navigated the steep learning curve for our specific archaeological application of remote sensing gear and shipboard logistic issues related to their operation. Second, and probably more importantly, during the course of our fieldwork, interested third parties and researchers have provided highly useful data on their experiences in and near the study area as well as information on similar inundated environments elsewhere and other types of remote sensing equipment that will facilitate this kind of research in the future. This information specifically included locational data on an infilled sinkhole at the mouth of the Suwannee River and, further afield, locational and descriptive data on clusters of intact tree stumps off the coasts of Alabama, Virginia, North Carolina, and Key West, Florida. A "forest", offshore of the Alabama coast in 60 feet of water, was uncovered by Hurricane Katrina and we are negotiating to be involved in future exploration and attempts to date the intact trees and the Paleo-landscape they are perched upon. An additional pair of interesting historical examples of prehistoric inundated landscapes were reported by Sellards in the original Vero man site report (1916:126). Sellards reports that dredge Captain O. N. Bie found a cypress tree swamp in 20ft of water seven miles below Tampa in Tampa Bay, further, he notes the discovery of swamp and peat deposits at 20ft down off the Keys on the Atlantic side of the state (Sellards 1916:126).

This kind of information—supplemented with additional data provided by scholars from the University of South Florida, the Florida Institute of Oceanography, the Florida Geological Survey, and the Florida Bureau of Archaeological Research—underscores the enormous potential of investigations of inundated late Pleistocene landscapes and has provided a valuable impetus for future work in the northeastern Gulf of Mexico.

Pre-Clovis and Clovis Archaeology in Florida

The cultural historical sequence of Florida is not clearly understood prior to the Early Archaic-Bolen age material of circa 10,000 rcybp. The relative Clovis-Simpson-Suwannee-Bolen sequence is widely accepted and may well be correct but is, alas, not yet chronologically or properly stratigraphically verified at this time. A Clovis ivory tool fragment has been dated to 11,050+/-50 at Sloth Hole and in situ Bolen points at Page-Ladson date 10,100 and 10,300 rcybp (Dunbar 2006, Hemmings 2004). There are additional well dated cultural contexts in Florida that lack diagnostic artifacts that may eventually help solidify this chronology. One particularly inscrutable lipped bifacial thinning flake was recovered from an indisputable closed context at Latvis-Simpson in mastodon digesta. A single Cucurbito pepo seed from this digesta sample was dated 31,550 +/-240 in 1995.

The relative cultural chronology of early Florida is tantalizing but also very poorly documented. A Bolen was found 30cm above a Page-Ladson point on land at Page-Ladson. At the Bolen Bluff site Clovis, Suwannee, Bolen, and a potential Pre-Clovis lozenge form discussed below were found in the vertical exposure in 1929 (Bullen 1959). Harney Flats contained Simpson and Suwannee in layers where it was not possible to distinguish a stratigraphic separation from Bolen material (Daniel and Wisenbaker 1987). There are a handful of other sites in Florida that may still contain stratified sequences of diagnostic Paleoindian and Pre-Clovis material that are receiving renewed interest at present await more research.

The eastern third of the country is producing a considerable amount of formal tool types that, for a number of intriguing reasons, suggest multiple technological traditions existed prior to Clovis (for example see Adovasio 1990, Adovasio and Pedler 2003, Dunbar 2006, Lowery 2009, and MacAvory and

MacAvoy 1997). Three technological strands appear to exist prior to Clovis based on current data: 1- Bipointed bifaces of two sizes that sort into large knives and smaller points (Cinmar is a larger form)- two of these are known from Florida; 2- small, typically less than 8cm long, lozenge or triangular points like the Miller point from Meadowcroft, Cactus Hill, and Miles Point, MY- they are also found in Florida at Wakulla Spring, Harney Flats, and possibly Bolen Bluff; and 3- Page-Ladson (or Delmarva) unfluted lanceolates.

The known range of Clovis material is slowly extending further south in Florida as well. In the last few years individual points have been found in Vero Beach (Indian River County) and on the beach in St. Lucie County on the Atlantic coast. The known Clovis range extends to at least Hillsborough County on the Gulf Coast. The Vero Beach point' material does not appear to be from Florida which is very rare to see in Florida Clovis. Further the material and workmanship of this point is strikingly similar to a Sloth Hole Clovis point.

In Florida's indeterminate chronologic sequence and relative sequence the temporal and technological relationship between Pre-Clovis, Clovis, Simpson, Suwannee, and other less common Paleoindian tool forms is unresolved.

It is thought that Simpson and Suwannee both post-date Clovis though there is better evidence for this being the case for Suwannee than there is for Simpsons in relation to Clovis (Dunbar and Hemmings 2004, Thulman 2009). Simpson points are uncommon and no single component occupation or layer has been found to date. Technologically they are very difficult to understand, particularly due to the very few known preforms or unfinished points that can inform on the bifacial reduction strategies employed to create them. Simpson should be considered a fluted point technology. On finished points the flutes are regularly reduced or even removed but on about half flutes or remnant scars remain. Like the bi-pointed Cinmar style bifaces Simpson's sort into two size and functional categories of very large knives (>14cm) and points (typically under 10cm). The large Simpson's have been found in South Carolina and in the Chipola and Peace Rivers of Florida. The consistent removal of flakes beyond midline or overshot flakes, widest and thinnest point above the haft, and exceptionally large size are traits most like the bipointed bifaces. The very narrow and sometimes waisted fishtail-like bases are not like any of the other Pre-Clovis materials. Morphologically smaller Simpson points are more like Clovis than anything else. Technologically the Simpson's could fit between Clovis and the Pre-Clovis forms. There is no chronological or stratigraphic evidence to support this idea, neither is it contradicted. A possible biface technological sequence is cautiously suggested only on the grounds that it attempts to account for the observed manufacturing processes and changes over time: Pre-Clovis: Bipointed, Page-Ladson, & Miller or Lozenges- Simpson – Clovis – Suwannee – Bolen (Early Archaic). Florida desperately needs a multicomponent Paleoindian site(s) that can be directly and relatively understood to solidify the sequence of technological change at the end of the Pleistocene.

The known cultural components in Florida, from Pre-Clovis through the Paleoindian, are still in a critical data gathering phase. It is important to note that Clovis and older material has been found North, East, and even Southeast of our research area, effectively showing widespread use of the still dry landscape by people at the moment they become archaeologically visible. As we search closer and closer to the LGM shore, in both time and space, it is not unreasonable to expect discovery of currently unknown cultural components as well. Particularly as they are likely to be exhibit an ever more aquatic and marine resource base adaptation.

Project Significance and Expected Results

It is confidently anticipated that the location, identification, and documentation of submerged archaeological sites on any part of the inundated landscape older than 10,000 radiocarbon years will have great significance on a variety of levels. First, and most obviously, the discovery of prehistoric archaeological material at the 40 m isobath will definitively confirm that human populations were present upon and utilized this paleo-landscape at least as early as any terrestrial counterparts. While confirmation of the use of nearshore submerged coastal margins has been provided for very limited portions of the Pacific littoral zone and, as noted earlier, for portions of the study area, direct in situ evidence for the use of earlier and more distant submerged coastlines has been lacking.

Second, depending on the type(s) of data to be recovered from our target loci, it should be possible to address a number of very specific issues about the nature, timing, and trends of the ancient human occupancy and use of this portion of the Gulf of Mexico. Indeed, based on currently available sea level change data, it may well be that the mode(s) of human utilization of large portions of the now inundated Florida landscape were, for several compelling reasons, far different from their terrestrial counterparts and, in fact, far different than anyone has heretofore posited or even imagined.

In this regard, it should be stressed that current research (Balsillie and Donoghue 2004) conclusively indicates that the last several thousands years of relative sea level stability are not representative of the much greater volatility which characterized the period from the Last Glacial Maximum (LGM), ca. 22,000 years ago (18,000 14C yr BP), until modern shorelines are reached roughly 4,500 years ago. This, in turn, suggests to us that the fundamental relationship between people and the ocean would have been very different than that evidenced during the comparative sea level stability which has characterized the last four and a half millennia.

Put in different terms, it is a given that humans in coastal or nearshore environments must and do adjust to periodic rises (or falls!) in sea level. The consequences of these events are easy to identify and

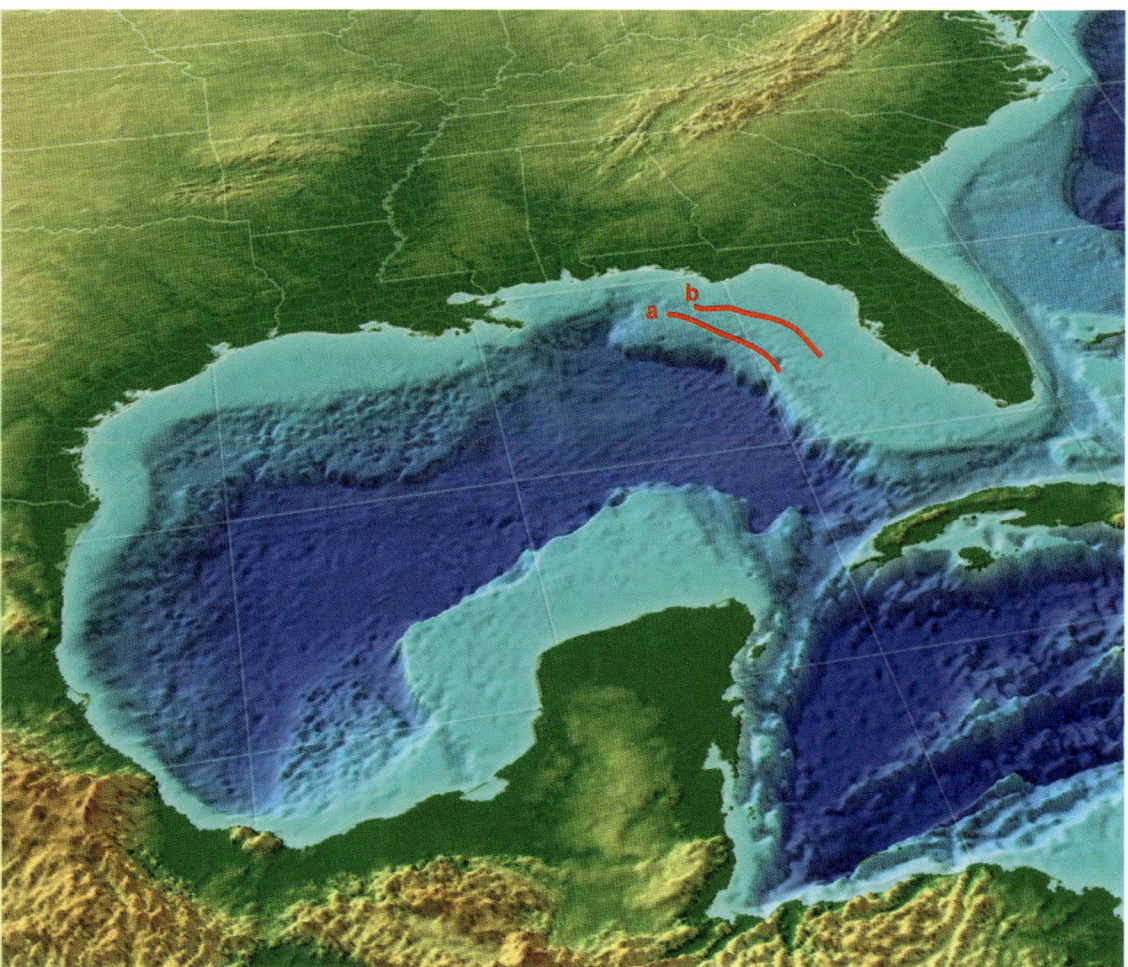

Figure 5. Approximate locations of inundated prehistoric shore lines in the study area. The western margin of line A represents the 18,000 14C yr BP LGM shore line submerged ca. 120 mbelow msl while the eastern margin of line A represents the 12,000 14C yr BP shoreline ca. 78–80m below msl. Line B represents the Clovis (~10,300 14C yr BP) shoreline submerged ca. 40 mbelow msl. Note the very rapid increase in the rate of land inundation over time.

include reductions (or increases) in the amount of potentially habitable portions of the subcoastal landscape, changes in the distribution of nearshore terrestrial resources (including fresh water), and concomitant alterations in the occurrences and density of exploitable marine biota. What has not been considered to date are the changing rates of sea level rise and, especially, the consequences of the escalating rapidity of these changes on the lifestyles of prehistoric coastal populations.

Before discussing some of these changes and their potential consequences, we conclude that any discussion of sea level rise or fall and its consequences is confounded by several variables. First, charts, maps, and many publications freely move between units rounded to the nearest minute, nautical mile, kilometer, terrestrial mile and/or fathom, foot, or meter. Further, the 2008–2009 data we collected routinely documents differences of up to 2 m from available charts. Additionally, tides, seasonality, wind, global warming, differential gravity (sensu GRCE Project), and a host of other factors introduce yet more imprecision. With all of these caveats in mind, we are confident that the figures cited below are internally consistent and reliable to within ± 5 m.

Constant citation and uncritical repetition of average sea level rise and fall in the extant literature often obscures the nuances of the process under consideration. By applying data on meltwater pulses and sea level stands to a contoured landscape, a much clearer picture emerges of this complicated series of events.

West of Tampa-St. Petersburg and the modern Florida coast, the LGM shore is 225 km away and now 120–130 m deep. If you average sea level rise over the 14,000 radiocarbon years from the LGM to the stabilization of the modern shoreline (18,000–4,000 14C yr BP), the average annual vertical increase is 8.6 mm. When considered in terms of calibrated calendar years (22,000–4500 BP), the sea level rise is only 6.9 mm per year. However, the vertical change is only part of the story (Figure 5).

Current data (Table 1) indicates that the LGM shoreline in project Subarea 3 (inside the Steamboat Lumps Marine Sanctuary) was initially inundated at ca. 18,000 14C yr BP and, further, that sea level rose 42 m vertically over the next 6,000 years. Because of the comparative steepness of the continental slope in this segment of the Florida coastline, the relatively high sea level rise translated to a coastal incursion of only 6 km. On average, only 1 km of land was lost every 1,000 years. In other words, for a very long time the results of sea level rise were slow, steady, and undramatic in terms of consequences for the distribution of terrestrial and offshore biota, water table fluctuation, availability of freshwater, erosional processes and, probably, human adjustments thereunto—if any humans were around. By the time incoming sea water is flooding the noticeably flatter plateau of the inundated continental shelf at ca. 12,000 14C yr BP, humans are present on the landscape of the New World at large, and in Florida specifically, while the seashore is still 220 km away.

Our initial examination of additional published data and charts for the Gulf of Mexico, western North America, and the unglaciated portions of the Atlantic continental shelf and slope indicates that the horizontal landscape covered in the 6,000 years after the LGM is not likely to exceed 20 km. In fact, for much of the west coast, and occasional Atlantic examples such as Miami, the entire inner continental shelf is less than 20 km wide.

As noted above, there are well-documented pre-Clovis occupations at Page Ladson and Sloth Hole in the immediate vicinity (Dunbar 2006; Hemmings 2004) and also at locations much farther afield at Monte Verde II in Chile (Dillehay 2000); Meadowcroft in Pennsylvania (Adovasio et al. 1975, 1977a, 1977b, 1978, 1989, 1990, 1992; Carlisle and Adovasio 1982); the Cinmar site off the coast of Virginia (Lowery 2009); Gault in Texas (Collins 2002); and the Paisley Caves in Oregon (Gilbert et al. 2008).

By 12,000–11,000 14C yr BP, sea level rise data is more abundant but somewhat contradictory, in part due to the inconsistency in calibrating "older" radiocarbon dates with more recent assays. During the Clovis interval (ca. 11,050–10,800 14C yr BP) using the short chronology of Waters and Stafford (2007), water level rose 75–50 m or 65–40 m below msl (depending on whose curve is employed [Balsillie and Donoghue 2004; Thulman 2009]). The much-cited Clovis shoreline of 40 m (Faught 2004:276) should be more accurately labeled the post-Clovis shoreline, which remains intact until the end of the Younger Dryas.

During the Clovis interval, the rate of vertical sea level rise is rapid, but not extraordinary when compared to earlier rates. As Table 2 indicates, however, the change in horizontal distance inundated by

Table 1. Depth, distance from shore, slope-to-shore, and approximate date of inundation for features in the greater study area.

Location	Date of Inundation		Water Depth (m below MSL)	Distance from Shore (km)	Slope to Shore (cm per km)
	14C yr BP	Calendar			
Steamboat Lump (Sub Area 3 LGM)	18,000	22,000	120-78[b]	220	700
Sub Area 3 (middle)	12,000	13,928	78-72	205	67
MWP-1A Pre-Clovis	12,400	14,500-11,500[a]	95	210[a]	45.2
Early Clovis Sub Area 2 A	11,000	13,000[a]	60	140	43
Sub Area 2 B	10,982	12,982[a]	54	110	49
Younger Dryas start	11,017	12,840	65-60	150[a]	43
Younger Dryas end	10,029	11,450	34	130[a]	26
FMG islands	9,360	10,514	35-30[c]	150[a]	—
Thor's Elbow	10,300	12,000[a]	40	148.8	27
MWP-1B	9,500	10,700[a]	39[a]	140[a]	28
Near Modern	4,000	—	2-1	<5	minimal

Notes: a, Estimation from Balsille and Donoghue (2004:1-55) raw data; b, Depth at 6 km; c. Maximum.

Table 2. Location of now-submerged features in the greater study area in relation to modern shore-line features.

Modern Shore Feature	Distance (km)	Direction	Slope-to-shore (cm per km)
Southeastern corner of Florida Middle Grounds (40 m below MSL)			
Aucilla River	213.9	Due North	18.7
Tarpon Springs	130	Due East	30.8
Thor's Elbow/Paleo-Suwannee Intersection (40 m below MSL)			
Aucilla River	167.4	20km E of N	24
Suwannee	151	NE 45 degrees	26.5
Tarpon Springs	148.8	N of E	27
Southwest Bay Peninsula, a Clovis-era Bay (60 m below MSL)			
Apalachicola	141	15 km W; 140km N	—

the incursion of the sea is staggering. Because the landscape is now much flatter, the average Clovis period sea level rise covers 330 m per year or roughly 1 km every three years! Unlike the period between the inundation of the LGM and the late pre-Clovis (ca. 18,000–12,000 14C yr BP, respectively), when attendant changes to sea level rise are minimal, the consequences of sea level rise in Clovis times are profound.

Emblematic of the scale of such changes is modern Tampa Bay and its surrounding peninsulas, bays, and islands. Tampa Bay averages less than 4 m of water depth and would not have existed as a bay until around 6000 14C yr BP (Balsillie and Donoghue 2004; Thulman 2009). By this time, sea level is within 6 m of modern and oscillates to near modern conditions with a high sea level stand of ca. 1 m within the last 3,000 years. Depending on the data used, sea level has been within 2 m of the modern coast for much of the last 4,500 years and all of the last 3,500 years (Balsillie and Donoghue 2004:19–21). It is during this post-6000 14C yr BP interval that Tampa Bay is largely in-filled, covering an area of 1,000 km2 (400 mi2) with an average depth of 4 m of saltwater. This, of course, has profound consequences for biotic communities and their human exploiters. Such as the Cypress tree swamp mentioned by Sellards earlier (Sellards 1916:126).

Space and temporal constraints prevent an in-depth elaboration of the consequences of individual events like the creation of the saltwater Tampa Bay, and similar now-inundated bodies further offshore, but that episode clearly reflects much broader processes operational over a very long period of time. In brief, incoming seawater drove biotic communities of plants, animals, and eventually people inland from the initial inundation of the LGM beaches until modern coastlines are formed in the last 4,500 years (cf. Balsillie and Donoghue 2004). This time frame subsumes the initial percolations of humans into the Americas sometime well before 13,000 14C yr BP, all of the Paleoindian period, and nearly two-thirds of the Archaic period. Indeed, much of our submerged study area would have been a habitable landscape to Native Americans until sometime in the Middle Archaic, prior to 6,000 years ago. This is true for much of the Atlantic continental shelf as well.

When discussing sea level rise using average rates over millennia, it is important to recall that at a human scale of individual lifetimes, years, or days, the story is much more nuanced and subtle with brief periods of coastal stability punctuated by rising and falling shorelines occurring in no predictable pattern. These oscillations probably occurred many times in a single person's life span. With that caveat, the long-term trend is clearly of regularly incoming water and near constant rearrangement of biotic communities being pushed inland with lag times varying with the speed of inundation—a movement which was acute during Clovis times. This, of course, produced a highly unusual situation for humans, especially in terms of seasonal scheduling, subsistence planning, and related activities. There are obviously a wide range of resources available to the aboriginal inhabitants, but they are never in the same place very long. This, in turn, necessitates careful consideration of the "costs" of movement as they relate to accessing and efficiently using these resources. The concept of this moving shoreline as a fluid "coastal oasis" was elaborated for this area by Thulman (2009). He observed that a great many of the first-order freshwater springs of modern Florida are within 10 km of the coast precisely because of the denser sea water pushing against the fresh aquifer water, which bubbles to the surface where it can. Thus surficial freshwater, plants, and animals were drawn to this watered "oasis" even while sea level rise pushed them ever landward.

The incursion of the sea meant that preferred site locales near the coast were continually lost from the LGM until the modern coastline stabilized. Furthermore, the change from a dynamic to a stable coast obviously profoundly altered the lifestyles of the populations inhabiting this portion of Florida. While we have a reasonably good picture of what more recent lifeways anchored in a stable coastline looked like, we have little real data regarding what kind of lifestyles were pursued during the longer period of coastal instability.

Operating at a finer resolution than gross sea level rise, there are also other processes effecting fresh water availability, coastal morphology, and biotic communities, particularly in sensitive shallow waters like the rapidly filled area in Subarea 2. Of particular interest within the overall survey area is the shoreline along the various Florida Middle Ground islands and mainland coast, which underwent rapid large scale changes for roughly 17,000 years and has been comparatively stable for most of the past 5,000 years.

Though recent work by Thulman (2009) focused on North Central Florida during Middle Paleoindian times, it nonetheless strongly supports long- and short-term processes in operation since the LGM throughout the state, specifically, including the northern half our research area. His analysis considers only the modern shoreline of Florida while noting that the landscape under consideration was largely xeric with some rare surficial water available. The surficial aquifer is very limited and the Floridian aquifer is only available along low-lying drainages through porous karst regions. At a regional scale, these variations are important but not as critical as when viewed within a single drainage such as Tampa Bay, where a change of even 1 m can cover or create islands and drown or desiccate the vegetation on the land.

Minor fluctuations in the fresh water table along the moving edge of the "Coastal Oasis" can be critically important to biotic communities and the humans that rely on them. Though the trends discussed here primarily focus on incoming seawater periods of lowered sea level, drought can destroy mangroves and overall wetland communities as well. A water table drop of only 30 cm desiccates wetlands to the point they will not rehydrate (Faure et al. 2002; Thulman 2009:251). Furthermore, a 75 cm drop is sufficient to "change a plant community from wetland to xeric" (see Thulman 2009:251–252). Again, because long-term oscillations include the rise and fall of water levels, it is important to keep in mind that a large shallow bay is particularly sensitive to even such seemingly minor changes in water availability.

As we continue to explore our primary site locations, it is worth noting that in addition to potentially illuminating the timing of, use, and lifestyles of the earliest dwellers of the eastern Gulf of Mexico coastline, the recovery of technologically diagnostic cultural material may help to elucidate the possible cultural affinities of these populations. Obviously, the recovery of Clovis-related material would dramatically enhance the known distribution of that cultural horizon, while the recovery of non-Clovis material would contribute to the understanding of prehistoric cultural diversity and the possible antecedents of Clovis in this part of southeastern North America.

Furthermore, from a purely paleoenvironmental perspective, it is anticipated that the continued research will recover direct evidence confirming the early subaerial exposure of this portion of the continental shelf. The two oldest pieces of direct evidence of an inundated terrestrial surface in the eastern Gulf of Mexico are an oak stump identified at the J&J Hunt site in the paleo-Aucilla river drainage, dated to 7,240 ± 100 14C yr BP (Basillie and Donoghue 2004:42), and a pine stump identified off Key West dating to ca. 8,400 14C yr BP (Lidz 2004). These stumps provide a minimum age for the inundation of these portions of the eastern Gulf of Mexico. Any similar finds at our sites will, at a minimum, be older than 10,000 14C yr BP and thus highly informative of sea-level rise rates.

Finally, our completed and proposed work fills a void in the NOAA Office of Natural Marine Sanctuaries Plan to create marine protection areas. Cultural resources—particularly shipwrecks and other maritime heritage sites, are mentioned as important—but because of a severe lack of hard data, the inundated landforms are not described or discussed in detail. This is precisely the kind of information our offshore work is in the process of gathering and analyzing. In this regard, we stress that the direct observation, instrumental, and photographic documentation and assessment of the LGM and later shorelines are of great significance of and by themselves. These inundated geological features, including the intersecting paleochannels, dune ridges, sand ripples, and possible barrier islands as well as the juxtaposed infilled karst features have not previously been the subject of systematic scrutiny. Their continued study affords the potential for not only better understanding the dynamics of late Pleistocene and early post-Pleistocene sea level rise, but also provides a clearer perspective on subaqueous sedimentation and erosion-related issues.

For all of the reasons stated here, the arena of Paleoindian studies in North America is currently in the most rapid state of flux since the seminal discoveries at Folsom and Clovis in the 1920s and 1930s. As the old Clovis-first paradigm has eroded, all of its fundamental postulates have been questioned, extensively modified, or discarded outright. In its place, a wide variety of new models have been proposed, none of which has been fully articulated, let alone broadly accepted.

The explorations detailed here and planned for 2013 promise to add significant new—and, frankly, exciting—data to the theoretical ferment which is currently brewing in the study of early humans in North America. If successful, it will at the very least place human beings on a deeply submerged shoreline of the New World when, at least for our species, that world was at its newest. At best, it stands to elucidate the timing of the arrival, lifestyles, and movements of humans into and through that world.

Acknowledgments

The research reported in this paper was funded by the National Oceanic and Atmospheric Administration, Mercyhurst University, The Gault School of Archaeological Research, and the Texas Archeological Research Laboratory of the University of Texas at Austin.

References Cited

Adovasio, J. M., J. Donahue, and R. Stuckenrath
1990   The Meadowcroft Rockshelter Radiocarbon Chronology 1975–1990. *American Antiquity* 55:348–354.
1992   Never Say Never Again: Some Thoughts on Could Haves and Might Have Beens. *American Antiquity* 57:327–331.

Adovasio, J. M., J. Donahue, R. Stuckenrath, and R. C. Carlisle
1989   The Meadowcroft Radiocarbon Chronology 1975–1989: Some Ruminations. Paper presented at the First World Summit Conference on the Peopling of the Americas, University of Maine, Orono.

Adovasio, J. M., J. D. Gunn, J. Donahue, and R. Stuckenrath
1975   Excavations at Meadowcroft Rockshelter. 1973-1974: A Progress Report. *Pennsylvania Archaeologist* 45(3):1–30.
1977a   Meadowcroft Rockshelter: Retrospect 1976. *Pennsylvania Archaeologist* 47(2–3):1–93.
1977b   Progress Report on the Meadowcroft Rockshelter-A 16,000 Year Chronicle. In *Amerinds and Their Paleoenvironments in Northeastern North America*, edited by W. S. Newman and B. Salwen. pp. 37-159. Annals of the New York Academy of Sciences 228.
1978   Meadowcroft Rockshelter 1977: An Overview. *American Antiquity* 43(4):632–651.

Adovasio, J. M., and D. R. Pedler
1996   Pioneer Populations in the New World: The View from Meadowcroft Rockshelter. Paper presented at the XIII International Congress of Prehistoric and Protohistoric Sciences, Forlì, Italy.
2000   A Long View of Deep Time at Meadowcroft Rockshelter. Paper presented at the 65th Annual Meeting of the Society for American Archaeology, Philadelphia, Pennsylvania.
2003   Pre-Clovis Sites and their Implications for Human Occupation Before the Last Glacial Maximum. In *Entering America: Northeast Asia and Beringia before the Last Glacial Maximum*, edited by D. B. Madsen, pp. 139–158. University of Utah Press, Salt Lake City.

Balsillie, J. H., and J. F. Donoghue
2004   *High Resolution Sea-level History for the Gulf of Mexico Since the Last Glacial Maximum.* Report of Investigations No. 103. Florida Geological Survey, Tallahassee.

Bullen, Ripley P.
1958   *The Bolen Bluff Site on Payne's Prairie, Florida.* Contributions to the Florida State Museum in Social Sciences No. 4. Gainesville.

Carlisle, R. C., and J. M. Adovasio (editors)
1982   Meadowcroft: Collected Papers on the Archaeology of Meadowcroft Rockshelter and the Cross Creek Drainage. Prepared for the Symposium "The Meadowcroft Rockshelter Rolling Thunder Review: Last Act". Forty-Seventh Annual Meeting of the Society for American Archaeology, Minneapolis, Minnesota April 14–17, 1982.

Collins, M. B.
2002   The Gault Site, Texas, and Clovis Research. *Athena Review* 3(2).

Daniel I. R. and M. Wisenbaker
1987   *Harney Flats: A Florida Paleo-Indian Site*. Baywood. Farmingdale, NY

Dillehay, T. D.
1989   *The Paleoenvironmental Context. Monte Verde: A Late Pleistocene Settlement in Chile, vol. 1.* Smithsonian Institution Press, Washington, D.C.
1997   *The Archaeological Context and Interpretation. Monte Verde: A Late Pleistocene Settlement in Chile, vol. 2.* Smithsonian Institution Press, Washington, D.C. 2000 The Settlement of the Americas : A New Prehistory. Thomas D. Dillehay. Basic Books, New York.

Dunbar, J. S.
1991   The Resource Orientation of Clovis and Suwannee Age Paleoindian Sites in Florida. In C*lovis: Origins and Adaptations*, edited by R. Bonnichsen and C. Turnmire, pp. 185–213. Center for the
Study   of the First Americans, Corvallis, Oregon.
2006   Paleoindian Land Use. In *First Floridians and Last Mastodons: The Page-Ladson Site in the Aucilla River*, edited by D. S. Webb, pp. 525–544. Springer Publishing Company, New York.

Dunbar, J. S., and C. A. Hemmings
2004   Florida Paleoindian Points and Knives. In New Directions on the First Americans, ed. B. T. Lepper and R. Bonnichsen, 65-72. College Station, Texas A & M Press.

Faught, M. K.
2004   The Underwater Archaeology of Paleolandscapes, Apalachee Bay, Florida. *American Antiquity* 69(2):235–249.

Faure, H., R.C. Walter, and D. R. Grant
2002   The Coastal Oasis: Ice Age Springs on Emerged Continental Shelves. *Global And Planetary Change* 33:47–56.

Gilbert, M. T. P., D. L. Jenkins, A. Gotherstrom, N. Naveran, J. J. Sanches, M. Hofreiter, P. F. Thomsen, J. Binladen, T. F. G. Higham, R. M. Yohe II, R. Parr, L. S. Cummings, E. Willerslev
2008   DNA From Pre-Clovis Human Coprolites in Oregon, North America. *Science* 320(5877):786–789.

Hemmings, C. A.
2004   The Organic Clovis: A Single, Continent-wide Cultural Adaptation. PhD dissertation, Department of Anthropology, University of Florida.
2005   An Update of Recent Work at Sloth Hole (8JE121), Jefferson County, Florida. *Current Research in
the   Pleistocene* 22:47–49.

Lidz, B. H.
2004   Coral Reef Complexes at an Atypical Windward Platform Margin: Late Quaternary, Southeast Flordia.   *GSA Bulletin* 116(7–8):974–988.

McAvoy, J. M., J. C. Baker, J. K. Feathers, R. L. Hodges, L. J. McWeeney, and T. R. Whyte
2000 *Summary of Research at the Cactus Hill Archaeological Site, 44SX202, Sussex County, Virginia.* Report to the National Geographic Society in Compliance with Stipulations of Grant #6345-98. Nottaway River Survey, Sandston, Virginia.

McAvoy, J. M., and L. D. McAvoy
1997 *Archaeological Investigations of Site 44SX202, Cactus Hill, Sussex County Virginia.* Research Report Series No. 8. Commonwealth of Virginia Department of Historic Resources, Richmond.

Sellards, Elias H.
1916 *Human Remains and Associated Fossils from the Pleistocene of Florida.* 8th Annual Report of the Florida Geological Survey, Tallahassee, FL.

Thulman, D. K.
2009 Freshwater Availability as the Constraining Factor in the Middle Paleoindian Occupation of North-Central Florida. *Geoarchaeology* 24(3):243–276.

Waters, Michael and Thomas Stafford
2007 Redefining the Age of Clovis: Implications for the Peopling of the Americas. *Science* 315(5815):1122–1126.

Wright, E. E., A. C. Hine, S. L. Goodbred, Jr., and S. D. Locker
2005 Effect of Sea-Level and Climate Change on the Development of a mixed Siliciclastic-Carbonate Deltaic Coastline: Suwannee River, Florida, U.S.A. *Journal of Sedimentary Research* 75:621–635.

Loess, Landscape Evolution, and Pre-Clovis on the Delmarva Peninsula

John S. Wah, Ph.D.
Matapeake Soil & Environmental Consultants, P.O. Box 186, Shippensburg, PA 17257
matapeake.soil@gmail.com

Darrin L. Lowery, Ph.D.
Chesapeake Watershed Archaeological Research Foundation, 8949 High Banks Drive, Easton, MD 21601
darrinlowery@yahoo.com

Daniel P. Wagner, Ph.D.
Geo-Sci Consultants, 4410 Van Buren Street, University Park, 20782
danwagner@juno.com

Introduction

Pre-Clovis assemblages lack tools with distinct morphological features that make them easily recognizable. Unlike fluted Clovis projectile points, stunted Pre-Clovis lanceolate bifaces and blade tools can be misidentified and attributed to later time periods (eg. Middle Woodland Jacks Reef or Pee Dee pentagonals or Fox Creek lanceolates). In mixed, multi-component sites or in upland settings where the same living surface has been available for occupation to everyone from the earliest inhabitants of North America to the present day, this problem is especially acute. Stratified settings provide the best potential for recognition of Pre-Clovis artifacts and sites. Not surprisingly, the more widely recognized sites with Pre-Clovis components in the eastern United States, including Meadowcroft Rockshelter (Adavasio, et al., 1975; Stuckenrath, et al., 1982), Cactus Hill (McAvoy et al., 2003; Wagner and McAvoy, 2004) and Miles Point (Lowery, et al., 2010), to name a few, are all stratified.

The Delmarva Peninsula in Maryland, particularly the western Delmarva along the Chesapeake Bay, offers an ideal setting for the preservation, and excellent opportunity for recognition and recovery, of Pre-Clovis materials and sites. Two relatively thin loess deposits from the Late Pleistocene have been identified along with a buried and preserved Late Pleistocene soil and landscape (Figure 1). The surface horizons of the paleosols buried by the more recent loess have been widely dated and shed considerable light upon Late Pleistocene climatic and environmental conditions and changes. Landscape evolution on the Delmarva, however, is fairly complex with regional consistencies but local variability in deposition, erosion, and landscape stability.

Miles Point Loess

The older loess deposit, the Miles Point Loess, covers an area of approximately 540 km^2 from Queen Annes County to northern Dorchester County, Maryland. The original extent of this loess is unclear; it is most notable on flat landscapes on the mid-Delmarva and largely absent on the more rolling topography of the northern peninsula, however, it has been tentatively identified as far north as Cecil County. Thickness of the Miles Point Loess deposit at sites sampled ranges from 63 to 118 cm with more than 56 percent silt (2-50 μm) sized particles. The Miles Point Loess is buried by the younger Paw Paw Loess. Figures 2 and 3 show the Delmarva Peninsula with sites discussed in the text and the extent of Miles Point and Paw Paw loesses, respectively.

A soil with strongly expressed morphologic features, the Tilghman Paleosol, has developed in the Miles Point Loess with a fragipan and extremely coarse prismatic structure. The surface horizon of the Tilghman Paleosol is a readily apparent stratigraphic marker at the contact of the earlier Miles Point and more recent Paw Paw loess deposits. Mineralogy of the fine silt sized (2-

Figure 1. Bank exposure showing the Paw Paw Loess over the Miles Point Loess. The surface horizon of the buried Tilghman Island Paleosol formed in the Miles Point Loess is clearly visible at the contact of the two loess deposits.

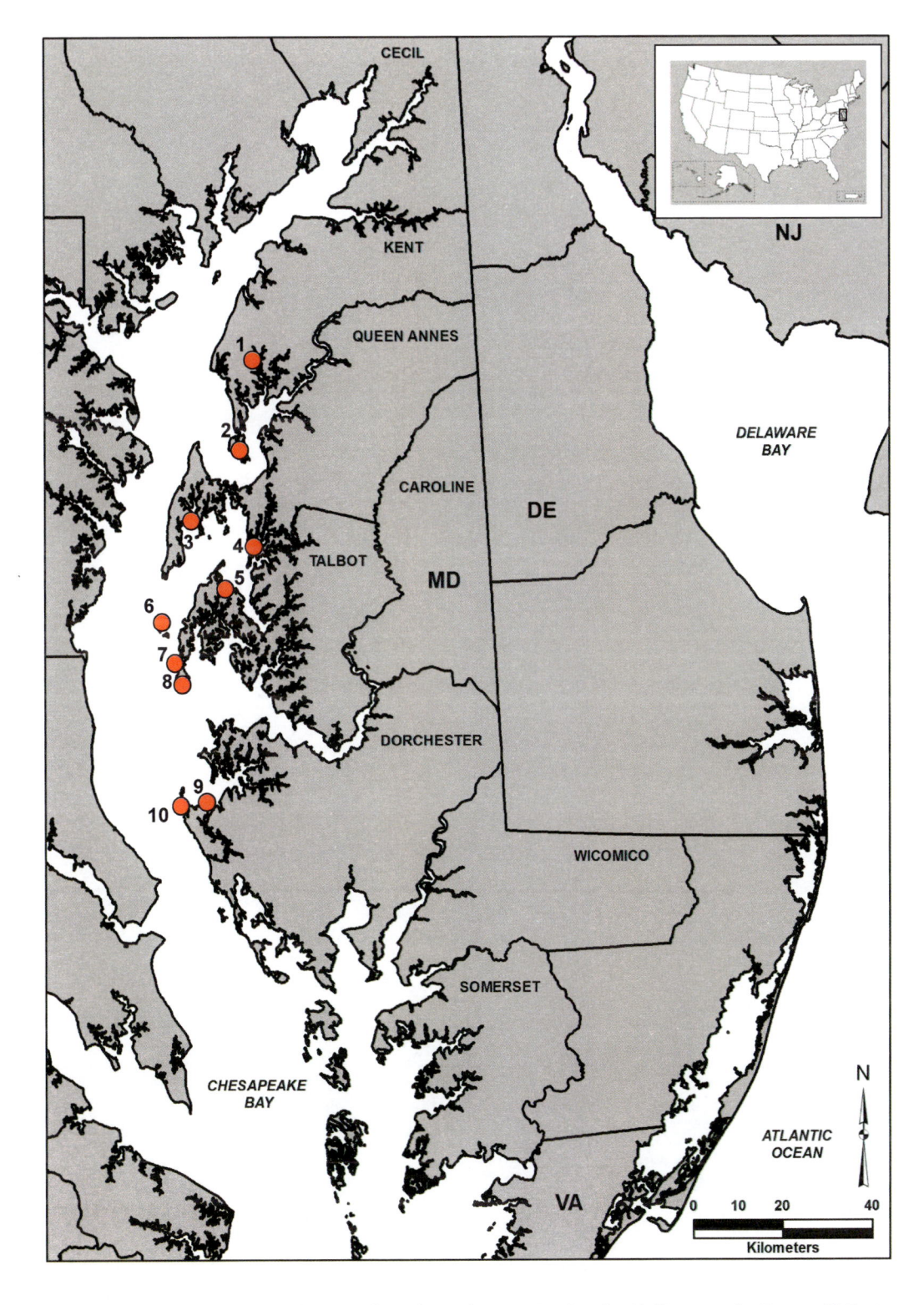

Figure 2. Location of the sites discussed on the Delmarva Peninsula: 1) Chesapeake Farms, 2) Eastern Neck Island, 3) Barnstable Farm, 4) Wye Island, 5) Miles Point, 6) Jefferson Island, 7) Paw Paw Cove, 8) Black Walnut Point, 9) Cators Cove, and 10) Oyster Cove.

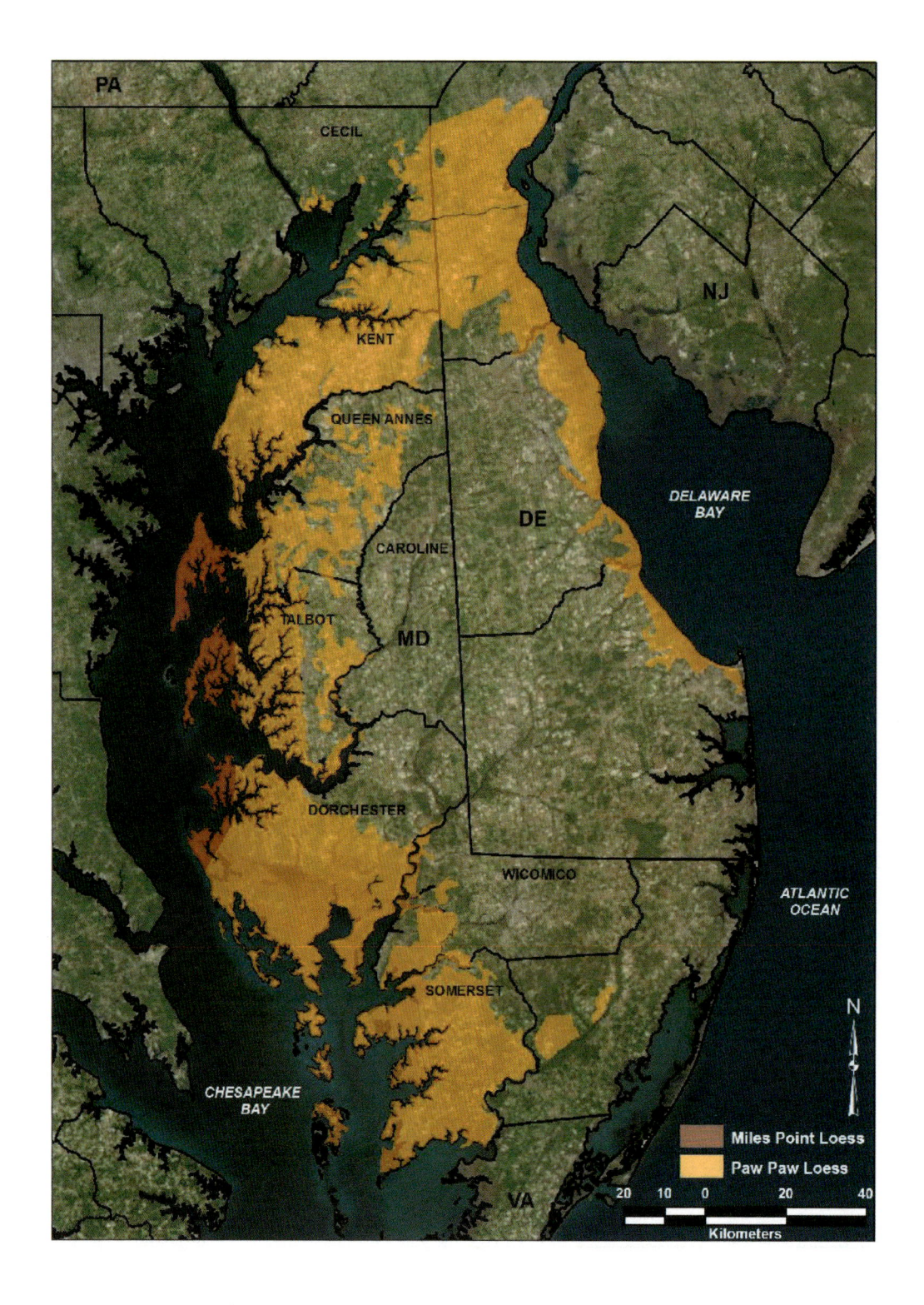

Figure 3. Distribution of Miles Point Loess and overlying Paw Paw Loess with the soils formed in eolian silts compiled from the USDA-NRCS STATSGO dataset (Soil Survey Staff, 2013).

20 μm) fraction is quartz dominated with trace amounts of muscovite, kaolinite, and feldspar. Clay mineralogy (<2 μm) is vermiculite and kaolinite, with lesser amounts of muscovite, chlorite, hydroxyl-interlayered vermiculite, and clay sized quartz and feldspar particles. There are no smectites that might shrink and swell and contribute to artifact movement down the soil profile (Wah, 2003). Figure 4 shows profiles of soils formed in the Miles Point and Paw Paw loesses.

Paw Paw Loess

The more recent loess, the Paw Paw Loess, is much more widespread than the Miles Point Loess, covering an area of approximately 5,000 km^2 from Cecil County in the north to the Virginia eastern shore in the south. This deposit is likely to have formed a fairly continuous mantle at one time but has largely been eroded from sloping landscapes (Foss et al., 1978). Paw Paw Loess is present at elevations that range from sea level to at least 90 feet above mean sea level and buries paleosols formed in sandy Coastal Plain sediments as well as the Miles Point Loess. The thickness of the deposit is generally less than 1.8 meters with the deposit thinning with distance from ancestral channel of the Susquehanna River, the presumed source of sediments, and from the north to south. Notable exceptions to the thinning trends are to the east of confluences of larger tributaries and the ancestral Susquehanna and east of meander bends in the larger tributaries. Silt contents generally range between 60 and 80 percent, however, there is some variability with one site sampled having as little as 44 percent silt sized particles but 23 percent very fine sands and 11 percent fine sands. As with deviations in deposit thickness, sites with relatively low silt contents and increased very fine and fine sand contents tend to be east of confluences and meander bends.

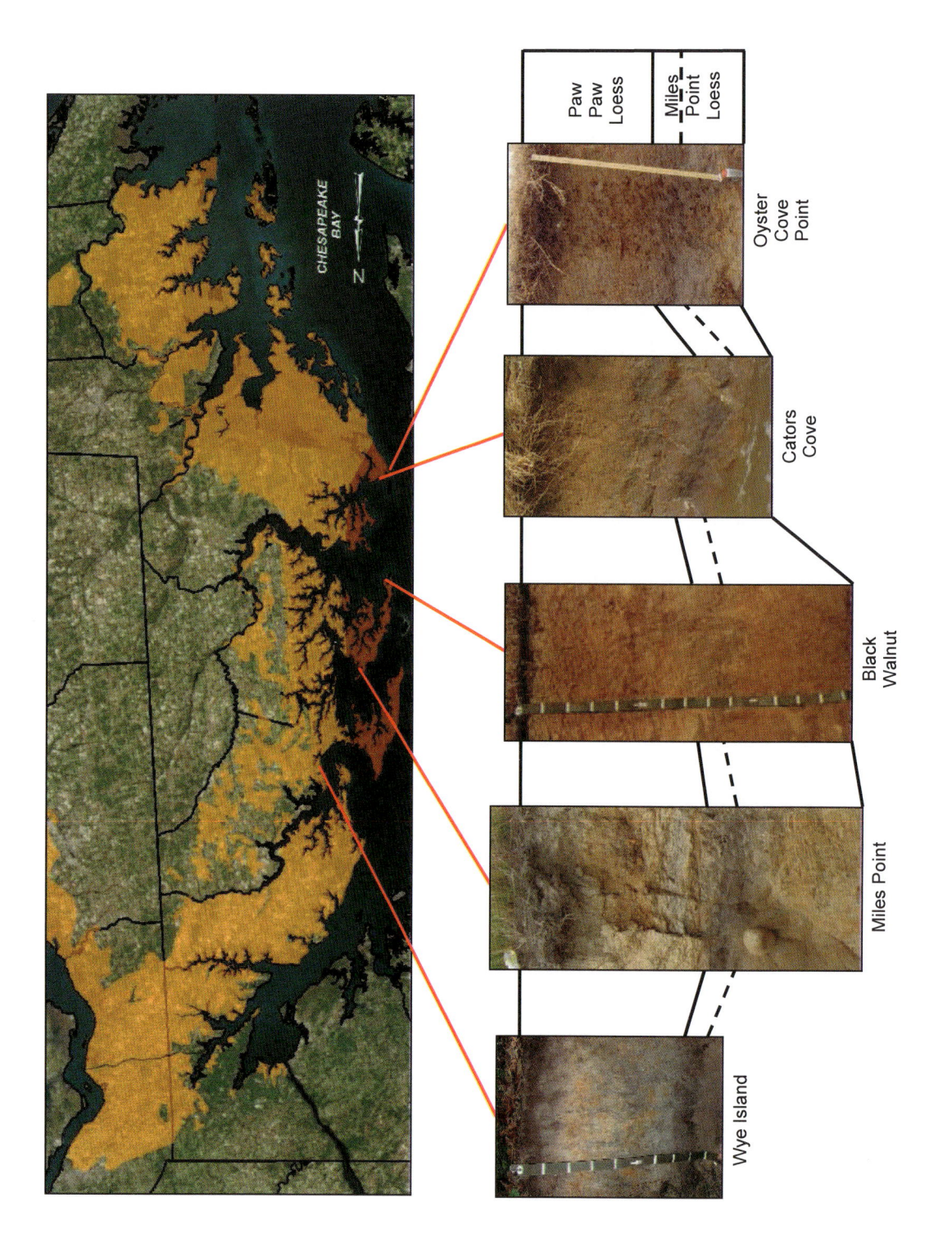

Figure 4. Soils developed in the Paw Paw Loess burying paleosols. At Wye Island the paleosol is formed in sandy Coastal Plain sediments, while at the remaining sites the paleosol is formed in the Miles Point Loess. The buried surface horizon is clear at the contact of the paleosol and overlying loess.

Soil development in the Paw Paw Loess is less strongly expressed than in the Miles Point Loess. Argillic horizons are present and at some locations, fragipans are described. Mineralogy of soils formed in the Paw Paw Loess is similar to that of the Miles Point Loess paleosol with predominately quartz with small quantities of muscovite, kaolinite, and feldspar in the fine silt (2-20 μm) fraction and kaolinite and vermiculite the more abundant minerals with muscovite, chlorite, hydroxyl-interlayered vermiculite, quartz, and feldspar in the clay (< 2 μm) size fraction (Wah, 2003).

Landscape evolution, environment, and potential for early occupation

^{14}C and OSL dates and cultural materials define the timing of events on the western Delmarva Peninsula. At Miles Point, OSL dates of 40,895 +/-2,370 BP at the base of the Miles Point Loess, above the contact with underlying sandy sediments, and 27,940 +/-1635 BP in the buried surface horizon of the soil formed in the Miles Point Loess place the deposition of sediments in Late Wisconsin, MIS3. ^{14}C dates from the surface horizon of the paleosol at Miles Point range from 21,940+/-140 BP to 27,240+/-230 BP, from charcoal, with a bulk soil ^{14}C date of 25,670+/-160 BP. Elsewhere on the Delmarva, the surface horizon of the Tilghman Paleosol has produced ^{14}C dates of 22,050+/-100 BP at Cators Cove, 25,800+/-120 at Oyster Cove Point, 20,850+/-90 BP at Black Walnut Point, and 19,433+/-90 at Barnstable Farm. The surface horizon of paleosols formed in sandy sediments and buried by the Paw Paw Loess has produced dates of 17,070+/-180 BP from charcoal at Wye Island and 24,280+/-150 at Chesapeake Farms on the northern Delmarva. Diagnostic Clovis-age projectile points recovered from the surface of the buried paleosol at Paw Paw Cove (Lowery, 2002; Lowery et al., 2010) date the onset of younger, overlying Paw Paw Loess deposition to approximately 13,000 years BP, coinciding with the onset of the Younger Dryas. Diagnostic Early Archaic projectile points at the base of the plowed surface horizon of the modern soil formed in the Paw Paw Loess suggest the termination of loess deposition at the end of the Pleistocene/beginning of the Holocene (Lowery et al., 2010).

Soils and sediments on the Delmarva show evidence of both long term stability across the region and episodes of erosion and deposition. The well developed paleosol formed in the Late Wisconsin Miles Point Loess is a function of extended landscape stability and pedogenesis through the Last Glacial Maximum. Sandy paleosols with intact surface horizons buried by the Younger Dryas Paw Paw Loess and a buried, upland freshwater wetland with ^{14}C dates of 20,020+/-80 BP and 20,400+/-90 BP on the southern Delmarva, also point to LGM stability. Conversely, the relatively limited distribution of the Miles Point Loess might be a result of severe erosional episodes prior to Paw Paw Loess deposition. Truncated sandy paleosol equivalents of the Tilghman Paleosol, from which the upper portions of the soil profile have been removed but argillic horizons preserved, buried by the Paw Paw Loess are certainly evidence of erosion prior to Younger Dryas loess deposition (Figure 5).

The Paw Paw Loess and the modern soil formed in it, like the Miles Point Loess and Tilghman Paleosol, demonstrate both the landscape stability of the region and erosional episodes. The well developed soil is a function of prolonged stability and pedogenesis through the Holocene. The extent and distribution of the loess deposit, its presence at all elevations but absence on sloping landscapes, point to significant reworking of sediments. Colman et al. (1999) in a core of Chesapeake Bay sediments taken off of southern Kent Island, note more than 17 meters of sediment above a bivalve with a ^{14}C date of 7,100+/- 45 BP, some sediments perhaps transported from the north by the Susquehanna River but the rest likely eroded from the Delmarva upland settings.

Figure 5. Soil profile and micrograph from Eastern Neck Island showing the truncated sandy paleosol buried by the Younger Dryas Paw Paw Loess. Note the absence of an organic rich surface horizon atop the paleosol. The micrograph shows the contact of the overlying loess and the sands with a distinct increase in particle size in the sands and also illuviated clay in the argillic horizon of the paleosol (courtesy of Philip Zurheide).

The modern landscape and elevation above mean sea level veil, for some, the origin of loesses and the landscape at the time at which the silts were deposited. During the Late Wisconsin, MIS3, the settings at which the Miles Point Loess was deposited - Miles Point, Paw Paw Cove, Black Walnut Point, Oyster Cove Point, Cators Cove, etc. - were upland landscapes with sea level in the Mid-Atlantic some 30 meters lower than present (Wright et al., 2009) and the sites around seven kilometers from the Susquehanna paleochannel. Similarly, around the time of the Paw Paw Loess deposition, sea level was between 55 meters (Odale, 1991) and 36 meters lower than present (Horton, 2009).

Foss et al. (1978) attributed the silts to the ancestral channel of the Susquehanna River draining and transporting sediments from glaciated areas of New York and northern Pennsylvania. While this is undoubtedly correct at least in part, it is also evident that there were more local sources providing sediment for wind transportation, particularly for the more recent Paw Paw Loess. For the Paw Paw Loess, it is not clear how much of the source sediment was derived directly from the Susquehanna paleochannel and how much of it was silts eroded from upland landscapes - silts that were potentially reworked Miles Point Loess initially derived from glaciated New York and Pennsylvania. Sites with abnormally thick loess deposits and/or high very fine and fine sand contents east of confluences and meanders demonstrate that at least some of the parent sediment to be redeposited by wind was from local sources.

The environment at the time of loess deposition was less hospitable than present. For loess deposition to be possible, sediments had to be available without the protection of anchoring vegetation. At the initiation of Miles Point Loess deposition, around 40,000 BP, temperatures at the GISP2 ice core were some 17 degrees Celsius colder than present. At the beginning of the

Younger Dryas and Paw Paw Loess deposition, the temperature at GISP2 was 18 degrees Celsius colder than present (Alley, 2004). Prevailing winds were from the west transporting the bulk of the available silt to the Delmarva, however, Late Pleistocene loess is also common on the western shore of the Chesapeake Bay and on the Delmarva west of the Delaware Bay. Carbonized organics recovered from the surface horizon of the buried paleosol at Miles Point are from krummholz yellow birch (Betula alleghaniensis), red spruce (Picea rubens), and balsam fir (Abies balsamea) (Lowery et al., 2010). Preliminary phytolith analysis from the Miles Point and Paw Paw loess at three sites (Paw Paw Cove, Black Walnut Point, and Oyster Cove Point) show a relative decrease in phytoliths from C3 grasses and relative increase in phytoliths from C4 grasses from the paleosol to the modern soil suggesting a shift to a warmer climate with more available soil moisture.

The potential for stratified archaeological sites with cultural materials from the early inhabitants of North America is particularly high on the western Delmarva Peninsula along the Chesapeake Bay. The timing of loess deposition and the thickness of the loess deposits provide vertical separation of materials of interest while the relatively thin overlying Paw Paw Loess allows reasonable hope for testing and recovery. At Miles Point, a Pre-Clovis artifact assemblage has been recovered from below the Miles Point Loess paleosol surface horizon. Similarly, a squat lanceolate projectile point and quartz cobble were found in the buried surface of the Tilghman Paleosol at Oyster Cove. At Paw Paw Cove and Jefferson Island, Clovis-age assemblages have been recovered from atop the paleosol surface buried by 62 centimeters and 75 cm of Younger Dryas loess, respectively (Figure 6). At Crane Point, diagnostic Early Archaic materials have been recovered from the base of the plowed surface of the modern soil formed in the Paw Paw Loess (Lowery, et al., 2010). Moving east from the Chesapeake Bay across the Delmarva, the Paw Paw loess thins until it becomes incorporated entirely in the modern soil surface or the deposit pinches out entirely. At that point on the landscape, the same living surface will have been available to people from all time periods. In Caroline County on the central Delmarva Peninsula, Brown (1979) has recovered three fluted projectile points from a plowed surface horizon context.

Summary

The record of early Native Americans on the Delmarva Peninsula is a very rich one. Not only have hundreds of Clovis-age fluted projectile points been recovered from a spectrum of former environmental settings across the peninsula, but the body of evidence for even earlier settlers is ever growing. The abundant evidence of early Americans here is a function of both a desirably complex suite of landscape econiches late in the Pleistocene as well as geologic and pedologic processes that have favored preservation of archaeological remains.

Respectively bounded on its west and east by the ancestral Susquehanna and Delaware Rivers, and prior to marine transgression rising some 50 m or more above them, the gently undulating plain that was to become the Delmarva Peninsula was also laced by numerous other lesser drainages. Between these local valleys broad interfluves contained pockets of boreal forest interspersed with swamps and meadows that supported a wide variety of desirable resources. And yet this seeming Eden of the Late Pleistocene collapsed. The burial of Pre-Clovis artifacts at Miles Point is a hint of at least a brief interval of instability perhaps near the LGM. The subsequent prolonged period of stability evinced by the extensive and consistently dated Tilghman Paleosol came to an abrupt end with the sudden onset of the Younger Dryas.

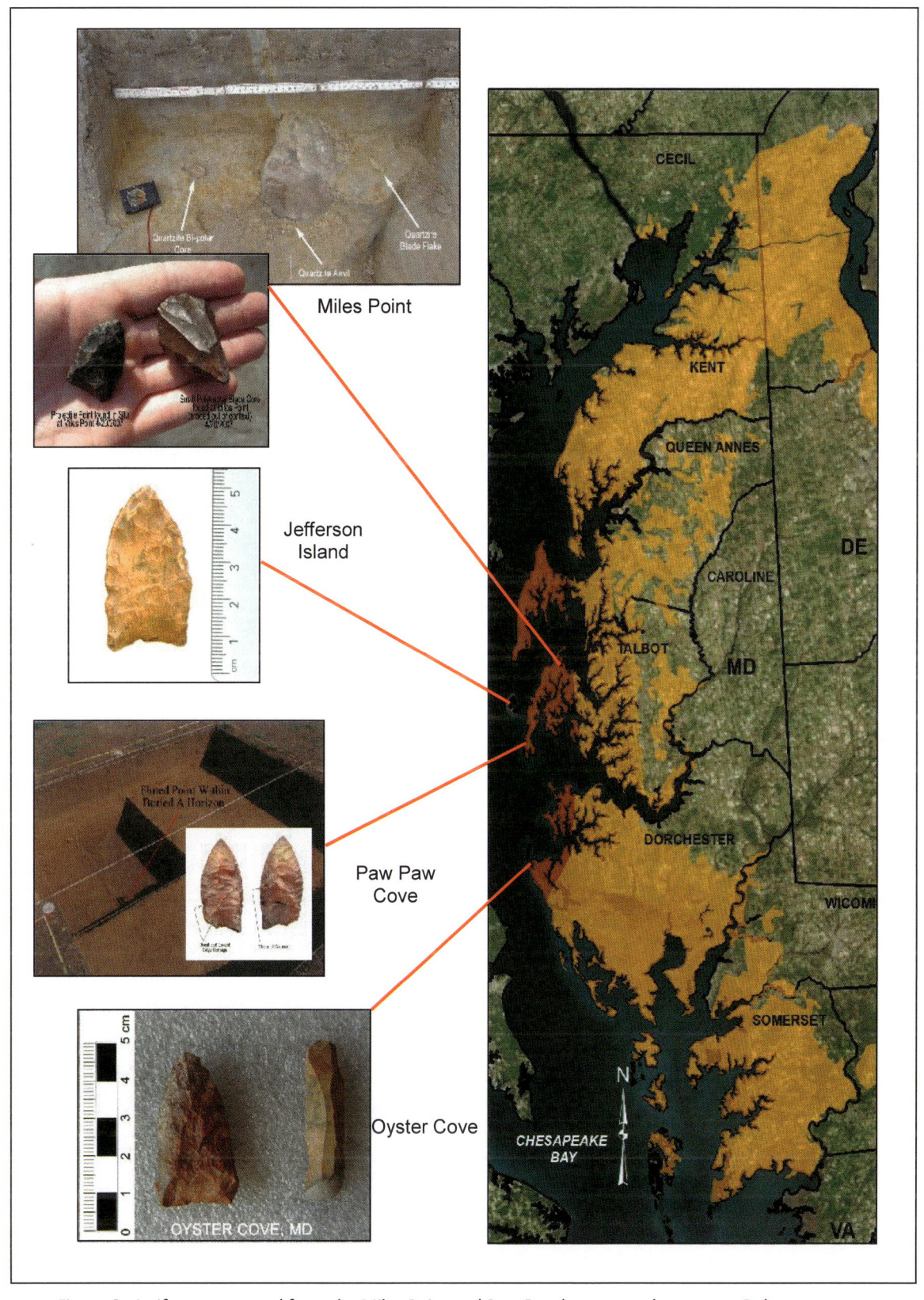

Figure 6. Artifacts recovered from the Miles Point and Paw Paw loesses on the western Delmarva Peninsula.

The rapid drop in temperature with subsequent loss of vegetation and loess deposition that marked the Younger Dryas was an environmental catastrophe to populations living on the Delmarva at the time. It has, however, become a boon for present day researchers. The widespread Paw Paw Loess provides an effective seal for a vast former landscape of the Late Pleistocene, and where present has preserved all Paleoamerican deposits beneath its well developed and dense subsoil. Nevertheless, from a geographic perspective it should also be recognized that although some of an ancient landscape still lies preserved, much has also been destroyed through the many kilometers of land retreat forced by sea level rise and marine transgression. Indeed, the remnant Paleoamerican materials still existing today are likely paltry in comparison to those once present on former upland landscapes and in river valleys long since inundated by the sea and estuaries.

References

Adovasio, J.M., J.D. Donahue, J. Gunn, and R. Stuckenrath. 1977. Progress Report on Meadowcroft Rockshelter - A 16,000 Year Chronicle. p. 137-159. *In* W.S. Newman and B. Salwen (eds.) Amerinds and Their Paleoenvironments in Northeastern North America. Annals of the New York Academy of Sciences 228.

Alley, R.B. 2004. GISP2 Ice Core Temperature and Accumulation Data. IGBP PAGES/World Data Center for Paleoclimatology Data Contribution Series #2004-013. NOAA/NGDC Paleoclimatology Program, Boulder CO, USA.

Brown, L. 1979. Fluted Projectile Points in Maryland. Manuscript on File. Maryland Historical Trust, Crownsville, MD.

Colman, S.M., J.F. Bratton, and P.C. Baucom. 2000. Radiocarbon Dating of Marion-Dufresne Cores MD99-2204, -2207, and -2209, Chesapeake Bay. *In* T.M. Cronin (ed.) Initial Report on the IMAGES V Cruise of the Marion-Dufresne to the Chesapeake Bay, June 20-22, 1999. U.S. Geological Survey, Open-File Report00-306 Online Version 1.0.

Foss, J.E., D.S. Fanning, F.P. Miller, and D.P. Wagner. 1978. Loess deposits of the Eastern Shore of Maryland. Soil Science Society of America Journal 42:329-334.

Horton, B. P., W. R. Peltier, S. J. Culver, R. Drummond, S. E. Engelhart, A. C. Kemp, D. Mallinson, et al. 2009. Holocene sea-level changes along the North Carolina coastline and their implications for glacial isostatic adjustment models. *Quaternary Science Reviews* 28: 1725–1736.

Lowery, D.L., M.A. O'Neal, J.S. Wah, D.P. Wagner, and D.J. Stanford. 2010. Late Pleistocene upland stratigraphy of the western Delmarva Peninsula, USA. Quaternary Science Reviews 29:1472-1480.

Lowery, D.L., 2002. A Time of Dust: Archaeological and Geomorphological Investigations at the Paw Paw Cove Paleo-Indian Site Complex in Talbot County, Maryland. Maryland Historical Trust, Crownsville, Maryland, 224 pp.

McAvoy, J.M., J.K. Feathers, R.I. Macphail, and D.P. Wagner. 2003. Verification of the Age and Integrity of the pre-Clovis occupation level at the Cactus Hill archaeological site, Status Report for Grant 7069-01. National Geographic Society, Washington, D.C.

Oldale, R.N., S. M. Colman, and G. A. Jones. 1991. Radiocarbon ages from two submerged strandline features in the Western Gulf of Maine and a sea level curve for the Northeastern Massachusetts coastal region. *Quaternary Research* 40:38–45.

Soil Survey Staff, Natural Resources Conservation Service, United States Department of Agriculture. U.S. General Soil Map (STATSGO2)[Maryland and Delaware]. Available online at http://soildatamart.nrcs.usda.gov. Accessed [4/30/2013].

Stuckenrath, R.J., J.M. Adovasio, J. Donahue, and R.C. Carlisle. 1982. The Stratigraphy, Cultural Features and Chronology at Meadowcroft Rockshelter, Washington County, Southwest, Pennsylvania. p. 69-90. *In* R.C. Carlisle and J.M. Adovasio (eds.) Meadowcroft: Collected Papers on the Archaeology of Meadowcroft Rockshelter and the Cross Cree Drainage. University of Pittsburgh Press, Pittsburgh.

Wagner, D.P. and J.M. McAvoy. 2004. Pedoarchaeology of Cactus Hill, a Sandy Paleoindian Site in Southeastern Virginia, U.S.A. Geoarchaeology 19:297-322.

Wah, J.S. 2003. The Origin and Pedogenic History of Quaternary Silts on the Delmarva Peninsula in Maryland. Ph.D. Dissertation. University of Maryland, College Park, MD.

Wright, J.D., R.E. Sheridan, K.G. Miller, J. Uptegrove, B.S. Cramer, and J.V. Browning. 2009. Late Pleistocene sea level on the New Jersey Margin: implications to eustasy and deep-sea temperature. Global and Planetary Change 66: 93-99.

Regional Variability in Latest Pleistocene and Holocene Sea-Level Rise Across the California-Oregon-Washington and Bering Sea Continental Shelves

Jorie Clark[a,*], Jerry X. Mitrovica[b], Jay Alder[c]

[a] College of Earth, Ocean, and Atmospheric Sciences, Oregon State University, Corvallis, OR 97331, USA

[b] Department of Earth and Planetary Sciences, Harvard University, Cambridge, MA 02138, USA

[c] U.S. Geological Survey, Oregon State University, Corvallis, OR 97331, USA

*Corresponding author. Tel.: +1 (541) 737 1575.
E-mail address: clarkjc@geo.oregonstate.edu (J. Clark).

Abstract

Sea-level rise during the last deglaciation was influenced by isostatic, gravitational, and rotational effects that led to significant regional departures from eustasy. Deglacial sea-level rise would have been particularly variable spatially in areas adjacent to the Cordilleran and Laurentide Ice Sheets. Here we predict relative sea-level (RSL) change across the California-Oregon-Washington and Bering Sea continental shelves using a state-of-the-art theory that incorporates time-varying shoreline geometry. Our results clearly demonstrate that deglacial sea-level rise across these continental shelves was non-uniform spatially and significantly different from eustatic. Such regional variability in sea level is important to consider when investigating potential coastal migration pathways used by early Americans. An improved understanding of regional sea-level rise may also be used for predictive modeling of archaeological sites that are currently submerged.

Introduction

One of the prevailing paradigms in the "Clovis First" model posits that the initial colonization of the Americas from Asia was by migration through the ice-free corridor that developed between the Laurentide and Cordilleran ice sheets during the last deglaciation. Once arriving to the interior of the continent south of the ice sheets, these peoples are thought to have slowly migrated towards coastal regions, where they adapted to the local environment (Erlandson et al., 2008). There is now compelling evidence, however, for pre-Clovis occupation of the Americas as early as ~15.5 ka (Erlandson et al., 2008; Kennett et al., 2008; Waters et al., 2011), which precedes opening of the ice-free corridor by several hundred to a thousand years (Dyke, 2004), bolstering arguments that support the hypothesis that the first migration was instead along the western coast of North America (Fladmark, 1979). Additional support for a coastal route has come from the evidence for older ages of human settlement and maritime activity along the Pacific Coast (Erlandson et al., 2008; Erlandson et al., 2011), and that areas along the Alaskan and British Columbian coasts were habitable before the opening of the ice-free corridor (Mandryk et al., 2001; Mann and Hamilton, 1995).

A key challenge in assessing the coastal migration hypothesis is the fact that most former occupation sites lie below current sea level. Paleogeography of coastal regions is thus important for understanding the archaeological record, including correctly predicting where sites may occur that are now below sea level. A common approach to assess such paleogeography is to uniformly lower sea level by the amount suggested from far-field (e.g., Barbados) or global mean sea-level records (e.g., Manley, 2002 (http://instaar.colorado.edu/qgisl/bering_land_bridge/blb_overview.html) (Anderson et al., 2010; Kennett et al., 2008). This uniform, global mean sea-level change is often referred to as eustatic change. Sea-level rise during the last deglaciation, however, was influenced by isostatic, gravitational, and rotational effects associated with the exchange of mass between ice sheets and oceans that, for the majority of the ocean basins, led to significant regional departures from eustasy (Milne and Mitrovica, 2008). Deglacial sea-level rise across the California-Oregon-Washington and Bering Sea continental shelves would have been particularly variable spatially in response to their relative proximity to the Cordilleran and Laurentide Ice Sheets. Here we predict relative sea-level (RSL) change across the Oregon-Washington and Bering Sea continental shelves as well as around the Channel Islands, California, using a state-of-the-art theory that incorporates time-varying shoreline geometry. We compare our predictions to sea-level changes that would be associated with commonly assumed eustatic rise.

Sea-Level Modeling

The ice-age sea-level calculations were performed using a gravitationally self-consistent sea-level algorithm (Kendall et al., 2005; Mitrovica and Milne, 2003) that accounts for the deformation of the solid Earth, as well as perturbations to the Earth's gravitational field and rotational state. In this regard, we adopt a spherically symmetric, self-gravitating, linear (Maxwell) viscoelastic Earth model (Peltier, 1974) and the rotational stability theory derived by Mitrovica et al. (2005). The sea-level algorithm also accounts for the migration of shorelines due to local sea-level variations and changes in the perimeter of grounded, marine-based ice sheets (Mitrovica and Milne, 2003). The former requires iterative improvement until the present-day topography matches the observed topography (Kendall et al., 2005). All calculations are based on a pseudo-spectral solver truncated at spherical harmonic degree 256, which provides a surface spatial resolution of approximately 100 km. A truncation at higher degree would not alter the results. Changes in shorelines presented in this study are illustrated by superimposing the sea-level predictions on the high-resolution ETOPO01 topography grid (Amante and Eakins, 2009).

The sea-level algorithm requires, on input, models for the Earth's viscoelastic structure and the global geometry of ice sheets over the last glacial cycle. The elastic and density structure of the Earth model are prescribed from the seismic model PREM (Dziewonski and Anderson, 1981). Moreover, we adopt the so-called VM2 radial profile of mantle viscosity, which is coupled to the ICE-5G ice history since the last interglacial (Peltier, 2004). The VM2 model (Peltier, 2004) is characterized by a relatively moderate, factor of 5, increase in viscosity from the base of a 90 km high-viscosity (effectively elastic) lithosphere to the core-mantle-boundary. In future work, we will explore the sensitivity of our predictions to a change in this profile, including the adoption of viscosity models with a higher increase with depth, as inferred in a number of studies of ice-age sea-level observations (Mitrovica and Forte, 2004; Nakada and Lambeck, 1989).

Results

We first show examples of relative sea-level (RSL) curves from the three general regions of interest, and how they compare to the global eustatic curve. The RSL curve from the area of the Channel Islands (Figure 1A) shows that during the Last Glacial Maximum (LGM) ~21 ka, RSL was ~20 m higher than would have been predicted by assuming eustatic sea level. Following the LGM, both sea-level curves began to rise, with the eustatic curve rising at a faster rate so that at ~14 ka, the two curves intersected and during the remainder of the sea-level rise, RSL remained below what would have been predicted from eustatic, usually by ~10 m. Finally, we note that differences of many meters between the two curves that continue well into the late Holocene have implications for predicting sites in present estuarine regions.

The relation between the RSL and eustatic curves for the continental shelves off of Oregon and underlying much of the Bering Sea is broadly similar to that described above for the Channel Islands (Figure 1B, 1C). Closer to the edge of the Cordilleran Ice Sheet (CIS), however, this relation changes significantly as the large isostatic effects of the ice sheet begin to dominate. For example, comparison of the RSL and eustatic curves from the continental shelf off of Washington near the former CIS shows that at the LGM, RSL was within ~15 m of present, reflecting the large isostatic depression of the crust just beyond the ice margin (Figure 1B). By ~16 ka, RSL started to fall in response to the isostatic uplift of the coast that was occurring as the CIS deglaciated. Specifically, the land was isostatically uplifting (rising) faster than the sea-level rise from melting of global ice sheets, the net effect of which is a fall in RSL. At ~12 ka, RSL started to rise again and followed a similar trajectory as that further south on the Oregon shelf (Figure 1B).

We next show a series of maps of the three regions considered here: Channel Islands, Oregon-Washington continental shelf, and Bering Sea shelf. Figure 2A shows the paleogeography for the Channel Islands region from 20 ka to 6 ka with an assumption of a eustatic sea-level history. Figure 2B shows the difference (anomaly) between RSL and modern sea level from 20 ka to 6 ka. These maps demonstrate that sea-level fall at these times, relative to modern, was not uniform, with a gentle SW-NE gradient in RSL corresponding to greater sea-level fall relative to present in the southwestern portion of the region. Figure 2C shows the paleogeography for the Channel Islands region from 20 ka to 6 ka when accounting for the RSL history. As identified by Figure 1A, the differences between the two paleogeographic reconstructions (eustatic versus RSL) range from ~20 m during and shortly after the LGM, to ~10 m during much of the remainder of the deglaciation. These differences would become important when considering, for example, the proximity of a site on the Channel Islands now above sea level to an inferred paleoshoreline (e.g., Kennett et al., 2008; Erlandson et al., 2011), and to the duration that the now-separate islands may have remained connected by land.

Figure 3A shows the paleogeography for the Oregon-Washington continental shelf from 20 ka to 6 ka with an assumption of a eustatic sea-level history. Figure 3B identifies the difference (anomaly) between RSL and modern sea level from 20 ka to 6 ka. As with the southern California region, these maps demonstrate that sea-level fall at these times, relative to modern, was not uniform, but now there are much stronger gradients in RSL that are associated with the isostatic and gravitational effects of the CIS. In particular, there is a strong SSW-NNE gradient in RSL on the northern Washington shelf, especially from 20-15 ka. As also illustrated by Figure 1B, RSL in this northern region was ~15 m below present at the LGM, indicating that any sites found today at these water depths could conceivably date from 20-15 ka as well as at ~5 ka, when RSL again reaches that level. Figure 3C shows the paleogeography for the Oregon-Washington continental shelf from 20 ka to 6 ka when accounting for the RSL history. Given the bathymetry of this region, there are some notable differences between the two paleogeographic reconstructions (eustatic versus RSL). To illustrate, we focus on two locations at two times where there are pronounced differences that have implications for predictive modeling of offshore sites. The first case is for the central Oregon shelf at 14 ka, where the assumption of eustatic sea level would predict a series of small islands occurring at this time as the shelf was submerged (Figure 4A). In contrast,, the prediction of RSL shows instead that much of the shelf in this area remained emergent, with an extensive coastline (Figure 4B). The second case is for the area around and just north of the mouth of the Columbia River. At 6 ka, the assumption of eustatic sea level would predict that much of the estuarine regions along this part of the coast were already starting to be submerged at 6 ka (Figure 4C), whereas the prediction of RSL indicates that these estuaries were completely emergent, and thus would not have been likely sites for coastal occupation sites (Figure 4D).

Figure 5A shows the paleogeography for the continental shelf underlying the Bering Shelf from 15 ka to 6 ka with an assumption of a eustatic sea-level history (there are no substantial differences in the paleogeography of this region between 20 ka and 15 ka). Figure 5B shows the difference (anomaly) between RSL and modern sea level from 15 ka to 6 ka. Similar to the northern Washington shelf, there is a clear influence by the Cordilleran Ice Sheet that persists until ~9 ka, with sea level being near present in the Norton Sound region and falling, relative to present, to the west and north of Norton Sound.

Figure 5C shows the paleogeography for the Bering Sea continental shelf from 15 ka to 6 ka when accounting for the RSL history. As with the Oregon-Washington shelf region, there are some notable differences between the two paleogeographic (eustatic and RSL) reconstructions. In particular, we note three differences that have implications for predictive modeling of offshore sites and use of the Bering land bridge for migration. First, there is an elaborate archipelago to the south of the Bering land bridge that may have served as occupation sites, but with important differences between the two reconstructions that will need to be included in predictive modeling. Second, an assumption of eustasy would predict that Norton Sound remained largely emergent until ~10 ka, but the strong influence of the nearby Cordilleran Ice Sheet caused much of the Sound to remain submerged since the LGM. Third, an assumption of eustasy would predict that the Bering land bridge did not become submerged until ~10 ka, whereas the prediction of RSL shows that it became submerged ~11 ka, with an earlier submergence history being supported by low-resolution sea-level records from the Chukchi Sea (Elias et al., 1996; Keigwin et al., 2006).

Conclusions

The evidence for pre-Clovis occupation of the Americas as early as ~15.5 ka provides additional support for the hypothesis that the first Americans migrated from Asia to lands south of the North American ice sheets by a route along the west coast of North America. However, understanding possible migration pathways and identifying potential early occupation sites that are now below sea level requires a correct representation of sea-level change during the last deglaciation. We have shown that departures from a uniform (eustatic) deglacial sea-level rise by isostatic, gravitational, and rotational effects associated with the exchange of mass between ice sheets and oceans during the last deglaciation led to substantial differences in the paleogeography across the California-Oregon-Washington and Bering Sea continental shelves. These departures must be accounted for in any paleogeographic reconstruction of regions influenced by deglacial sea-level rise in order to correctly assess coastal migration routes and predict underwater occupation sites.

Acknowledgments:

We thank Steve Hostetler for his helpful review.

References

Amante, C., and Eakins, B. W., 2009, ETOPO1 1 Arc-Minute Global Relief Model: Procedures, Data Sources and Analysis.

Anderson, D. G., Yerka, S. J., and Gillam, J. C., 2010, Employing high-resolution bathymetric data to infer possible migration routes of Pleistocene populations: Current Research in the Pleistocene, v. 27, p. 60-64.

Dyke, A. S., 2004, An outline of North American deglaciation with emphasis on central and northern Canada, in Ehlers, J., and Gibbard, P. L., eds., Quaternary Glaciations: Extent and Chronology, Volume 2b: Amsterdam, Elsevier, p. 373-424.

Dziewonski, A. M., and Anderson, D. L., 1981, Preliminary reference Earth model: Physics of the Earth and Planetary Interiors, v. 25, p. 297-356.

Elias, S. A., Short, S. K., Nelson, C. H., and Birks, H. H., 1996, Life and times of the Bering land bridge: Nature, v. 382, p. 60-63.

Erlandson, J. M., Moss, M. L., and Des Lauriers, M., 2008, Life on the edge: early maritime cultures of the Pacific Coast of North America: Quaternary Science Reviews, v. 27, p. 2232-2245.

Erlandson, J. M., Rick, T. C., Braje, T., Casperson, M., Culleton, B., Fulfrost, B., Garcia, T., Guthrie, D. A., Jew, N., Kennett, D. J., Moss, M. L., Reeder, L., Skinner, C., Watts, J., and Willis, L., 2011, Paleoindian seafaring, maritime technologies, and coastal foraging on California's Channel Islands: Science, v. 331, p. 1181-1185.

Fladmark, K. R., 1979, Routes: alternate migration corridors for Early Man in North America: American Antiquity, v. 44, p. 55-69.

Keigwin, L. D., Donnelly, J. P., Cook, M. S., Driscoll, N. W., and Brigham-Grette, J. P., 2006, Rapid sea-level rise and Holocene climate in the Chukchi Sea: Geology, v. 34, p. 861-864.

Kendall, R. A., Mitrovica, J. X., and Milne, G. A., 2005, On post-glacial sea level - II. Numerical formulation and comparative results on spherically symmetric models: Geophysical Journal International, v. 161, no. 3, p. 679-706.

Kennett, D. J., Kennett, J. P., West, G. J., Erlandson, J. M., Johnson, J. R., Hendy, I. L., Westg, A., Culleton, B. J., Jones, T. L., and Stafford, J., T.W., 2008, Wildfire and abrupt ecosystem disruption on California's Northern Channel Islands at the Ållerød–Younger Dryas boundary (13.0–12.9 ka): Quaternary Science Reviews, v. 27, p. 2530-2545.

Mandryk, C. A. S., Josenhans, H., Fedje, D. W., and Mathewes, R. W., 2001, Late Quaternary paleoenvironments of Northwestern North America: implications for inland versus coastal migration routes: Quaternary Science Reviews, v. 20, p. 301-314.

Mann, D. H., and Hamilton, A. C., 1995, Late Pleistocene and Holocene paleoenvironments of the North Pacific Coast: Quaternary Science Reviews, v. 14, p. 449–471.

Milne, G. A., and Mitrovica, J. X., 2008, Searching for eustasy in deglacial sea-level histories: Quaternary Science Reviews, v. 27, no. 25-26, p. 2292-2302.

Mitrovica, J. X., and Forte, A. M., 2004, A new inference of mantle viscosity based upon joint inversion of convection and glacial isostatic adjustment data: Earth and Planetary Science Letters, v. 225, no. 1-2, p. 177-189.

Mitrovica, J. X., and Milne, G. A., 2003, On post-glacial sea level: I. General theory: Geophysical Journal International, v. 154, no. 2, p. 253-267.

Mitrovica, J. X., Wahr, J., Matsuyama, I., and Paulson, A., 2005, The rotational stability of an ice-age earth: Geophysical Journal International, v. 161, no. 2, p. 491-506.

Nakada, M., and Lambeck, K., 1989, Late Pleistocene and Holocene sea-level change in the Australian region and mantle rheology: Geophysical Journal International, v. 96, p. 497-517.

Peltier, W. R., 1974, Impulse response of a Maxwell Earth: Reviews of Geophysics, v. 12, p. 649-669.

Peltier, W. R., 2004, Global glacial isostasy and the surface of the ice-age earth: The ice-5G (VM2) model and grace: Annual Review of Earth and Planetary Sciences, v. 32, p. 111-149.

Waters, M. R., Forman, S. L., Jennings, T. A., Nordt, L. C., Driese, S. G., Feinberg, J. M., Keene, J. L., Halligan, J., Lindquist, A., Pierson, J., Hallmark, C. T., Collins, M. B., and Wiederhold, J. E., 2011, The Buttermilk Creek Complex and the origins of Clovis at the Debra L. Friedkin site, Texas: Science, v. 331, p. 1599-1603.

Figure captions

Figure 1. Relative sea level (RSL) curves (blue) as compared to a eustatic sea level curve (red) for the three regions considered here. The RSL curves are representative for each region, but there will be variations in RSL across each region. (A) Channel Islands region, CA. (B) The Oregon-Washington continental shelves. The lighter blue RSL curve corresponds to the southern Oregon shelf (~41.5°N), whereas the darker blue RSL curve corresponds to the northern Washington shelf (~48.5°N). (C) The continental shelf underlying the Bering Sea.

Figure 2. Paleogeography and sea-level predictions for the Channel Islands region, CA. (A) The paleogeography for the Channel Islands region from 20 ka to 6 ka with an assumption of a eustatic sea-level history. (B) The difference (anomaly) between RSL and modern sea level from

20 ka to 6 ka. (C) The paleogeography for the Channel Islands region from 20 ka to 6 ka when accounting for the RSL history.

Figure 3. Paleogeography and sea-level predictions for the Oregon-Washington continental shelves. (A) The paleogeography of the Oregon-Washington continental shelves from 20 ka to 6 ka with an assumption of a eustatic sea-level history. (B) The difference (anomaly) between RSL and modern sea level from 20 ka to 6 ka. (C) The paleogeography of the Oregon-Washington continental shelves from 20 ka to 6 ka when accounting for the RSL history.

Figure 4. (A) Paleogeography of the central Oregon continental shelf at 14 ka assuming a eustatic sea-level history. (B) Paleogeography of the central Oregon continental shelf at 14 ka when accounting for the RSL history. (C) Paleogeography of the continental shelf near the mouth of the Columbia River at 6 ka assuming a eustatic sea-level history. (D) Paleogeography of the continental shelf near the mouth of the Columbia River at 6 ka when accounting for the RSL history.

Figure 5. Paleogeography and sea-level predictions for the Bering Sea continental shelf. (A) The paleogeography of the Bering Sea continental shelf from 15 ka to 6 ka with an assumption of a eustatic sea-level history. (B) The difference (anomaly) between RSL and modern sea level from 15 ka to 6 ka. (C) The paleogeography of the Bering Sea continental shelf from 15 ka to 6 ka when accounting for the RSL history.

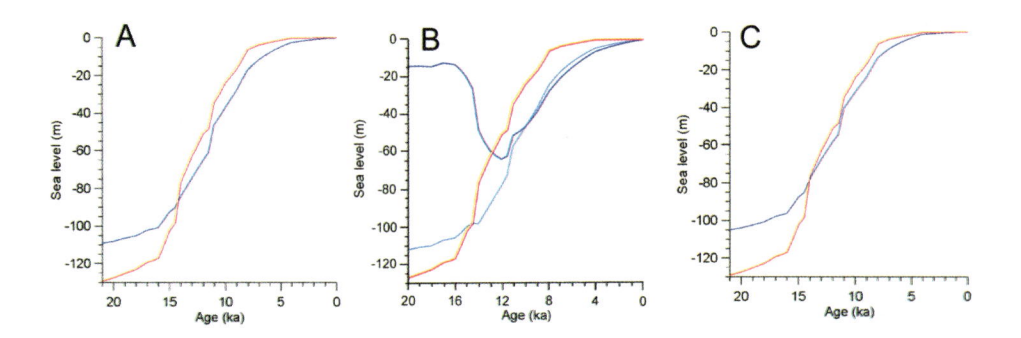

Figure 1.

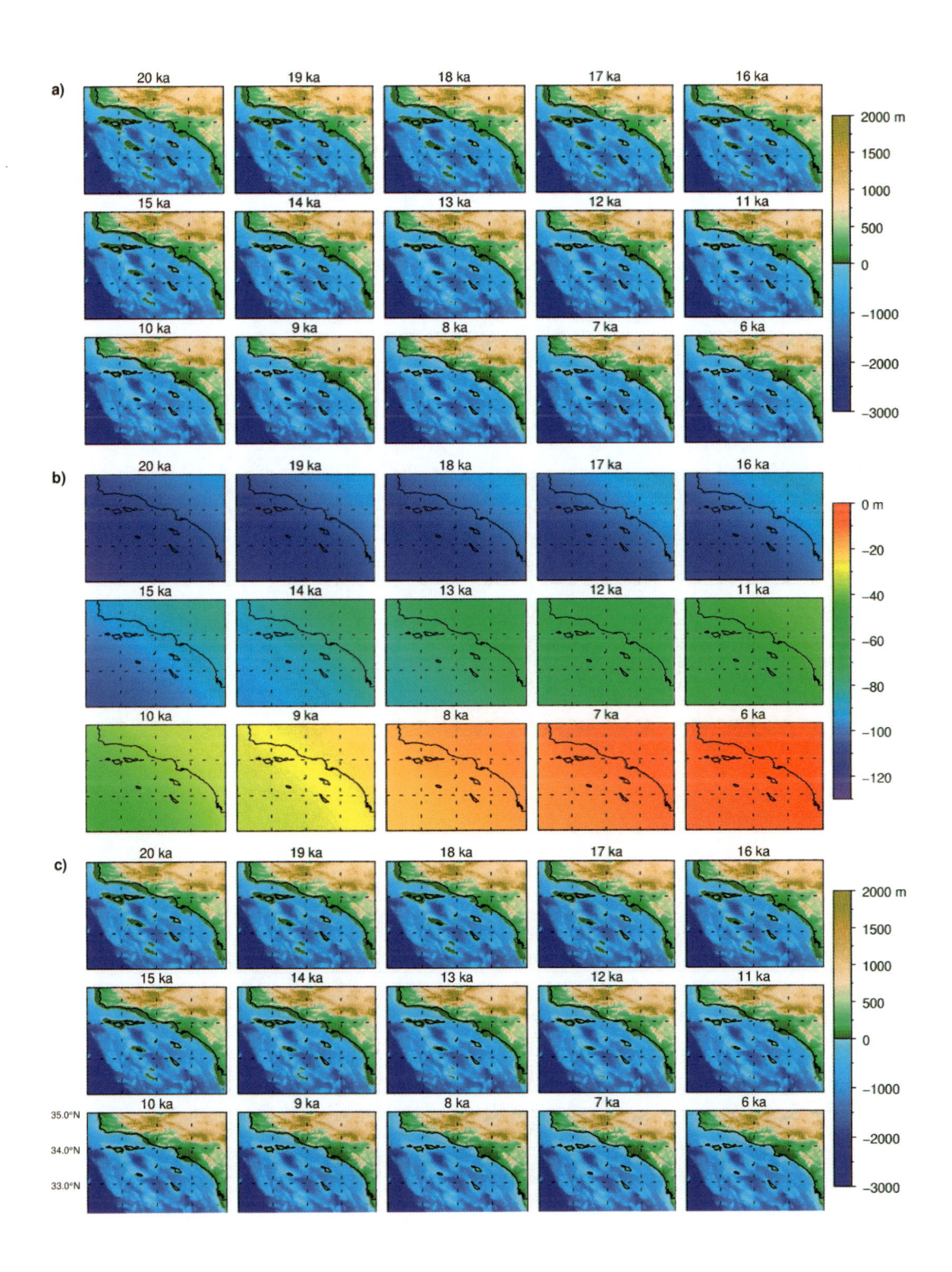

Figure 2.

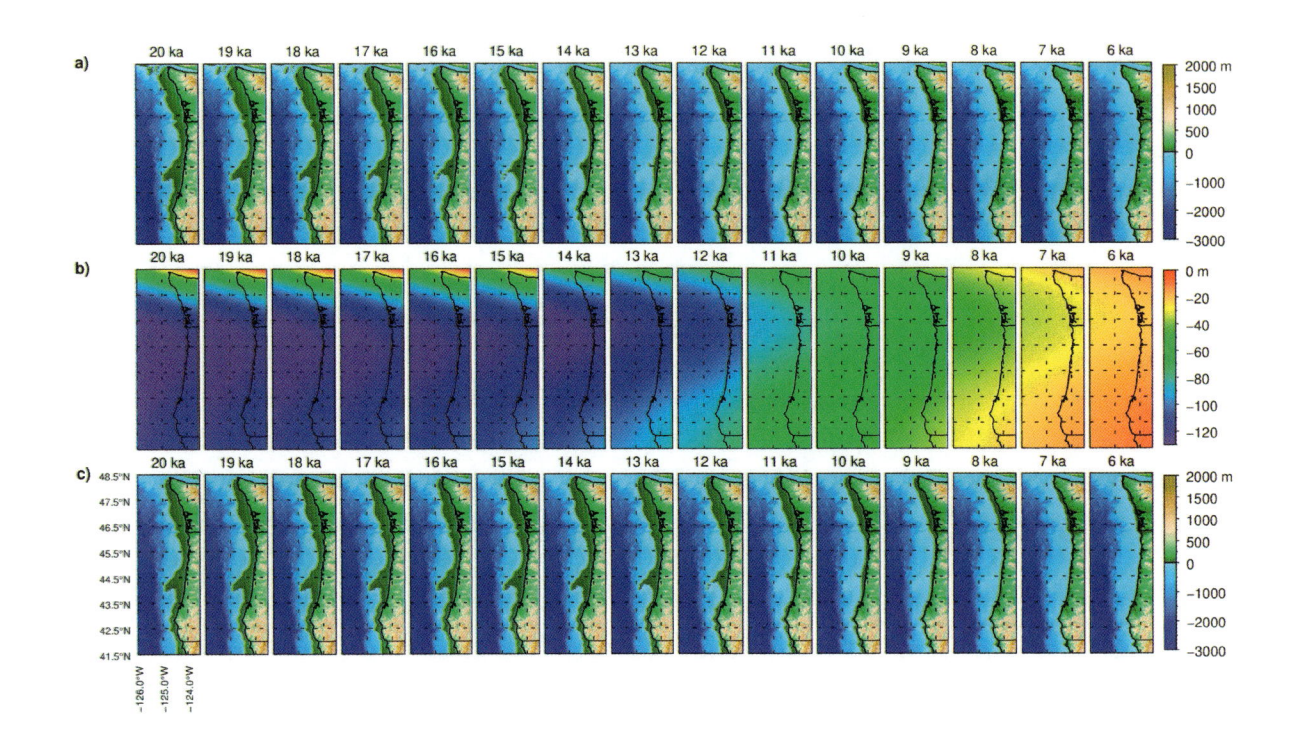

Figure 3.

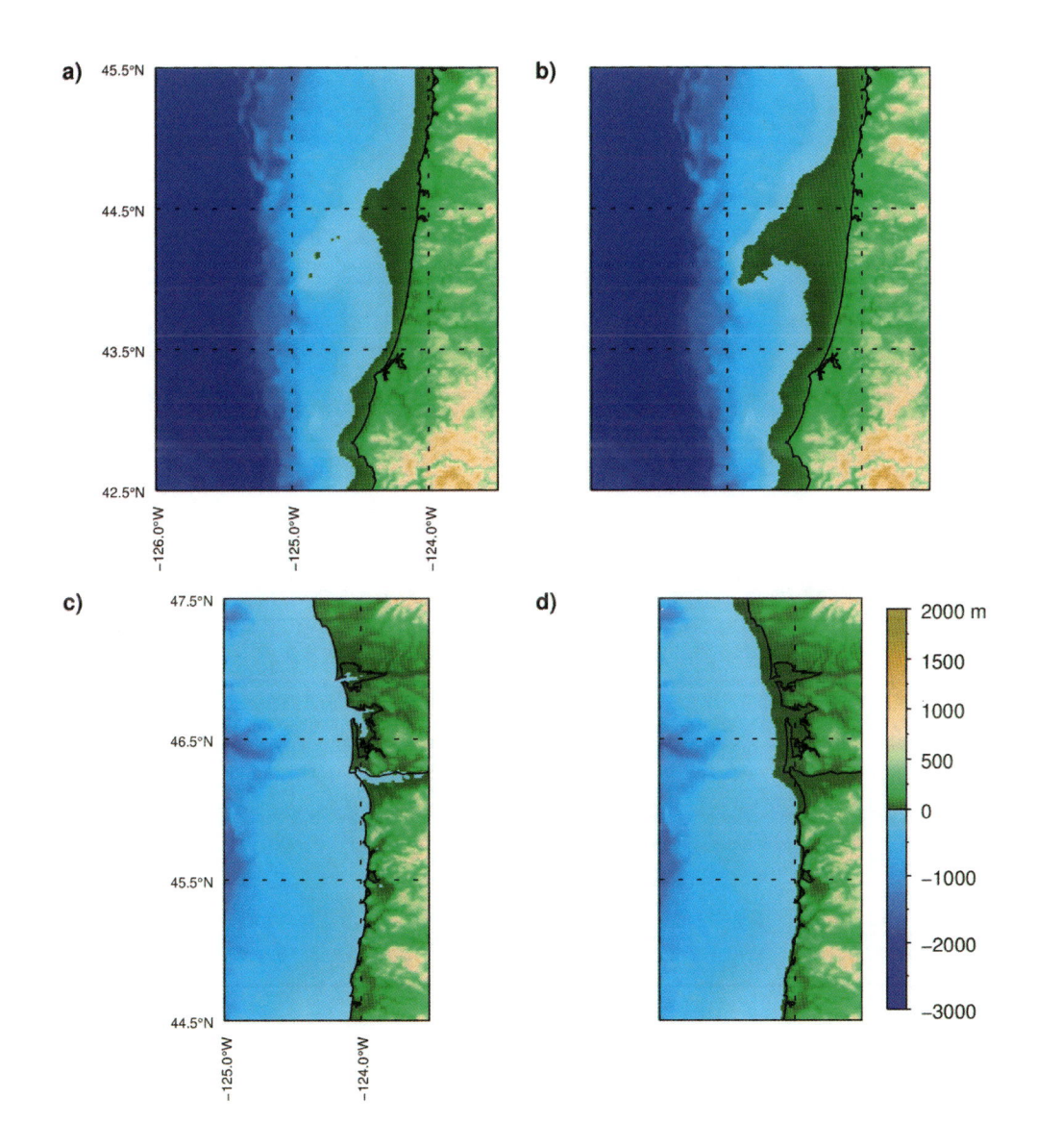

Figure 4.

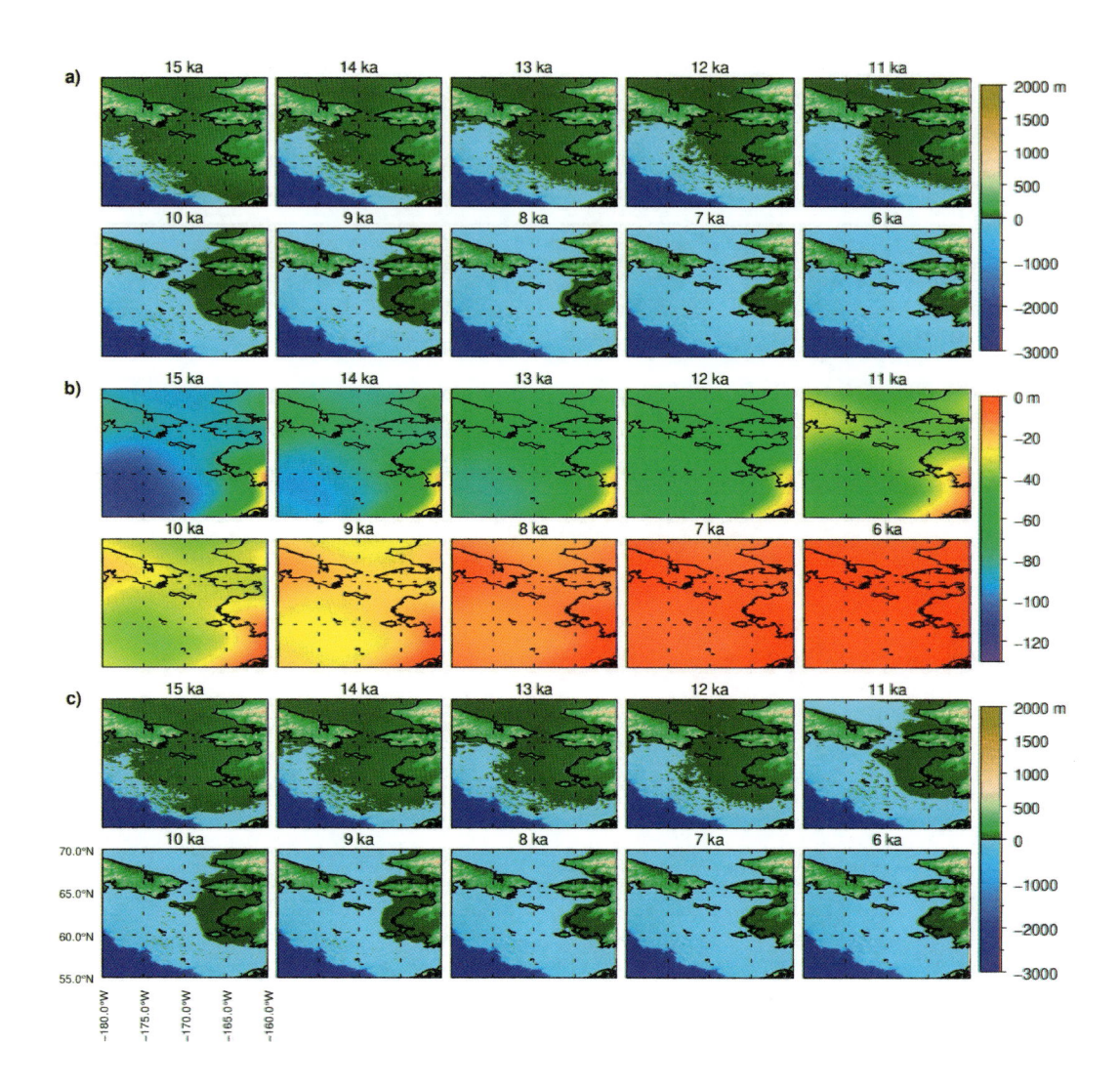

Figure 5.

Meadowcroft Rockshelter: Retrospect 2012

J. M. Adovasio and David R. Pedler

Mercyhurst Archaeological Institute
Mercyhurst University
Erie, Pennsylvania 16546

Abstract

With the publication of the first radiocarbon sequence from Meadowcroft Rockshelter in 1974, the site has become and remains the most controversial North American locality ever advanced for early occupation of the New World since Abbott's excavation in the Trenton Gravels. The salient features of this site, including the stratigraphy, cultural features, artifactual suite, and floral and faunal collections, are summarized from the perspective of 39 years of research. Additionally, the first calibrated radiocarbon sequence from the site is presented. We stress that despite the now nearly extinguished debate over the earliest occupation of this site, the excavation and documentation protocols employed there are still considered even by its severest critics as the state-of-the-art. With the benefit of nearly 40 years of hindsight, we conclude that perhaps the most enduring non-methodological contribution of the Meadowcroft research is the delineation of an occupational sequence that spanned at least 14 millennia and occurred against an ever and often subtly shifting backdrop of environmental change.

Presented at the Conference "Pre-Clovis in the Americas"
Smithsonian Institution, Washington DC
9–10 November 2012

Next year marks the fortieth anniversary of the beginning of the Meadowcroft/Cross Creek Archaeological Project, a multidisciplinary multiyear research project that underwent its most intensive phase over the years 1973–1984 and has continued to the present. When the project began, the pictures of the Paleo-Indian lifeway and chronology that were available to North American archaeologists were very much different from those of today. In terms of chronology, the hemisphere's first colonists were thought to have been unable to cross from Beringia to North America any earlier than 15,000 BC, when the Wisconsin ice sheet apparently began to retreat and ultimately open the Ice-Free Corridor. The hemisphere's first cultural manifestation, called Clovis, was thought to have occurred no earlier than 10,500 BC, the age of that entity's earliest recorded component. The Clovis lifeway was then uniformly characterized as a rapidly moving hunter-gatherers focused on the hunting of late Pleistocene megafauna. The initial publication of radiocarbon dates from Meadowcroft Rockshelter (Adovasio et al. 1975) and subsequent secondary publications (e.g., Dragoo 1976; Funk 1978) posed the first widely published credible challenge to this paradigm, at least for the Northeast, with the release of dates several millennia older than Clovis. Counter challenges soon appeared in the academic press and have continued to this day.

Much has changed since then, both in terms of previously held scholarly assessments of the Paleo-Indian lifeway and chronology, and—perhaps more importantly—in terms of a dramatic rethinking of virtually all aspects of the scientific disciplines employed to explore them. This change has resulted in varying levels of resolution and consensus on one hand, and continued (if not intensifying) intractability and irresolution on the other. It is generally agreed, for example, that the earliest inhabitants of the Americas were at first Siberians. Resolution and consensus begin to break down, however, once one starts speaking about the timing (pre- versus post-Late Glacial Maximum [LGM]) and routes of the migration (coastal versus interior, or both?). Moreover, considerable lacunae continue to exist in our understanding of late Pleistocene lifeways in the Americas, hinging as it does on quite variable data from far-flung sites that are often relatively small, poorly preserved, and ephemeral occupations compared to later (and sometimes overlapping) cultural manifestations.

There are, by recent reckoning (Pedler and Adovasio 2011:65), at least five—and perhaps as many as twenty—archaeological sites in the Americas that are substantially older than Clovis. With very few exceptions, the technologies represented at these localities are significantly different both from each other as well as from Clovis. What follows is a brief but hopefully comprehensive summary of a site that has been referred to as "the most long-standing case for archaeological evidence of pre-Clovis (earlier than 11,500 RCYBP) humans in the eastern United States" (Goodyear 2005:103), along with a discussion of its impact on our understanding of Paleo-Indian lifeways and the peopling of the Americas.

Site Context

Meadowcroft Rockshelter is a deeply stratified, multicomponent site located 4 km (2.5 mi) northwest of the town of Avella in Washington County, Pennsylvania, and 47 km (29 mi) southwest of Pittsburgh (Figure 1). It is situated on the north bank of Cross Creek, ca. 12 km (7.6 mi) west of this small tributary's confluence with the Ohio River (Figure 2). Topographically, the region surrounding Meadowcroft is maturely dissected. More than 50 percent of the 14,164 ha (35,000 acre) Cross Creek watershed is in valley slopes with upland and valley bottom settings in the minority. Maximum elevations in the Cross Creek drainage are generally higher than 396 m (1,299 ft) above mean sea level (msl), while Meadowcroft is considerably lower at 360 m (853 ft) above msl.

Within the Cross Creek watershed, the main stem of Cross Creek flows west for some 32 km (19 mi). The maximum north-south width of the watershed is ca. 15 km (19 mi). The prevailing stream pattern is dendritic, with numerous small creeks and runs supplying the main stem of Cross Creek. The drainage trends generally west toward the West Virginia–Ohio border and the Ohio River. Cross Creek exhibits a markedly asymmetric drainage pattern, with the northern tributaries significantly longer than those on the south. Consequently, the drainage area to the north of Cross Creek is much larger than its counterpart to the south. This condition is probably the result of a drainage pattern superimposed on the 3–5° southeast-trending regional dip of the study area.

The region's present topography developed during the later part of the Pleistocene, when increased precipitation and runoff caused extensive down-cutting. The area was also unaffected by glacial ice, as the

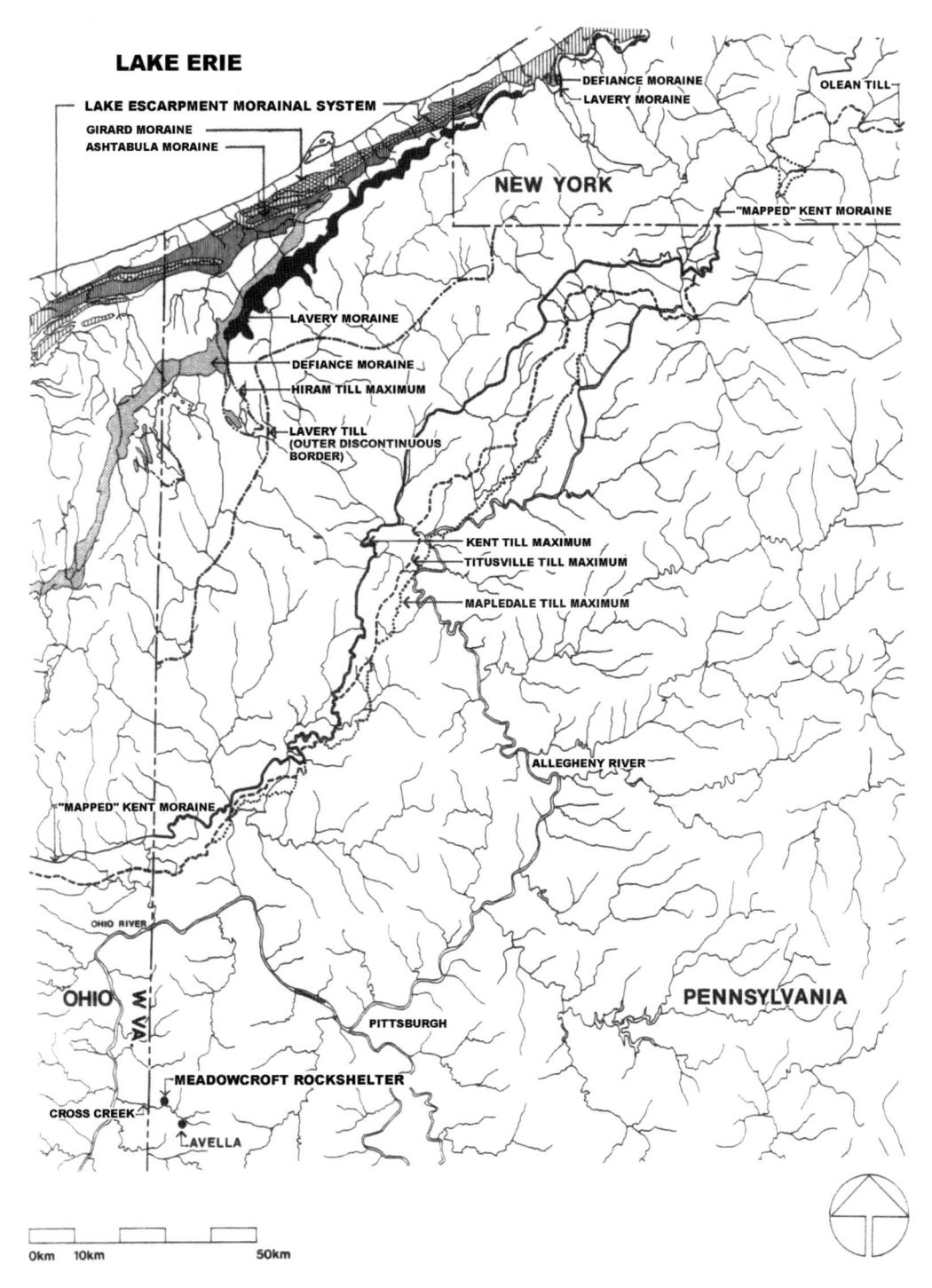

Figure 1. Location of Meadowcroft Rockshelter in relation to various glacial features and watersheds.

late Pleistocene maximum glacial boundary (see Figure 1) extends only as far south as northern Beaver County, ca. 83 km (52 mi) north of the Cross Creek drainage. By the time the watershed was initially occupied by humans, the ice front was forming the Lavery terminal moraine more than 150 km (93 mi) north of Meadowcroft.

Figure 2. View of Meadowcroft Rockshelter from the opposite bank of Cross Creek in the late autumn.

Stratigraphy and Chronology

The excavations at Meadowcroft Rockshelter distinguished 11 natural strata (Figure 3), labeled numerically from the oldest and deepest (Stratum I) to the latest and uppermost (Stratum XI). Achieving a maximum excavated thickness of over 4 m (13 ft), each of these depositional units varies considerably in thickness, composition, and texture—among other parameters—and many include numerous, often very thin, microstrata (Figure 4), the great majority of which reflect discrete occupation/visitation events. The stratigraphic sequence, as detailed in Stuckenrath et al. (1982), significantly, manifests no depositional hiatuses or disconformities.

Meadowcroft's stratigraphic sequence is anchored by 52 radiocarbon dates that were processed by four different laboratories (Table 1). The radiocarbon sequence is consistent with the observed stratigraphy, with only four low-order, statistically irrelevant reversals (i.e., SI-2363, SI-2049, SI-1681, and SI-2056 [see Table 1]) during the Late Archaic and Woodland periods. The calibrated ages for these assays (Figure 5) indicate a Woodland period ascription for Strata XI–IV(upper), an Archaic ascription for Strata IV(middle)–IIb, and a predominantly Paleo-Indian ascription for Stratum IIa. The Woodland and Archaic ages are uncontroversial and conform to the accepted ranges those cultural periods recognized for Pennsylvania and the upper Ohio River watershed. The lower culture-bearing strata (Middle and Lower IIa) and their attendant radiocarbon dates, however, have been the subject of four decades of vigorous debate and controversy.

The newly calibrated radiocarbon ages from the site presented here indicate that the upper reaches of Stratum IIa (see Table 1) are consistent with determinations that are commonly accepted for Early Archaic and Paleo-Indian sites throughout the Northeast and greater North America. The middle and lower portions of that Stratum, however, register ages whose ranges indicate a human presence that pre-dates the calibrated age of Clovis manifestations by several millennia. For example, the averaged radiocarbon age (11,570 ± 70 BP [sample size of two assays]) calculated by Faught (2008:Table 1) for the Aubrey site in Texas, the earliest securely dated Clovis fluted point locality in his sample, when calibrated using OxCal 4.1 (IntCal 09 curve) yields an age of 11,671–11,321 cal BC (1σ). It is very important to note that Faught (2008:Table 3) rejects the older Meadowcroft dates due to "possible contaminated age estimate," but his calculated mean (sample size of four assays) of 13,540 ± 272 for lower Stratum IIa at Meadowcroft nonetheless calibrates to 15,217–13,328 cal BC (1σ). The most recent limits of both age ranges do not overlap, and suggest that Meadowcroft is at least 2,000 calendar years older than the earliest Clovis manifestation.

As Table 1 indicates, however, the middle (one assay) and lower (nine assays) reaches of Stratum IIa appear to be even older, indeed considerably so, with calibrated ages ranging stratigraphically and chronologically from 13,606–9311 cal BC (1σ) to 24,690–22,098 cal BC (1σ). Numerous critiques of

Figure 3. General view of east-face profile at Meadowcroft.

Figure 4. Close-up view of east-face profile at Meadowcroft, showing microstratigraphic units.

the radiocarbon chronology from the site have been by turns offered (e.g., Haynes 1980, 1991, 2005; Mead 1980; Tankersley and Munson 1992) and refuted (Adovasio et al. 1980, 1992, 1998; Goldberg and Arpin 1999; Adovasio and Pedler 2005), with many of the questions focusing on contamination and questionable cultural associations, particularly with the older, stratigraphically lower dates. But even if one dismisses the radiocarbon dates associated with possible basketry (SI-2060), following the objection of Grayson (2004:381), or those not directly associated with fire pit cultural features (SI-2062, DIC-2187), the considerably narrowed range of calibrated dates is nonetheless much older than Clovis, ranging from 13,606–9311 cal BC (1σ) to 20,210–15,586 cal BC (1σ). Furthermore, should one also discard radiocarbon dates with rather large sigmas of error, following Faught›s (2008:672) recommendation, and average lower Stratum IIa›s two assays (i.e., SI-2488 and SI-1686 [see table 1]) with the smallest sigma values, the result of 15,293–14,896 cal BC (1σ) is still much older than Clovis. In any case, the occupational sequences at Meadowcroft currently remains the longest documented for the entire New World.

Table 1. Radiocarbon Chronology for Meadowcroft Rockshelter.

Lab #	Material	Context	14C Date BP	Age Cal AD/BC (1σ)[a]	Comment
Stratum XI					
SI-3013	charcoal	fire pit	175 ± 50	cal AD 1649–1954	middle third of unit
Stratum IX					
SI-2363	charcoal	fire pit	685 ± 80	cal AD 1187–1421	upper third of unit
Stratum VIII					
SI-3023	charcoal	fire pit	630 ± 100	cal AD 1186–1454	
Stratum VII					
SI-2047	charcoal	fire pit	925 ± 65	cal AD 993–1225	middle third of unit
SI-3026	charcoal	fire pit	1290 ± 60	cal AD 649–878	middle third of unit
Stratum V					
SI-3024	charcoal	fire pit	1665 ± 65	cal AD 236–543	upper third of unit
SI-3027	charcoal	fire pit	1790 ± 60	cal AD 85–386	upper third of unit
SI-3022	charcoal	fire pit	1880 ± 65	38 cal BC– cal AD 320	upper third of unit
SI-2362	charcoal	fire pit	2075 ± 125	396 cal BC– cal AD 210	upper third of unit
SI-2487	charcoal	fire pit	2155 ± 65	379–48 cal BC	upper third of unit
Stratum IV					
SI-2051	charcoal	fire pit	2290 ± 90	750–110 cal BC	upper third of unit
SI-1674	charcoal	fire pit	2325 ± 75	751–199 cal BC	upper third of unit
SI-2359	charcoal	fire pit	2485 ± 350	1491 cal BC– cal AD 215	upper third of unit
SI-3031	charcoal	fire pit	2655 ± 120	1112–414 cal BC	upper third of unit
SI-1665	charcoal	fire floor	2815 ± 80	1211–812 cal BC	middle third of unit
SI-1668	charcoal	fire pit	2820 ± 75	1210–818 cal BC	middle third of unit
SI-1660	charcoal	fire pit/fire floor	2860 ± 80	1266–836 cal BC	middle third of unit
SI-2049	charcoal	fire pit/fire floor	3050 ± 85	1494–1053 cal BC	middle third of unit
Stratum III					
SI-2066	charcoal	fire pit	2930 ± 75	1378–927 cal BC	upper third of unit
SI-1664	charcoal	fire pit	3065 ± 80	1500–1057 cal BC	upper third of unit
SI-2053	charcoal	fire pit	3090 ± 115	1612–1024 cal BC	upper third of unit
SI-3030	charcoal	fire pit	3100 ± 90	1606–1092 cal BC	upper third of unit
SI-2046	charcoal	fire pit	3115 ± 70	1528–1133 cal BC	upper third of unit
SI-1679	charcoal	fire pit	3255 ± 115	1876–1267 cal BC	middle third of unit
Stratum IIb					
SI-1681	charcoal	fire pit	3210 ± 95	1735–1266 cal BC	upper third of unit
SI-1680	carbonized basket fragment		3770 ± 90	2466–1962 cal BC	upper third of unit
SI-2063	charcoal	fire pit	3950 ± 240	3097–1770 cal BC	middle third of unit
SI-2058	charcoal	fire pit	3970 ± 85	2857–2206 cal BC	middle third of unit
SI-2054	charcoal	fire pit	4005 ± 85	2867–2291 cal BC	middle third of unit
SI-2356	charcoal	fire pit	4380 ± 500	4313–1773 cal BC	middle third of unit

Lab #	Material	Context	14C Date BP	Age Cal AD/BC (1σ)[a]	Comment
SI-1685	charcoal	fire pit	4820 ± 85	3778–3373 cal BC	middle third of unit
SI-2358	charcoal	fire pit	6290 ± 355	5901–4457 cal BC	middle third of unit
PITT-122	charcoal	fire pit	6315 ± 280	5766–4606 cal BC	lower third of unit, small sample, diluted
PITT-292	charcoal	fire pit	6630 ± 70	5668–5474 cal BC	lower third of unit, Juglans nigra
SI-2055	charcoal	fire floor	6670 ± 140	5873–5344 cal BC	lower third of unit
SI-2056	charcoal	fire pit	5300 ± 130	4438–3800 cal BC	lower third of unit
	Stratum IIa				
SI-2064	charcoal	fire pit	8010 ± 110	7297–6641 cal BC	upper third of unit
SI-2061	charcoal	fire pit	9075 ± 115	8621–7944 cal BC	upper third of unit
SI-2491	charcoal	fire pit/floor	11,300 ± 700	13,606–9311 cal BC	middle third of unit
SI-2489	charcoal	fire pit	12,800 ± 870	15,860–11286 cal BC	lower third of unit
SI-2065	charcoal	fire pit	13,240 ± 1010	16,631–11480 cal BC	lower third of unit
SI-2488	charcoal	fire pit	13,270 ± 340	15,075–13,021 cal BC	lower third of unit
SI-1872	charcoal	fire pit	14,925 ± 620	17,461–14,950 cal BC	lower third of unit
SI-1686	charcoal	fire pit	15,120 ± 165	16,731–16,004 cal BC	lower third of unit
SI-2354	charcoal	fire pit	16,175 ± 975	20,210–15,586 cal BC	lower third of unit
SI-2062	charcoal	charcoal concentration	19,100 ± 810	23,119–18,652 cal BC	lowest level in unit
SI-2060	carbonized fragment of cut bark-like material, possible basketry		19,600 ± 2400	31,329–16,688 cal BC	lowest level in unit
DIC-2187	charcoal	charcoal concentration	21,070 ± 475	24,690–22,098 cal BC	base of unit, directly above Strata IIa/I interface
		I/II interface			
SI-2121	charcoal	lens	21,380 ± 800	25,987–21,918 cal BC	—
SI-1687	charcoal	lens	30,710 ± 1140	36,632–31,151 cal BC	—
OxA-363	charcoal	lens	31,400 ± 1200	37,419–31,672 cal BC	solid fraction
OxA-364	charcoal	lens	30,900 ± 1100	36,620–31,416 cal BC	soluble fraction

Note: All ages calibrated using OxCal 4.1 (IntCal 09 curve).

General Character of Site Utilization

Subsistence and Seasonality

Despite the long record of aboriginal visitation to Meadowcroft Rockshelter, it is clear from a variety of data sets that the primary function of the site remained essentially constant through time. Specifically, the data recovered from Meadowcroft to date strongly argue that throughout its history the site served as a temporary locus for broad-spectrum hunting, collecting, and food processing activities. The predominance of utilized flakes along with projectile points showing multifunctional use—in combination with limited numbers of bifaces and unifaces, the relative abundance of food bone, voluminous edible plant remains and, at certain times, invertebrate resources plus the presence of perishable artifacts used in the acquisition, transport, or processing of these remains—all support this conclusion. In marked contrast, the general absence of extensive in situ manufacture of lithic, ceramic, or shell artifacts as well as other categories of evidence militate strongly against a more long term or permanent occupation of the site.

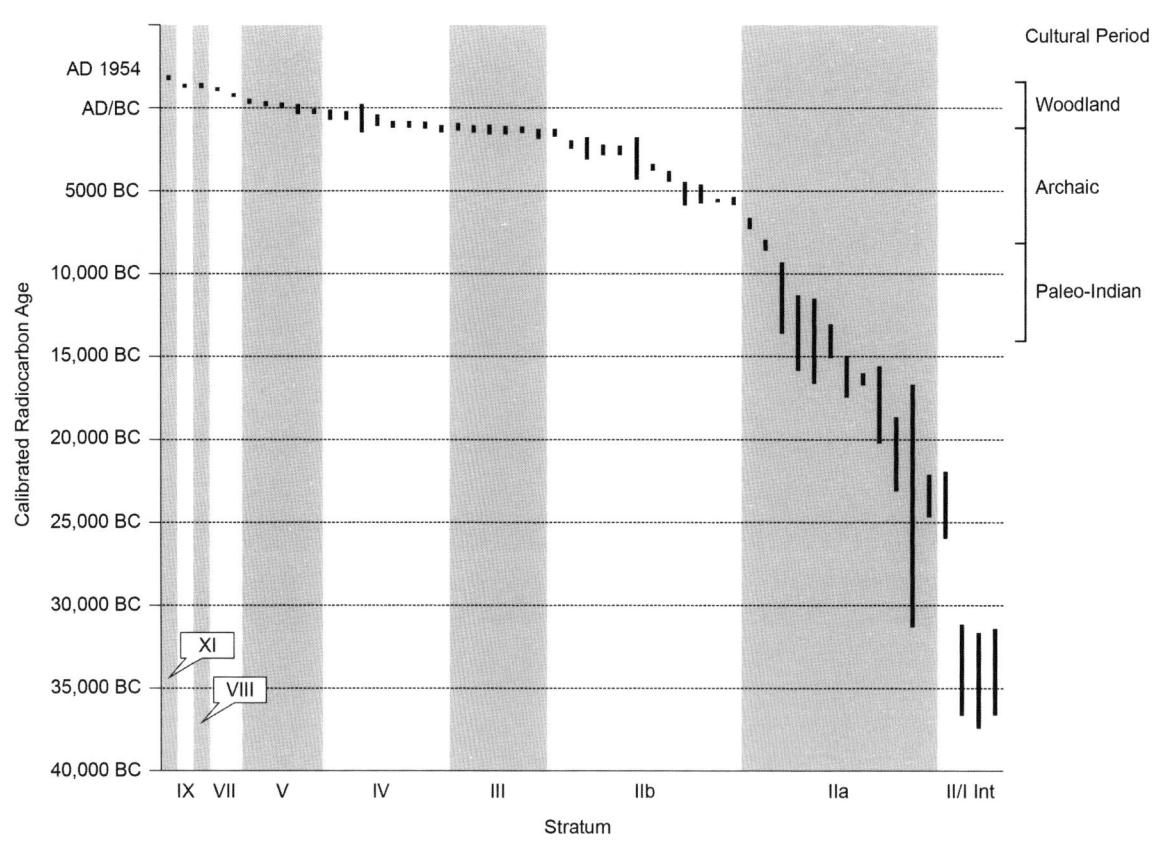

Figure 5. Calibrated radiocarbon chronology from Meadowcroft Rockshelter.

Although 115,166 identifiable bones and bone fragments representing more than 140 species have been recovered from the site, only 11 identifiable specimens and less than 11.9 grams of plant remains derive from middle and lower Stratum IIa. If all of the identified faunal remains from these levels represent food remains, which is highly unlikely, then the earliest visitors to the site exploited white-tailed deer and perhaps much smaller game. The meager floral assemblage suggests the possible gathering of hickory, walnut, and hackberry (*Celtis* sp.). It is conceivable and likely that these populations might have exploited now-extinct Pleistocene big game animals, notably mastodon, but no evidence of such predation was observed at the site. The later site deposits directly indicate that the primary subsistence modes of the various post-10,600 B.P. Meadowcroft populations included the hunting of deer, elk, and turkey augmented by the taking of a wide range of smaller game and, at various times, by the exploitation of riverine fauna, notably mussels. The intensive collection of hackberries, nuts, and a variety of other fruits and seeds is also consistently indicated.

Examination of the constituents of the dietary modes of the sequent Meadowcroft populations reveals some interesting trends. The hunting mode seems to have been more or less constant throughout the occupational sequence, while the other subsistence modes mirror a number of potentially significant changes. Hackberry exploitation sharply diminished after 925 ± 65 B.P. (cal AD 993–1225 [1σ]) along with the collection of *Rubus* sp. and *Vaccinium* sp. Conversely, the gathering of other nuts and seeds remained more or less constant. Similarly, exploitation of riverine resources, relatively insignificant before the Late Archaic, became markedly important immediately thereafter and remained so until ca. 1665 ± 65 B.P. (cal AD 236–543 [1σ]), when it virtually ceased. Finally, the addition of cultigens toward the end of the long Meadowcroft sequence does appear to correlate with a diminution in collection of certain wild plants, though it does not seem to have altered the basic character or function of site use.

Based on the availability of the principal plant food exploited at Meadowcroft, the principal time of site visitation was late summer through middle to late autumn. A small quantity of bird eggs in cultural features also suggests brief episodes of spring visitation during Archaic and Woodland times.

Intensity of Site Use

Despite the fact that the primary function of Meadowcroft remained consistent through time, the intensity of site use appears to vary significantly. It should be emphasized that intensity of site use is a difficult parameter to define precisely. It may refer to the length of the aboriginal visitation interval measured in days or weeks, the frequency or timing between separate visitation events, the number of persons per visit, or any combination of the foregoing. Whatever its defining properties, however, several discrete and complementary proxies may be used—with caution—to measure this phenomenon. These include: (1) the number and kind of cultural features per stratum or attendant time period, (2) artifacts of various compositional classes or types by stratum or chronological interlude, (3) frequency and density of ecofactual materials through time, and (4) the spatial and temporal distribution of selected constituents of constant volume samples (CVS) selectively collected from across the site and throughout the deposits.

The frequency and types of cultural features at Meadowcroft are plotted by stratum and cultural period in Table 2. As that table indicates, the densest concentration of cultural features occurs in Strata III (Late Archaic/Transitional) and IV (Early Woodland), with reduced or markedly lower concentrations before and after these periods. Significantly, the greatest concentration of more carefully prepared (i.e., lined, rimmed, etc.) fire pits, re-used fire features, specialized activity areas, thick ash and charcoal lenses, and extensive burned areas or fire-floors—which may represent the intentional incineration of trash by its aboriginal inhabitants—also occur during this time segment.

Interestingly, the incidence of discrete microstrata which may represent individual seasonal visitations is far more numerous during the emplacement of Strata III and IV than at any other time in the site's occupational trajectory. Not coincidentally, this index of differential visitation is also reflected in a wide array of other proxy intensity indices. Though space precludes a reproduction of the complex tables which detail those distributions, suffice to note that the occurrence of temporally diagnostic projectile points (cf. Fitzgibbons 1982:Table 2), all other classes of lithic artifacts (cf. Fitzgibbons 1982:Table 1), floral remains (e.g. Cushman 1982:Tables 1–4) as well as vertebrate and invertebrate faunal remains (Guilday and Parmalee 1982; Lord 1982) mirror the concentration of cultural features noted above.

While all of these distributions clearly indicate that the moments of most intensive site use fall into the Late Archaic or Transitional through Early Woodland periods, this conclusion is also etched in yet sharper relief by scrutiny of the CVS flotation data from Meadowcroft. Six vertical columns of heavy and light fractions from the flotation sampling have been sorted and their components identified and quantified (Skirboll 1982). Three of these columns derive from inside the modern dripline and three from outside. As detailed by Skirboll (1982:224–228), two of the heavy fraction columns from inside the modern dripline span every excavated stratum, microstratum, and level from the modern ground surface to the base of the excavation in culturally sterile Stratum I. Examination of the distribution of select components from these columns—which are principally attributable to anthropogenic agency such as lithic flakes, charcoal, burned bone, mussel shell, fish scales, and thermally altered hackberry seeds—indicates that all of these materials are differentially distributed through time and exhibit a marked tendency to co-cluster in different time periods.

A relatively low frequency of anthropogenically introduced material is found in Paleo-Indian Stratum IIa which is late Pleistocene in age. However, the amounts of all classes of flotation components attributable to human origin increase during the Early and Middle Archaic represented in Stratum IIb. Significantly, all classes of material are most abundant in Strata III and IV, which encompass the Late Archaic or Transitional and Early Woodland period occupations at the site. Moreover, with few exceptions, the occurrence of these intensity proxies fall off markedly on either side of this ca. 1400 radiocarbon year interval (Figure 6).

Earlier publications (Adovasio et al. 1977; Skirboll 1982:224) noted that there have been some questions as to whether the large amounts of bones, eggshells, and (especially) hackberries recovered at the rockshelter are attributable to human origin transport or some other non-human vector. While it is certain that numerous birds (explicitly including raptors) and other animal predators doubtless occupied the site

Table 2. Frequency of Cultural Features at Meadowcroft Rockshelter, by Stratum.

Stratum (Field Designation)	Fire Pits	Refuse/ Storage Pits	Roasting Pits	Fire Floors	Ash/ Charcoal Lenses	Burials	Specialized Activity Areas	Total
XI (F3, F8)	4	–	–	–	–	1[a]	–	5
X (F25)	1	–	–	–	–	–	–	1
IX (F9)	2	–	–	–	–	–	–	2
VIII (F12)	1	–	–	–	–	–	–	1
VII (F13)	9	–	–	–	1	1[b]	1	12
VI (F63, F129)	9	–	–	1	2	–	–	12
V–IX (F4) (outside dripline)	5	1	–	2	2	–	–	10
V (F14)	20	1	1	2	6	–	4	34
IV (F16)	35	9	3	13	15	–	3	78
III (F18)	26	2	–	8	17	–	1	54
IIb (F46)	6	3	–	6	8	–	2	25
IIa (F46)	26	5	1	1	1	–	4	38
I/IIa interface (F85)	3	–	–	–	–	–	1	4
I (F99)	–	–	–	–	–	–	–	–
Total	147	21	5	33	52	2	16	276

Source: Stuckenrath, Adovasio, Donahue, and Carlisle (1982:Table 1).
Notes: (a) dog burial, (b) human burial.

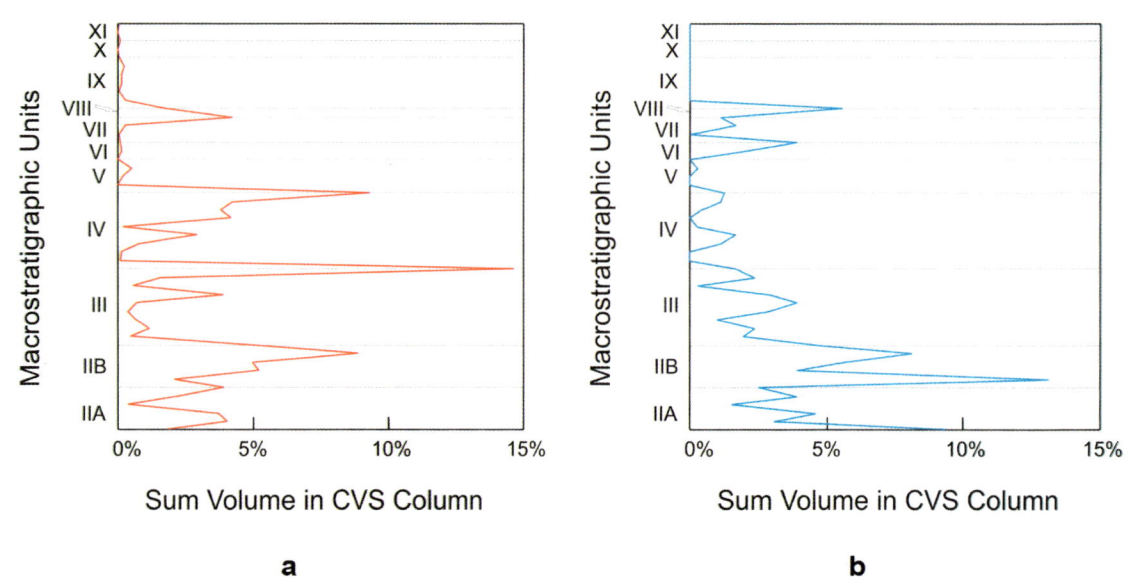

a

b

Figure 6. Distribution of material recovered from four CVS columns at Meadowcroft: (a) anthropogenic material, (b) and sandstone. (Adapted from Adovasio et al. [1982] and Skirboll [1982].)

throughout its long history, there is strong evidence to support the view that at the very least the burned materials were associated with human occupation.

Again, as detailed by Skirboll (1982:228), there is a positive correlation between burned bone, hackberries, eggshells, and other classes of material certainly or highly likely to have been anthropogenically deposited at the site. The incidence of tertiary flakes, crustacean shell, and fish scales correlate congruently with burned eggshell, bone, and thermally altered hackberries. Additionally, and significantly, the unburned fraction of the large hackberry sample does not correlate well with either the burned specimens or with the lithic flakes, crustacean shell or fish scales. (For this reason, unburned hackberries are not included in Figure 6.)

Unburned hackberries are most abundant in Paleo-Indian Stratum IIa while burned hackberries are most common in Late Archaic/Transitional Stratum III. Finally, while unburned hackberries could represent materials introduced by animals, their condition may also reflect differential food processing procedures by humans.

Overview

When the initial set of radiocarbon dates from Meadowcroft Rockshelter were published in 1975 (Adovasio et al. 1975), there were no other Meadowcrofts. Hence, the apparently pre-Clovis chronometric determinations from the site were viewed with more than casual skepticism. Indeed, it then appeared that Meadowcroft was simply another of the 500 or so localities to lay claim to putative pre-Clovis credentials since 1933. Unlike them, however, for better or worse, Meadowcroft would not enjoy its Warholesque 15 minutes of fame and then go away. As is well known to all of the participants in this symposium, since 1975 other Meadowcrofts have been excavated and, to varying degrees, published.

Most notable, of course, were the excavations and attendant analysis at Monte Verde, Chile (Dillehay 1989, 1997). This very carefully excavated and extensively reported site is widely considered to be the first to unequivocally shatter the Clovis barrier—this despite persistent and, frankly, convoluted efforts to discredit it (e.g., Fiedel 1999). Significantly, though the apparent subsistence strategy at Monte Verde parallels the broad-spectrum pattern evidenced at Meadowcroft, the durable technology represented is radically different.

Closer to home, the multi-year excavations at Cactus Hill in Virginia (McAvoy and McAvoy 1997; Johnson 1997) and the ongoing research at Miles Point in Maryland (Lowery 2007; Lowery et al. 2010) have produced lithic assemblages much closer in technological affinity to those recovered from middle and lower Stratum IIa at Meadowcroft. These notably include unfluted lanceolate to sub-triangular points, small blades or blade flakes, and polyhedral cores. Indeed, except for the raw material represented, all three of these assemblages appear to represent the same basic lithic technology (cf. Stanford and Bradley 2012). Interestingly, the subsistence regime posited for at least one of these sites, Cactus Hill, is, again, broad-spectrum foraging. Somewhat further afield in Florida, stratified inundated sites like Page Ladson (Webb 2006) and Sloth Hole (Hemmings 2004, 2005) as well as terrestrial loci like Harney Flats (Daniel and Wisonbaker 1987) suggest that an unfluted, lanceolate to lozenge-shaped projectile point horizon may underlie and be locally ancestral to the Clovis phenomenon in the far Southeast (James Dunbar, personal communication 2012).

Of course, these are by no means the only notable pre-Clovis manifestations in the Americas. Other papers in this conference address the striking discoveries at the Gault (Collins 1999, 2002) and directly adjacent Debra L. Friedkin (Waters et a. 2011) sites in Texas as well as the Paisley Cave site complex in Oregon (Gilbert et al. 2008), to name but a few examples of what is now—except to the obdurately witless—an incontrovertible reality. The tipping point in the pre-Clovis versus Clovis debate clearly was passed some time ago. Indeed, as Brian Fagan noted to first author Adovasio while perusing the stratigraphy at Meadowcroft, the remaining arguments for the Clovis-first position are "background noise" (Brain Fagan, personal communication 2011).

To our thinking, with the collapse of Clovis-first as with unraveling of any long-established paradigm in any field, a new and fresher urgency is accorded to the interrelated issues of the first Americans' lifeway(s), point of departure, means of travel, and—most vexatiously—the timing of their arrival or arrivals. It is clear from just the sites briefly mentioned here that there is no one model that provides a single set of answers to the foregoing questions. It is also clear that, unfettered as we presently are by the restrictions of a constraining paradigm, there has never been a greater opportunity to refocus on the process of the peopling of the New World. In this renewed dynamic and fluid intellectual climate, there is ample room to consider multiple ancestral homelands (e.g., Iberia rather Siberia), multiple sequential and/or roughly contemporaneous peopling pulses (including via multiple entry routes) by populations with different genetic profiles, linguistic backgrounds, technologies, and subsistence orientations.

We suggest in closing, that this welcome ferment was, in no small way, first stirred, if not catalyzed, by discoveries at what Dave Meltzer (2004:376) agreed is "arguably the most controversial site" of it time—Meadowcroft Rockshelter.

Acknowledgment

The authors wish to thank Jeffrey Illingworth for his assistance in producing this paper.

References Cited

Adovasio, J. M., Jack Donahue, and Robert E. Stuckenrath
    1980   Yes, Virginia, it really is that old: a reply to Haynes and Mead. *American Antiquity* 45:588–95.
    1992   Never Say Never Again: Some Thoughts on Could Haves and Might Have Beens. *American Antiquity* 57:327–31.

Adovasio, J. M., Jack D. Donahue, Joel Gunn, and Robert Stuckenrath
    1977   Progress Report of Meadowcroft Rockshelter – A 16,000 Year Chronicle. In *Amerinds and Their Paleoenvironments in Northeastern North America*, edited by W. S. Newman and B. Salwen, pp. 137–159. Annals of the New York Academy of Sciences 228.

Adovasio, J. M., Robert Fryman, Allen G. Quinn, Dennis C. Dirkmaat, and David R. Pedler
    1998   The Archaic West of the Allegheny Mountains: A View from the Cross Creek Drainage, Washington County, Pennsylvania. In *The Archaic Period in Pennsylvania: Hunter-Gatherers of the Early and Middle Holocene Period*, edited by Paul A. Raber, Patricia E. Miller, and Sara M. Neusius, pp. 1–28. Pennsylvania Historical and Museum Commission. Harrisburg.

Adovasio, J. M., Joel D. Gunn, Jack Donahue, and Robert Stuckenrath
    1975 Excavations at Meadowcroft Rockshelter, 1973–1974: A Progress Report. *Pennsylvania Archaeologist* 45(3): 1–30.

Collins, Michael B.
    1999   Clovis Blade Technology. University of Texas Press, Austin.
    2002   The Gault Site, Texas, and Clovis Research. *Athena Review* 3(2).

Cushman, Kathy A.
    1982   Floral Remains from Meadowcroft Rockshelter, Washington County, Southwestern Pennsylvania. In *Meadowcroft: Collected Papers on the Archaeology of Meadowcroft Rockshelter and the Cross Creek Drainage*, edited by Ronald C. Carlisle and J. M. Adovasio, pp. 207–220. Department of Anthropology, University of Pittsburgh, Pittsburgh.

Daniel, I. Randolph, and Michael Wisenbaker
    1987   Harney Flats: A Florida Paleo-Indian Site. Baywood Publishing, Amityville, New York.

Dillehay, Tom D.
    1989. *Monte Verde: a Late Pleistocene Settlement in Chile*, vol. 1. Smithsonian Institution Press, Washington, D.C.
    1997   *Monte Verde: a Late Pleistocene Settlement in Chile*, vol. 2. Smithsonian Institution Press, Washington, D.C.

Dragoo, Don W.
    1976   Some Aspects of Eastern North American Prehistory: A Review 1975. *American Antiquity* 41(1):3–27.

Faught, Michael K.
    2008   Archaeological Roots of Human Diversity in the New World. *American Antiquity* 73(4):670–698.

Fiedel, Stewart J.
    1992   *Prehistory of the Americas*. Second ed. Cambridge University Press, Cambridge.

Fitzgibbons, Phillip T.
    1982   Lithic Artifacts from Meadowcroft Rockshelter and the Cross Creek Drainage. In *Meadowcroft: Collected Papers on the Archaeology of Meadowcroft Rockshelter and the Cross Creek Drainage*, edited by Ronald C. Carlisle and J.M. Adovasio, pp. 91–111. Department of Anthropology, University of Pittsburgh, Pittsburgh.

Funk, Robert E.
    1978   Post-Pleistocene Adaptations. In *Northeast,* edited by B. G. Trigger, pp. 16–27. Handbook of North American Indians, vol. 15, W. C. Sturtevant, general editor. Smithsonian Institution, Washington, D.C.

Gilbert, M. T. P., D. L. Jenkins, A. Gotherstrom, N. Naveran, J. J. Sanches, M. Hofreiter, P. F. Thomsen, J. Binladen, T. F. G. Higham, R. M. Yohe II, R. Parr, L. S. Cummings, E. Willerslev
    2008   DNA From Pre-Clovis Human Coprolites in Oregon, North America. *Science* 320(5877):786–789.

Goldberg, Paul, and Tina L. Arpin
    1999   Micromorphological Analysis of Sediments from Meadowcroft Rockshelter, Pennsylvania: Implications for Radiocarbon Dating. *Journal of Field Archaeology* 26:325–42.

Goodyear, Albert C.
    2005   Evidence for Pre-Clovis Sites in the Eastern United States. In *Paleoamerican Origins: Beyond Clovis*, edited by Robson Bonnischsen, Bradley T. Lepper, Dennis Stanford, and Michael R. Waters, pp. 103–112. Center for the Study of First Americans, Department of Anthropology, Texas A&M University. Texas A&M University Press, College Station.

Grayson, Donald K.
    2004   Monte Verde, Field Archaeology, and the Human Colonization of the Americas. In *Entering America: Northeast Asia and Beringia Before the Last Glacial Maximum*, edited by David B. Madsen, pp. 379–387. University of Utah Press, Salt Lake City.

Guilday, John E., and Paul W. Parmalee
    1982   Vertebrate Faunal Remains from Meadowcroft Rockshelter, Washington County, Pennsylvania: Summary and Interpretation. In *Meadowcroft: Collected Papers on the Archaeology of Meadowcroft Rockshelter and the Cross Creek Drainage,* edited by Ronald C. Carlisle and J. M. Adovasio, pp. 163–174. Department of Anthropology, University of Pittsburgh, Pittsburgh.

Haynes, C. Vance
    1980   Paleoindian charcoal from Meadowcroft Rockshelter: Is Contamination a Problem? *American Antiquity* 45:582–7.
    1991   More on Meadowcroft Radiocarbon Chronology. *The Review of Archaeology* 12:8–14.
    2005   Clovis, Pre-Clovis, Climate Change, and Extinction. In *Paleoamerican Origins: Beyond Clovis*, edited by Robson Bonnischsen, Bradley T. Lepper, Dennis Stanford, and Michael R. Waters, pp. 113–132. Center for the Study of First Americans, Department of Anthropology, Texas A&M University. Texas A&M University Press, College Station.

Hemmings, C. Andrew
    2004   *The Organic Clovis: A Single, Continent-wide Cultural Adaptation*. Ph.D. dissertation, Department of Anthropology, University of Florida, Gainesville.
    2005   An Update of Recent Work at Sloth Hole (8JE121), Jefferson County, Florida. *Current Research in the Pleistocene* 22:47–49.

Johnson, Michael F.
    1997   Additional Research at Cactus Hill: Preliminary Description of Northern Virginia Chapter—ASV's 1993 and 1995 Excavations. In *Archaeological Investigations of Site 44SX202, Cactus Hill, Sussex County, Virginia*, edited by James M. McAvoy and Lynn D. McAvoy, Appendix G. Research Report Series No. 8, Virginia Department of Historic Resources, Richmond.

Lord, Kenneth
    1982   Invertebrate Faunal Remains from Meadowcroft Rockshelter, Washington County, Southwestern Pennsylvania. In *Meadowcroft: Collected Papers on the Archaeology of Meadowcroft Rockshelter and the Cross Creek Drainage*, edited by Ronald C. Carlisle and J. M. Adovasio, pp. 186–206. Department of Anthropology, University of Pittsburgh, Pittsburgh.

McAvoy, James M., and Lynn D. McAvoy
    1997   *Archaeological Investigations of Site 44SX202, Cactus Hill, Sussex County Virginia*. Research Report Series No. 8. Virginia Department of Historic Resources, Richmond.

Mead, Jim I.
    1980   Is It Really that Old? A Comment about the Meadowcroft Rockshelter "Overview." *American Antiquity* 45(3):579–582.

Meltzer, David J.
    2004   On Possibilities, Prospects, and Patterns: Thinking About a Pre-LGM Human Presence in the Americas. In *Entering America: Northeast Asia and Beringia Before the Last Glacial Maximum*, edited by David B. Madsen, pp. 359–377. University of Utah Press, Salt Lake City.

OXCAL 2012
    OXCAL Version 4.1, Oxford Radiocarbon Accelerator Unit. Electronic document available at https://c14.arch.ox.ac.uk/oxcal/OxCal.html, accessed September 2012.

Pedler, David R., and J. M. Adovasio
    2011   The Peopling of the Americas, in *Peuplements et Pr*éhistoire *en Amériques*, edited by Denis Vialou, pp. 55–67. Collection Document Préhistoriques No. 28, Comité des Travaux Historic et Scientifiques. CTHS, Paris.

Skirboll, Esther
    1982   Analysis of Constant Volume Samples from Meadowcroft Rockshelter, Washington County, Southwestern Pennsylvania. In *Meadowcroft: Collected Papers on the Archaeology of Meadowcroft Rockshelter and the Cross Creek Drainage*, edited by Ronald C. Carlisle and J. M. Adovasio, pp. 221–240. Department of Anthropology, University of Pittsburgh, Pittsburgh.

Stanford, Dennis J., and Bruce A. Bradley
    2012   *Across Atlantic Ice: The Origin of America's Clovis Culture*. University of California Press, Berkeley.

Tankersley, Kenneth B, and Cheryl A. Munson
    1992   Comments on the Meadowcroft Rockshelter Radiocarbon Chronology and the Recognition of Coal Contaminants. *American Antiquity* 57:321–6.

Stuckenrath, Robert, J. M. Adovasio, Jack Donahue, and Ronal C. Carlisle
    1982   The Stratigraphy, Cultural Features and Chronology at Meadowcroft Rockshelter, Washington County, Southwestern Pennsylvania. In *Meadowcroft: Collected Papers on the Archaeology of Meadowcroft Rockshelter and the Cross Creek Drainage*, edited by Ronald C. Carlisle and J. M. Adovasio, pp. 69–90. Department of Anthropology, University of Pittsburgh, Pittsburgh.

Waters, Michael R., Thomas W. Stafford Jr., H. Gregory McDonald, Carl Gustafson, Morten Rasmussen, Enrico Cappellini, Jesper V. Olsen, Damian Szklarczyk, Lars Juhl Jensen, M. Thomas P. Gilbert, and Eske Willerslev
    2011   Pre-Clovis Mastodon Hunting 13,800 Years Ago at the Manis Site, Washington. *Science* 334:351–353.

Webb, David
    2006   *First Floridians and Last Mastodons: The Page-Ladson Site in the Aucilla River*. Springer, New York.

Modeling Cactus Hill (44SX202)

Michael F. Johnson

Fairfax County (Virginia) Senior Archeologist (retired), 2828 Cleave drive, Falls Church, Virginia, USA 22042

Corresponding author. Tel. 703-534-9479, E-mail address: mj44fx1@verizon.net

1.0 Introduction

During the closing stages of the Archeological Society of Virginia's nine year excavation at the Cactus Hill site (44SX202) in the Inner Coastal Plain portion of the Nottoway River Valley of southeastern Virginia (Slide 1) (Johnson, 1997), it became apparent that if the hypothesized pre-Clovis age occupation there were to stand up to normal scrutiny it would have to be replicated. The author's dissertation (Johnson, 2013), which details an attempt to replicate Cactus Hill in the Nottoway River Valley, forms the basis for this presentation.

The fact that a site of Cactus Hill's age could be found in an Atlantic Coastal Plain floodplain in the Middle Atlantic Region, today, is a product of differential impacts of Wisconsin tectonic uplift and post-glacial subsidence on various river basins. As an outgrowth of cooperation with a United State Geological Survey team conducting research on Late Pleistocene climate change as manifested in the recent geology of the Middle Atlantic Region (Pavich et al., 2008), the author has developed a model for how so few Coastal Plain floodplains and possibly other eastward flowing rivers have survived inundation caused by the combined factors of post-Wisconsin sea level rise, subsidence and differing melt-water discharge rates, during the last glacial maximum (LGM).

As can be seen from the map in Slide 2, Coastal Plain down-cutting and subsequent inundation in the Chesapeake Bay, and Albemarle and Pamlico Sounds of North Carolina has produced large bays, which are not evident south (*) of the hypothesized LGM forebulge. The bays south of Pamlico Sound (*) are south of the hypothesized forebulge and, therefore, were less impacted by LGM erosion and Holocene sea level rise.

However, note that Slide 3 shows the inundation of Coastal Plain watersheds is not as pronounced in the smaller drainages, like the Nottoway. These are likely due to excessive down-cutting of the major rivers with headwaters in the Appalachians and less erosion in those rivers that drain the Piedmont. With smaller drainage basins and less access to spring run-off from alpine glaciers the spring discharge rates of those rivers would have been dramatically reduced, especially when compared to the rivers with headwaters that penetrated the Appalachians.

As a result, the Coastal Plain portions of the Maherrin and Nottoway (Chowan), Pamunkey and Mattaponi (York), and Pautuxent still have extant floodplain terrace systems (Slide 3). Those systems in the Roanoke, James, Rappahannock, Potomac and Susquehanna (Chesapeake Bay) are currently under tidal estuaries. For those wanting to discover more Cactus Hill-like sites, drainages like the Nottoway offer excellent opportunities for successful predictive modeling, similar to that discussed below.

2.0 Cactus Hill Site (44SX202)

2.1 Lines of Evidence

This paper will not rehash the details, pros, and cons of the lines of evidence supporting a pre-Clovis age occupation at Cactus Hill. That would take a dissertation. They are presented here merely to demonstrate that the best working hypothesis is that Cactus Hill, located on the Nottoway River (Slide 4) in southeastern Virginia contains a stratified pre-Clovis age component. Details on each line of evidence are available from the associated references.

2.1.1 Geomorphology

Multiple studies of the site's geomorphology indicate that Cactus Hill has a consistent depositional history (Jones and Johnson, 1997; Lund, 1999; MacPhail and McAvoy, 2008; McAvoy and McAvoy, 1997; McAvoy, et al., 2000; Wagner and McAvoy, 2004; Perron, 1999).

2.1.2 Disturbances

Multiple intact hearth and pit features mapped by the two independent investigating teams support the hypothesis that disturbance factors, while present, have had minimal impact on the overall stratigraphic integrity of the site (Johnson, 1997: Appendix G, Figure 42; McAvoy and

McAvoy, 1997: Figure 5.67 and Addendum, Figures 17 and 1). This is supported by the geomorphological analysis (MacPhail and McAvoy, 2008: 691-692).

2.1.3 Stratigraphic consistency of points

One of the more striking lines of evidence is that in most instances the vertical/stratigraphic positions of diagnostic points and pottery are consistent with their chronological sequences (Johnson, 1997: G14 (Figure 7), G19 (Figure 10), G22 (Figure12); McAvoy and McAvoy, 1997: 177). This line of evidence is also consistent horizontally.

2.1.4 Stratigraphic consistency of other stone tools

In addition to the differences between points from the Clovis age and pre-Clovis age levels in areas A/B and B, McAvoy and McAvoy (1997) and McAvoy et al., (2000: 5) reported tool type differences between the Clovis and pre-Clovis age levels. The blade levels were also noteworthy. The hypothesized pre-Clovis age artifacts (Slide 4c-i), stratigraphically and horizontally associated with a cross-mended point (Slide 4 e/f) from Area A, consisted of four quartzite prismatic blade-like flakes (Slide 4d, g, h, and i) and an expended unifacial core (Slide 4c). These were eight inches deeper than the base of a quartz fluted-point preform (Johnson, 1997).

2.1.5 Stratigraphic consistency of artifact raw materials

McAvoy et al., (2000: 3-4) noted the striking differences between the stone artifacts in the Clovis age and pre-Clovis age occupation levels. Raw material differences were not evaluated in area A due to the small size of the sample.

2.1.6 Cross-mends within, not between levels

In both areas A and B (including area A/B) cross-mends were identified within but not between the Clovis and pre-Clovis age levels (Johnson, 1997: Addendum; McAvoy and McAvoy, 1997: 105, 108, Addendum-Figure 19).

2.1.7 C-14 date consistency

McAvoy's team obtained several radiocarbon dates on the deeper levels of areas A/B and B. These were consistent with the natural stratigraphy (McAvoy et al., 2000: 2, 5, 9).

2.1.8 Optically stimulated luminescence (OSL) dates

The OSL dates from areas A/B and B were in agreement with the C-14 chronology (Feathers et al., 2006: 185; McAvoy, 2000: 9).

2.1.9 Phytolith consistency

Analysis by Lucinda McWeeney's showed an increase in phytolith weight in levels of heavy occupation as compared with levels of lighter occupation and sterile zones, which is consistent with a culturally stratified deposition (McAvoy et al., 2000: 11-12).

2.1.10 Phosphate consistency

Although not as pronounced as the results from the phytolith analysis, the variations in phosphate content from occupation levels to sterile zones were still significant, demonstrating a "discernible anthropogenic modification of the soil in the form of increased phosphate in the occupation levels (McAvoy et al., 2000:11)."

2.1.11 Horizontal replication of vertical patterns between Areas A, A/B, B and D)

With respect to lines 1-7 above, where samples were recovered from more than one area, vertical consistencies were replicated from one area to another.

2.2 Discussion

Among the various lines of evidence the most noteworthy is that the point types shown in Slide 4 (a, b, e/f) were only recovered from below mapped and/or dated Clovis age levels. Had there been a "donor" level in or above the Clovis age levels, none were identified (Johnson, 1997; McAvoy and McAvoy, 1997).

Technologically, neither point type (a and b, and e/f) is consistent with Clovis age point technologies from the Middle Atlantic Region (Johnson, 1989, 1996). The two points (a and b) from areas B and A/B are consistent with a uniform, distinct technology. Both are clearly reworked and at the ends of their use lives. They are not preforms for fluted points as has been suggested.

During the ten years the author was the senior researcher for the McCary Virginia Fluted Point Survey (McCary, 1996), the author observed several full sized points that involved this technology and raw material. All were similar in size to Clovis age fluted points. However, none had lateral edge grinding; none were fluted; all were exceptionally thin; and all were made of greenish and gray meta-volcanic stone. Clovis age fluted points from this region are almost always ground along the lateral and basal edges in the hafting area; fluted on one or both faces, and bi-convex in cross section forward of the flute terminations. Additionally, and most importantly, fluting produced an "I-beam" cross section in the hafting are, which is not present on Cactus Hill-like points.

The two cross-mended pieces of the point mid-section (Slide 4e/f), is unlike any other point recovered from the excavations at Cactus Hill. It is excessively reworked and made of a metamorphosed-quartz-like material, which is a rare material on the site.

One of pre-Clovis's most ardent critics, Stuart Fiedel (2012: 659) concurs in the author's oft stated premise that the simplest explanations for the numerous lines of evidence from Cactus Hill are that there is a pre-Clovis age cultural component, albeit ephemeral, at Cactus Hill. It is also evident (Fiedel, 2012) that there are alternative explanations for each line of evidence. However, it is highly unlikely that the simplest explanations are all or even mostly in error.

The hypothesized pre-Clovis age component at Cactus Hill is, without question, the best working hypothesis to explain what was recovered there. However, it is incumbent on both the supporters and critics of that hypothesis to either find supporting or negating evidence. The effort underlying the following research was pursued in that vein. As such, it is a logical and necessary next step for the site's proponents and detractors.

3.0 The Barr Site (44SX319)

The Barr site testing was an early attempt to model Cactus Hill. It was made near the end of work on Cactus Hill. Local resident, Annette Barr, who reported the fluted points that resulted in the 1993 excavations at Cactus Hill, showed the author another sand quarry, approximately 11 miles downriver, that had produced artifacts, including Nottoway River chert, similar to that from the nearby Williamson Paleoamerican site (McCary, 1951, 1996).

The author led two testing sessions on the site. Slide 5 shows the excavation overview. The testing was based on surface landforms. Like Cactus Hill, the site was on the upstream edge of a point bar exposed to the north. The main part of the site also was on slight rise above the floodplain, which had produced potentially early artifacts.

The testing resulted in a very similar pattern to that which had been found at Area D at Cactus Hill. At Barr the deepest Early Archaic Palmer phase artifacts, items a-e in Slide 6, were underlain by sterile sediment as shown in Slide 7. The presence of a distinct fire cracked rock hearth with cross mends, shown in Slide 6, strongly reinforce the potential integrity of sites in Tarboro soil. One must control for the disturbances as one must with any site.

The key to understanding why the Barr site is not a prime candidate for pre-Younger–Dryas occupations is both the lack of a protective high clay bank and the site's surface elevation above the current Nottoway River channel. The current top of the landform is no more than 15 feet above the current river bed and is subject to flooding even today. When coupled with the stratigraphy discovered in Area D at Cactus Hill, Barr helped demonstrate the significance of both site surface elevation and the protection of older soils by underlying clay banks (elevated/truncated paleosols).

4.0 The Model

The purpose of the follow-on research was designed to locate buried pre-Younger Dryas age cultural occupations in the Nottoway drainage. Whether one accepts or remains skeptical about the hypothesized oldest occupations at Cactus Hill, such a line of research is the appropriate next step in evaluating the Cactus Hill pre-Clovis age hypothesis. Cactus Hill cannot and should not stand alone. It should have been part of a larger settlement pattern that should be replicable.

4.1 Criteria:

4.1.1 Well-drained living surface

Slide 8 shows a hypothesized, north-south, cross-section of the Nottoway River floodplain between Areas A, A/B, B, and D, which were separated by a sand quarry. The quarry removed a significant portion of the sites center. The top edge of the clay bank can be seen in the inset, which seems to have influenced the sand quarrying to its north. This clay bank was to prove pivotal in the final interpretation of the site.

Cactus Hill is on extensive, deep, well-drained loamy sand deposits. It would have provided a relatively dry camping area. Since Cactus Hill and most of the other Paleoamerican sites identified previously by McAvoy (1992) were located on Tarboro loamy sand, mapping the distribution of Tarboro soil was the first step used in targeting other potential sites (Slide 9).

4.1.2 Exposure to strong (north) winds

The second criterion was that Cactus Hill has a north aspect (Slide 8). As such it would have been open/exposed to prevailing north and northwest winds off the glacier, which would have made the site freer of flying, biting insects. As a result, the areas within Tarboro soil deposits would have to possess a northward aspect to fit. This was the basis for the assumption that Cactus Hill was occupied during warmer months. It is not likely Cactus Hill would have been occupied in the pre-Younger-Dryas winter.

4.1.3 Flash flood avoidance

The third criterion was that, although well-drained soil was important, it needed to be high enough not to have been exposed to flash flooding. Because of the results from Cactus Hill Area D; peripheral areas to Cactus Hill; and the testing at Barr, it was clear that Tarboro covered terraces below 15 feet high contained evidence of only post-Younger-Dryas cultural occupations. As a result, areas 20 feet and higher above the adjacent Nottoway River Channel, like Cactus Hill and Blueberry Hill, on Tarboro soil; and possessing a northerly aspect were considered prime testing areas. It should be noted that in one area - Chub Sandhill - the opportunity arose to test additional areas that did not fit this and other parts of the model. This additional control testing only added to the reinforcing results from Area D; from peripheral testing around Cactus Hill and Blueberry Hill; and on Barr.

4.1.4 Younger-Dryas scouring

The premise for the fourth and most critical criterion was that the immediate, ca. 12,900 BP, Younger-Dryas onset caused significant scouring of the Nottoway River floodplain. Therefore, one had to find residual pre-Younger-Dryas landforms to find evidence of pre-Younger-Dryas soils and cultural occupations.

Based on work at Cactus Hill, Barr and preliminary testing at nearby Blueberry Hill, the best places to find surviving pre-Younger-Dryas landforms was predicted to be in Tarboro deposits; with northerly aspects; >20 feet above the current Nottoway river; on top of deeply buried, horizontally truncated paleosols (clay banks) that parallel old river channels.

It appeared that the buried clay banks served to prevent older soils/sediments, immediately above them, from being scoured away by Younger-Dryas flooding, which was evident from control testing at Cactus Hill and Barr. At Cactus Hill, as can be seen in Slide 8, Area D is underlain by coarse sediment down to the level of the bottom of the current flood chute.

McAvoy's team excavated Area D and recovered artifacts down to approximately three feet deep, where they recovered several Early Archaic carbon dates, the earliest of which was associated with an Early Archaic Palmer phase occupation (McAvoy and McAvoy, 1997: 169-170). The author confirmed the natural stratigraphy later with a bank cut. This indicated that Area D was on a younger landform than Areas A, A/B and B, which brought the significance of the clay bank underlying Areas A, A/B and B into focus. Coupled with test excavations at the Barr site, this assessment led to the successful model testing discussed below.

5.0 Method/Methodology

Excavations at Cactus Hill involved large blocks with detailed horizontal and vertical control of data recovery (Johnson, 1997; McAvoy and McAvoy, 1997). Following the recovery by the author's team of a pre-Clovis age lithic feature in Block A, 5x5-foot sub-squares; 1/2-inch sub-levels; three dimensional piece plotting of all potential artifacts, including individual pieces of charcoal; and systematic partial 1/16-inch window screening were employed for maximum quality and quantity data retrieval.

Based on the results from the Cactus Hill excavations, the author developed the warm season model for the Paleoamerican occupations there. This model was enhanced by several landform and soil factors that were critical to its potential success (see above).

Since the Cactus Hill excavation methods would be too cumbersome to be used on a site identification survey, the follow-on research into other potential site locations involved methodological approaches more suited to locating and evaluating previously unidentified sites. However, the approach was complicated by the fact that the target site components were expected to be ephemeral and located in deeply buried contexts. Economically isolating ephemeral site in such contexts was the first major hurdle. Detecting them would require more than widely spaced, shallow shovel testing, employing 1/4-inch mesh, dry screen samples, all of which are commonly used in cultural resource management (CRM) surveys (government archeology).

5.1 Four Phases

5.1.1 First phase

This phase involved identifying landforms and macro-soil conditions in the Nottoway River floodplain, which were comparable to those at Cactus Hill.

5.1.2 Second phase

This phase involved identifying comparable geological features and micro-soil conditions within those landforms.

5.1.3 Third phase

This phase involved identifying the buried cultural stratigraphy of the targeted parts of the landforms.

5.1.4 Fourth phase

This phase involved test excavations of sites that could possibly produce Clovis or pre-Clovis age, Cactus Hill-like components. (Artifact producing areas/sites in the vicinity of those components that were predicted not to contain Clovis or pre-Clovis age components were also tested in an attempt to negate the model.)

5.1.5 Discussion

The choice to explore downstream from Cactus Hill was made to avoid conflicting with McAvoy's (1992; McAvoy and McAvoy 1997) primary research areas. Fortunately, the author was able to obtain rough copies of the field maps from a recent, unpublished, United States Department of Agriculture soil survey of Sussex County, Virginia, where Cactus Hill is centrally located (Slide 9). The soil survey was critical to identifying similar soils to those at Cactus Hill (phase 1 above).

With respect to phase 2, the author determined that the most efficient and economical approach, which was employed in 2001 in locating the nearby Blueberry Hill site, was to use three-inch diameter bucket auger sampling on a relatively tight interval grid. That could be used to economically obtain preliminary information on deeply buried soil features and landforms.

With respect to phase 3, the horizontal size differences between a one-foot square shovel test and a three-inch diameter auger test for recovering archaeological data was overcome by using 1/16-inch window screen to sift the auger samples, rather than the 1/4-inch hardware cloth commonly used with shovel testing. Although the comparison was not statistically evaluated, the author assumed it would suffice for an archaeological survey. The author has almost 40 years' experience at knapping stone, with a primary focus on Clovis age technology. The amounts of sub-1/4-inch shatter and debitage produced from lithic reduction are quantifiably overwhelming by comparison to debitage larger than1/4-inch.

Once auger testing detected a potential buried activity surface in the desired context, the phase 4 assessment would be employed. That involved selectively oriented 5x10-foot trenches, positioned as to optimize recovery of potential diagnostic artifacts from the targeted depths.

As mentioned above, auger testing had been used to locate the Blueberry Hill site, downstream from Cactus Hill. That site was also tested using 5x10-foot trenches. These trenches were initially excavated in four-inch levels from the bottom of the plow zone. However, that testing encountered a deeply buried, potential Paleoamerican, activity surface. The author thought that the method used there was not rigorous enough for the implications of the model being tested.

Therefore, with only minor exceptions, the author divided the 5x10-foot trenches into 5x5-foot sub-squares and 2-inch levels below the plow zone. In more sensitive areas the 2-inch levels were further split into 1-inch sub-levels.

Artifacts larger than a United States 25-cent piece (one-inch diameter) were two-dimensionally mapped in the levels. The soil/sediment matrix was sifted initially through 1/4-inch mesh. However, on Rubis-Pearsall and Blueberry Hill that was eventually changed to 1/8-inch mesh. The shift to 1/8-inch was done, because small (usually <1/4-inch) artifacts and charcoal have a tendency to percolate down in the soils/sediment contexts of Tarboro sites. The 1/8-inch mesh was chosen a check on that un-quantified assumption. It was later to prove prophetic, when a crinoid bead was recovered in the fine screen residue from the deepest recognized cultural level during testing on Blueberry Hill.

All artifacts/geo-facts were recovered by provenience and returned to the lab. Nothing but modern flora and fauna were discarded in the field. From 1996 onward, in the author's excavations at Cactus Hill and elsewhere in the Nottoway River Valley, maximum data recovery was employed within the context of the appropriate method (survey, testing and excavation).

Methodologically, phases 1 through 4 were adequate to demonstrate that the model did work, at least in part. Even in the areas where questions remained, the testing opened the possibilities that those questions could be answered by full-scale excavations of the remaining parts of the sites.

The discovery of one deeply buried Clovis age site and one possible deeply buried pre-Clovis age site on landforms where there were minimal indications that sites were even on the landforms, demonstrates beyond question the efficacy of the model and method.

6.0 Chub Sandhill

Although we tested the Blueberry Hill site in 2001/2, the results at that time proved to be controversial and had to be further tested in 2010 to resolve weaknesses. It will be discussed below as the third test of the model.

Slide 10 shows the section of Sussex County soil maps containing the Chub Sandhill testing area (upper left inset); the testing areas including the Koestline, Watlington, and Rubis-Pearsall sites (annotated map), and an example of how individual auger cores were laid out (lower right inset).

The east-west, 2,800-foot long by 100-foot interval auger transect was designed to get a preliminary assessment of the point bar feature's geomorphology. It involved only rough field analysis of the soil/sediment data. No efforts were made to recover archeological data from the auger samples.

Slide 11 shows a cross section of the results from the field analysis of the data recovered from the 26 cores evaluated along the east-west transect across the point bar. It revealed three terraces above the current meander channel with a flood chute behind the first terrace/levee. Of primary importance, we located an elevated paleosol (g) underlying the third terrace. This was separated from the paleosol under the first terrace and flood chute by an apparent old river channel, which appears to have been buried by sediment. Using 16 feet of auger we were never able to reach the bottom of the overburden, due to saturated sand at 14 feet deep. The results from this transect interval sampling served as the basis for subsequent auger testing and test excavations done on the point bar.

Later auger transects were run parallel to the outer levee, and second and third terraces. These transects were roughly aligned north-south from the approximate up-stream ends of the terraces toward the south. Each core sample from these cores was sifted through 1/16-inch window screen and all artifacts were recovered by level. The artifact recovery demonstrated which depths we could expect artifacts.

6.1 Koestline site (44SX332)

Based on expected results, test excavations were started at the Koestline site, located on the upstream (north) end of the levee (see Slide 10). It was the least likely location for early cultural horizons.

Slide 12 shows the diagnostic artifacts from the two 5x10-foot test trenches excavated on the site. The earliest artifacts were the small Morrow Mountain points (Slides 12k and 12p-q), which date to approximately 7,000 BP. It is noteworthy that all of the diagnostics and most of the other artifacts occurred from the surface down to approximately 21 inches deep. The cultural zone was underlain by coarse, upward fining, poorly sorted sand. This indicates that the occupation there probably began shortly after the 8,200 BP cold event.

6.2 Watlington site (44SX331)

Watlington, located on the second terrace (see Slide 10), was the second site tested. Much of the site had been destroyed by an old sand quarry. As a result, only the eastern edge away from the likely river channel (current flood chute) was testable.

Slides 13 and 14 show the diagnostic artifacts by level from the two block excavations, which involved six 5x10-foot trenches. Here the diagnostics began with the Middle Archaic Guilford phase (Slide 13s) and extended back as far as early, Early Archaic, Kirk corner notched phase (Slide 13y). Most artifacts occurred below 15 inches deep and above 33 inches deep. Almost no surface expression of a site was found. This indicates abandonment as a major occupation area after Guilford, possibly during the Middle Archaic.

The presence of artifacts as deep as 60 inches suggested the possibility that there are earlier components, which were not identified. However, most trenches were excavated to below 40 inches and two were excavated to 60 inches deep. Several very small flakes were recovered from the top of a lamellae zone at 60 inches deep, which was separated by 12 inches of relatively clay free sand from the next artifact producing level above. With the recognized propensity for small debitage to percolate through the profile, it is likely that the flakes were caught on top of a Bt trap formed by the less permeable lamellae.

6.3 Rubis-Pearsall site (44SX360)

Part of the reason the third terrace was excavated at the end was that the land was in private hands and we did not want to have it generally known that we were excavating for Paleoamerican sites in that area. The Virginia-North Carolina border is well known for its hard core artifact hunters, who do not merely surface collect but destroy stratified Paleoamerican sites in search of Clovis age points. This was a constant problem during the nine years of excavating on cactus Hill. This problem was resolved, when the Virginia state government acquired the property and added it to the Chub Sandhill preserve, where we were working.

Methodologically, it is noteworthy that the third terrace was initially tested with one auger core located immediately inside the state property boundary, which ran along the site's western edge. Only one very small quartzite flake was recovered from fine screening the entire core. That flake came from the 37-43-inch deep auger core sample. Otherwise no surface indication of a site was noted in either a forest road running along the property boundary or exposed surfaces on the site.

The initial testing involved three-inch bucket auger sampling on a 50-food grid with 25-foot in-fill samples within the site core. These were excavated in a similar manner to those done on Watlington. Each core was evaluated and then fine screened (1/16-inch mesh) by sample, back into the auger holes. Resulting artifact distributions were used to place test trenches.

Slide 15 shows the approximate edge of the buried clay bank; test trench locations, and major diagnostic artifacts most relevant to the Paleoamerican model. Initial testing (RP1-RP6) was begun off (west of) the clay bank to test for Paleoamerican components in the area not predicted to contain older cultural material. Those pits produced nothing older that Early Archaic Fort Nottoway points and no potential Clovis or pre-Clovis age material or contexts.

As predicted, the third terrace produced a Clovis age component (upper left inset) deeply buried in the terrace. This was in the first test trench (RP7) excavated on the clay bank. No other Clovis or pre-Clovis age evidence or contexts were recovered from RP8-RP10.

Although the test trench results from the site demonstrated severe disturbance in many areas, the cross-mend of a jasper flake onto its core (lower left inset) from the same level and square as the fluted point, suggests that there are areas of minimal disturbance. A quartzite adz was also discovered from the same level and square as the fluted point and core. The Early Archaic Palmer age point (lower right inset) was recovered from nearby trench 8, which (due to wall collapses) produced the only walls suitable for OSL sampling.

Milan Pavich (et al. 2008), a geologist with a U.S.G.S team studying late Pleistocene climate volunteered to run two OSL samples on the site as part of his team's study of Pleistocene dunes. As can be seen from Slide 16 the two dates, 20.5 ± 2.6 ka on the Early Archaic level and 18.8 ± 2.4 ka on the Clovis age level, were not in sync with the cultural stratigraphy. However, they supported one of the major discoveries from our excavations in the Nottoway.

Years earlier on the Lorton Town Center 1 site, Dan Wagner (1998: Personal communication) noted to the author that ants and worms do extensive mixing of the top 18 inches of any normal soil profile (Johnson and Anthony, 2002: 21). On that site the author also noted that 4,000 year old fire cracked rock filled platform hearts had migrated (as units) to approximately 5-6 inches below the surface. While it was apparent from Wagner's observation that ants were depositing soil on the surface, it was also clear that they were removing that soil from beneath the artifacts. This was not only causing features and artifacts to be slowly buried but also they were percolating slowly down through the surrounding soil as tunnels and chambers were collapsing under them.

Cactus Hill, because of the tight precision in which it was excavated, revealed that smaller artifacts percolated far faster than the larger artifacts. As a result, the multiple grain OSL data from Rubis-Pearsall indicated that the artifacts in Tarboro loamy sand, and probably most other soils, could be in soil that is older than the artifacts themselves. That would indicate that the site is, at least on the scale of medium to coarse sand grains, somewhat disturbed. This phenomenon has recently been studied in depth in Florida (Rink et al., 2012; Thulman, 2012).

However, it has been demonstrated repeatedly, as with the re-fitted jasper core and flake in Slide 15, that although smaller items migrate downward through the profile, the larger items (and sometimes even the smaller items) are in good context relative to each other. This appears also to apply to larger pieces of charcoal and also charcoal concentrations associated with hearths.

From the standpoint of the predictive model the anomalous dates were not as big a problem as they may seem. The purpose of the model was to predict and identify old landforms that would produce Paleoamerican cultural levels. Although Rubis-Pearsall did not produce clear evidence of pre-Clovis age artifacts, it produced Clovis age artifacts and a landform that was old enough to have been occupied during and prior to that time.

It is noteworthy that the auger testing behind (east of) the clay bank produced evidence of very deep artifacts. One auger core sample, 80-85 inches deep, produced a medium sized quartzite flake. Two other deep auger core samples produced very small flakes but these fell within the size range of the normal down-drifting small debitage. Considering the instability of the deeper Rubis-Pearsall sediments, it was not safe to take any of the test pits below 50 inches deep without incurring wall collapses.

Slide 17 shows an idealized series of transects showing the apparent migration of the main Nottoway River channel from east to west from the Younger-Dryas onset through today. The cultural data indicate that the occupations tended to migrate to the west, following the changes in the main river channel. This led to abandonment of older terraces in favor of those closer to the river. The earliest landform on the third terrace (Rubis-Pearsall) was used after the apparent abandonment, as is evident from the discovery of moderate Early Archaic and more ephemeral Middle Archaic and Woodland components in the upper levels. These could have been associated with the apparent buried spring (perched water) identified during auger sampling (see Slide 15).

Because of the Clovis age component and ca. 20,000 year old OSL dates, the Chub Sandhill part of the model testing was considered a success. Although, no definitive potential pre-Clovis age artifacts were identified, the site still has potential to produce cultural material from that period.

A fact that was not address was that the terraces further south would have been along the hypothesized pre-Younger-Dryas channel. It is possible that even more pristine pre-Younger-Dryas aged archeological sites could be buried in those terraces. The following discussion of the Blueberry Hill site offers strong support for that possibility.

7.0 Blueberry Hill Site (44SX327)

7.1 Preliminary testing

Testing at Blueberry Hill was initially the result of an attempt to define the eastern boundaries of Cactus Hill (Slide 18). Test trenches/pits E1, E2 and F1 were initial attempts to determine if there were other occupations associated, first, with the floodplain and, second, with the spring head at the upper end of a wetland adjacent to the eastern edge of Cactus Hill.

Test square F1 produced historic ceramics well below the surface and no positively identified prehistoric artifacts in poorly sorted coarse to very coarse sand, indicating a recent alluvial deposit. Test trenches E1 and E3 produced relatively shallow prehistoric occupations above a well-developed Bt soil.

However, test trench E1 produced a fire cracked rock hearth containing the two large bifaces shown in the inset in Slide 18. The feature and bifaces were in the level below the Middle Archaic Morrow Mountain II point also shown in the inset. This trench, unlike E2, was excavated through the Bt horizon, which produced the broken prismatic blade-like flake also shown. A trench along the northeastern wall og E1 was dug to 84 inches deep before hitting a water table, probably associated with the spring aquifer. These trenches were completed before the significance of the clay bank was discovered.

Once the importance of the clay bank was recognized as the key to protecting pre-Younger-Dryas landforms from scouring an effort was made to address the possibility that the pre-

Younger-Dryas, Nottoway River channel may have run west to east rather than its current south to north direction (Slide 19). This was begun in 2001, when Mike Waters of Texas A&M University, and the author made an effort to detect a buried clay bank (shown in Slides 19 and 20) under northern edge of the sand ridge that was to become Blueberry Hill.

Although not tested, it is possible that there is another channel or relic flood chute on the south side of the Cactus Hill/Blueberry Hill sand ridge The unique stratigraphy observed in test pits E1 and E2 suggests that there is either an old flood chute there or the sand ridge may cover a braid scar/chute between the two archeological sites. The small spring and wetland south of Cactus Hill, which drains toward the current Nottoway River channel, may be post Younger-Dryas.

7.2 2001-2002 testing

After Waters and Johnson auger tested the Blueberry Hill landform and discovered the buried clay bank, the author excavated two test trenches. These were located on the north side of the sand ridge (Slide 20).

7.2.1 Test trench BBH1

The first, a preliminary 5x10-foot test trench (BBH1), near Waters' and Johnson's auger transect, was excavated in 2001. The test produced only eight artifacts of which four were immediately below the plowzone and four were fire cracked rocks. Slide 21 shows the profile of the trench and the two deepest artifacts, which came from approximately two feet below the surface. One is an end thinned early stage quartzite bifaces and the other is an edge worn flake (inset). Ironically both were found together in the east wall of the trench. Had the trench been placed three inches to the west they would not have been found and no further testing would have been done on the site. The fact that they were almost touching suggested that the area had potentially excellent integrity in the deeper levels.

7.2.2 Test trenches BBH10, BBH11, and BBH12

In 2002, based on the results from the E1 and E2 testing, which indicated the clay bank was truncated between Blueberry Hill and Cactus Hill, an auger testing effort was made to identify the western end of the Blueberry Hill clay bank. Its possible location helped determine the location for a 5x30-foot long trench, which was divided into three, end-to-end 5x10-foot sub-trenches (Slides 21 and 22). These trenches were designated BBH10, BBH11 and BBH12, due to the fact that the site boundaries had not been defined and we were not sure if earlier test pits were on or off the site.

Slide 23 shows the stratigraphic locations of significant diagnostic artifacts, which were mapped and their stratigraphic locations pined in the adjacent east wall (see center photo in Slide 22). As is readily evident from Slide 23 the diagnostic artifacts lie in two planes. The upper artifacts are four, Late Archaic, Savannah River points. These are separated from the lower series of artifacts by a relatively sterile zone, approximately 10-15 inches thick.

The deeper artifacts represent a clear activity surface of large artifacts. They include (left to right insets) the distal end of a quartzite prismatic blade-like flake; a broken meta-volcanic bifaces with a burinated edge (right edge in photo); a quartzite broken sidescraper; a large sandstone abrader, and a quartzite Cactus Hill-like point base, the latter being mapped in BBH10. The tools were all mapped in the 30-34-inch depth and the point was mapped in the 34-36-inch depth. Based on the actual depths they were essentially from the same plane. The artifact in the lower right is a quartzite fluted point base that was recovered from the 14-18 inch depth in a screen. Its depth is problematic due to the fact that it was not mapped *in situ*, and there was a large disturbance in the center of the trench that, due to wall collapse issues, was not removed first. This was considered a test, and therefore was not as rigorously excavated as Cactus Hill.

Based on the results and particularly the nature of the deeper artifacts, it was assumed that the deep activity surface probably represented either a Paleoamerican or pre-Clovis age activity. However, the tools do not fit the formal Clovis tool kit. The only Clovis age–like artifact recovered from this and subsequent testing on the site is the quartzite fluted point base (Slide 24, right).

The lanceolate Cactus Hill-like point base (Slide 24, left) is made of quartzite but technologically resembles the two meta-volcanic Cactus Hill points from the deepest levels of Cactus Hill (see Slide 4). However, it also resembles the pre-Early Archaic, Hardaway blades, reported by Coe (1964) from beneath Late Paleoamerican Hardaway sidenotched points at the Hardaway site in North Carolina. Hardaway blades have not been dated and no Clovis age artifacts were recovered from Hardaway (Coe, 1964, Daniel, 1998).

7.3 2010 testing

Due to valid geomorphological questions raised by Wagner (2009: Personal communication), the author revisited the site in 2010. The main questions were about the age of the landform, which Wagner had auger tested near and contended did not appear to be old enough to contain pre-Younger-Dryas occupations. This was based on elevation, which he had been led to believe was lower than Cactus Hill, and the apparent lack of lamellae, which were prominent at Cactus Hill, indicating an older, more stable landform.

Considering the low level rigor with which the site had been tested in 2001 and 2002, the author considered the criticism potentially valid. However, the main reason the author had tested the Blueberry Hill landform was because the USGS 7.5 minute series, Sussex quadrangle had mapped the landform within on five-foot contour interval higher than Cactus Hill (see Slide 4). Still, the author felt it imperative to re-open the site to test the site's validity as containing a pre-Younger-Dryas occupation.

7.3.1 Transect interval sample

Subsequent logging had obliterated any surface evidence of the 2001 and 2002 test trenches, to the point that, even though they had been shot in with a transit from Area A at Cactus Hill, we were not able to re-locate them. As a result, a more rigorous testing strategy, similar to that used on Rubis Pearsall, was employed. It involved an initial auger testing program based on a 20-foot interval grid to determine the best locations for later test trenches (Slide 25). Ironically, this method led to the rediscovery of the 2002, 5x30-foot long trench and the re-location of the deep activity surface located by that trench (Slide 26).

7.3.2 Test excavations

It was decided from the outset that the level of rigor on this site should be greater than other testing methods used previously on Bar and Blueberry Hill. Like Chub Sandhill, trenches were divided into 5x5-foot sub-squares and levels were reduced to two inches thick with one-inch sub-levels being used in the lower levels. Also all artifacts larger than a quarter (25-cent piece) were mapped. All soil residues were sifted through 1/8-inch mesh. One methodological short-cut, which was to come back to haunt the results, was that the plowzone and first, two-inch level were sifted through 1/4-inch mesh, This was due to the heavily disturbed nature of the plowzone, which contained large amounts of dead vegetation from the recent tree harvest and re-planting and recent historic trash. The first level below the plowzone was a mixing zone, impacted by immediate sub-plowzone roots, and rodent and insect burrows.

Tarboro soil appears to be relatively free of deep root activity. This is likely due to the poor nutrient value of the loamy sands – there is not much for roots to eat below the plowzone. Hickory tap roots, which anchor the trees in the paleosols and covering Bt horizons, are a problem where they are prevalent.

7.3.2.a Test trench BBH2

The first test trench, BBH2, demonstrated the potential integrity of the site. Slide 27 shows the larger artifacts in stratigraphic sequence from the northern sub-square (BBH2B), which was located higher up on the sand ridge than the other test trenches (see Slide 26). Note the quartzite bifaces cross-mend between artifacts "27b" and "27c". The two fragments were horizontally overlapping. Artifact "27b" was at the bottom of level 11B, and "27c" was at the top of level 12A. Their depth also put them and the deeper items within the depth range of the deep activity surface identified in 2002.

7.3.2.b Test trench BBH3

Slides 27 and 28 show what appears to be a two-tiered artifact distribution in the area of the deep activity surface. The 2002 test trench was discovered 2.5 feet to the east of BBH3. In both sub-squares a significant artifact gap was identified between the bottom of the Late Archaic Savannah River component (ca. 4,000 BP) and the deep activity surface. It is 11 inches in BBH3A and 18 inches in BBH3B.

The large quartzite hammerstone (Slide 28f) was recovered from just below the north arrow in the floor plan shown in Slide 29. Slide 29 shows the locations of the quartzite hammerstone (29d) and unifacial sandstone (29f) and quartzite (29g), unifacial, split cobble choppers on the same level, five feet apart. No other artifacts like the two choppers have been recovered by the author in any of the work done in the Nottoway River valley. Of course, that does not date them to any particular time period. However, their clear stratigraphic association with the Cactus Hill-like point (see Slide 23) recovered from the same depth in BBH10, less than ten feet to the east is an excellent association.

7.3.2.c Test trench BBH7 (crinoid bead)

The probable crinoid bead (Slide 30d and inset) recovered in the lab from the 1/8-inch screen residue from the 31-32-inch depth in BBH7A is instructive from several standpoints. First, it is likely a bead that was transported to the site across the Blue Ridge Mountains from more than 125 miles to the west (David B. Spears, Virginia State Geologist, Personal communication 2011).

Second, it was not found in either troweling or in the dry screen residue – it was recovered in the lab during washing of the dry screen residues. Throughout the Cactus Hill and subsequent research, no items, other than obvious modern biota were discarded in the field. Everything was brought back to the lab and sorted under lab conditions. This was done to reduce inherent biases ubiquitously introduced by crew member differences and field conditions.

Third, the original depth of the bead is problematic. Although it is tempting to ascribe its original depth to the deep activity surface, which was its context when recovered, smaller items of that size are notorious for percolating down through the profiles. A case in point is a small lead shot recovered in the lab sort from 30-32 inches deep in BBH2. There is no doubting that the shot originated in the plowzone. This does point out the problem created by not having applied the same screen size to the plowzone and level 2 as the rest of the depths. Had the bead come from the plowzone and we had screened the plowzone through 1/8 rather than 1/4-inch mesh, at least there would have been an opportunity to recover other crinoid beads there. The plowzone did produce Woodland pottery (Slide 28a and b, and Slide 29a). All can be stated is that the small crinoid bead was recovered in that deep level but it is not clear that it originated there. Had no beads been recovered from the plowzone using the more rigorous method, at least one could state there were no similar beads recovered from above it. The lead shot is instructive.

7.3.2.d Test trench BBH6 (lamellae)

One of the potential problems raised by Wagner (2009: Personal communication) was that he had not identified lamellae in his auger testing in the area of the Blueberry Hill landform. Although we did not identify any strong lamellae in any of our auger tests along the 20-foot interval transects or our first nine test trenches, we did locate very strong lamellae in BBH6. The trench was located approximately 100 feet east of the main excavation. Although we had observed feint lamellae in the walls of BBH3 it is clear that strong Cactus Hill-like lamellae are not common on the site. However, they were restricted to a relatively narrow zone on Cactus Hill and appear to be similarly restricted on Rubis-Pearsall.

Lamellae have been demonstrated in a lab experiment to start forming in a saturated environment in 16 days (Bond, 1986). For lamellae to form in the Cactus Hill and Blueberry Hill loamy sands, the optimum porosity and iron-rich clay contents have to be there, and there would have to be some mechanism by which those landforms could be saturated for enough time to dissolve and move the clays through the sand matrix.

Wagner is correct that the strong lamellae are indicative of time and stability. However as demonstrated at Rubis-Pearsall their apparent absence does not preclude great antiquity and stability on the landform. There were lamellae on Rubis-Pearsall but they were not very well developed. The discovery of strong lamellae in BBH6 (Slide 31) merely supports the potential temporal comparability of the Blueberry Hill and Cactus Hill landforms.

The fact that strong lamellae formed on the highest landforms adjacent to the old channels indicates that, for the soil to have been saturated long enough to form such strong lamellae, the landforms had to be very old, as contended by Wagner. Hurricane Floyd saturated the lower levels at Cactus Hill for more than two weeks, which demonstrated a potential mechanism for lamellae formation over a long period of time.

7.3.2.e Test trench BBH6 (OSL dates)

In order to add another line of evidence to resolve the question of age, the author, with the assistance of David Thulman, contracted for six OSL dates from the western profile of BBH6A. They were recovered as shown in Slide 31. The resulting dates, run by the USGS in Denver and analyzed by Kevin Burdett of the Florida Geological Survey, are shown in Slide 32. They were on multiple grains. According to Rink (2012: Personal communication), due to bioturbation, at minimum they demonstrated that the age of the dune was more than 13,000 years old. From the age of landform perspective this was critical to the model.

The author decided to highlight the best quality dates just to see what would happen. Surprisingly, a distinct chronological pattern, consistent with the hypothesized geomorphology of the Cactus Hill part of the sand ridge, emerged. As can be seen in the table in Slide 32 the sequence of highest quality dates, beginning at the 20-inch depth is approximately ca. 13,000 BP. This is followed at 30 inches by a date of approximately ca. 15,800 BP. The closeness in quality of the ca. 11,900 BP date from that depth cannot be discounted. The 40-inch date of ca. 25,400 BP and 50-inch date of ca. 30,600 BP are also consistent. The oldest dates conform to a post-Heinrich III depositional history for the dune's formation (Pavich 2012: Personal communication).

The three diagnostic date markers are shown to the right in Slide 32. They, too, are coincidentally in proper date sequence with the highest quality OSL dates. However, they were excavated 100 feet to the west, so the association is merely relative and not direct. This indicates a need for the area of the deep activity surface to be re-tested, with the express purpose of taking and analyzing a comparable suite of OSL samples.

7.4 Discussion

Slide 33 shows the horizontal distribution of larger, functionally and temporally diagnostic artifacts associated with the deep activity surface. It shows that there are likely areas of the component that remain unexcavated. The circled area only represents a rough conservative estimate of site component boundaries, based on the excavation and auger test results. However, a 20-foot grid, using a 3-inch bucket auger, is inadequate to detect the internal structure of any, much less and ephemeral prehistoric site. It is likely that there are several more ephemeral activity surfaces of this age, buried in the Blueberry Hill and other sand ridges along the south bank of the pre-Younger-Dryas channel.

8.0 Cactus Hill Model Conclusions

Slide 33 shows the surface configuration of the Cactus Hill/Blueberry Hill sand ridge along the south bank of the hypothesized pre-Younger-Dryas ancestral Nottoway River channel. The dotted line shows either an alternative buried channel or flood chute. The two question marks ("?") are on adjacent rises along the sand ridge that have not been tested.

As with Rubis-Pearsall the Blueberry Hill/Cactus Hill situation demonstrates that to locate pre-Younger-Dryas landforms that are prime candidates for pre-Younger-Dryas human occupations, one must account for changes in the river channel. These changes can influence significant abandonment of formerly prime landform by human populations. Cactus Hill was a prolific site with occupations dating to the full temporal range of local cultural periods. It is likely due to the fact that the Cactus Hill part of the sand ridge was not abandoned by the main river channel, when it shifted course, probably during the Younger Dryas.

The early occupations at Cactus Hill are not dissimilar in artifact quantities from those at Blueberry Hill. However, those above are far, far greater than at Blueberry Hill. In fact only pottery from the plowzone and a shallow but distinct Late Archaic Savannah River component were the only obvious post-Younger-Dryas occupations identified.

Although no positive pre-Clovis age occupations were identified as a result of this test of the Cactus Hill model, it is clear that the model works as far as identifying appropriate aged landforms. In fact the Rubis-Pearsall site was positively identified as being occupied by Clovis age occupants and possibly earlier. By the same token, due to multiple lines of evidence, the deep activity surface at Blueberry Hill is a likely candidate for being of pre-Clovis age.

Both the deep and ephemeral Rubis-Pearsall and Blueberry Hill components are located in soils that are of comparable age and stratigraphic depths to Cactus Hill. Dennis Stanford (2012: Personal communication) noted how consistent the depths on the pre-Younger-Dryas occupation are from site to site, relative to later occupations. All appear to be below 30 inches deep. This might be expected if the landform geomorphologies were similar, which they appear to be.

9.0 What Next: Smith Mountain Gap

As a result of the above model testing the question remains: what next? How can the results of the above research, which took 19 years, be expanded? Current plans are to attempt replication of the Cactus Hill pre-Clovis age occupation outside the Nottoway Watershed but, more importantly, also to address the broader issue of interaction and possibly inter-regional communication during the Early Archaic through pre-Clovis age periods.

9.1 44PY152

Cactus Hill-like points (Slide 34a-d) from another Virginia site, 44PY152 below Smith Mountain Gap (SMG), were recovered from association with a stratified Clovis age assemblage (upper right inset) (William A. Childress, personal communication 2011). The Cactus Hill-like points are almost identical in morphology to those recovered from pre-Clovis age levels in Cactus Hill Areas A/B and B (Slide 34, upper left) and Blueberry Hill (Slide 34, lower left). Two of the points (c and d) from 44PY152 were made of greenish patinated meta-volcanic stone, similar to the material most commonly used for Cactus Hill-like points, recovered from the Nottoway Watershed.

Site 44PY 152 is located on the south bank (again) of the Roanoke River watershed approximately 150 miles west of Blueberry Hill. The lower left inset in Slide 34 is a view of the site looking toward the southern edge of SMG. It also shows the significant exposure of the Early Archaic and Paleoamerican levels which have resulted from hydro-electric power generation over more than 40 years. The north aspect is consistent with the Nottoway model.

9.2 Transportation funnel/macro-band rendezvous area model

Slide 35 shows that the Nottoway/Chowan and Roanoke Watersheds both drain into Albemarle Sound in eastern North Carolina. Prior to the current inundation of the Albemarle Sound, the Nottoway River was a tributary of the Roanoke. That means that the pre-Clovis age occupations at Cactus Hill and possibly Blueberry Hill would have been in the same watershed as that of PY152. They may represent two parts of the same macro-band settlement pattern. If that is the case, then the Early Archaic macro-band model for the Southeastern United States, proposed by Anderson (1996: 29-45) and Anderson and Hanson (1988: 267-272) may have had a far deeper heritage in the Middle Atlantic than originally proposed by the author 24 and 17 years ago (Johnson 1989: 127; 1996: 211).

This hypothesis is further supported by significant differences between the points, tool-like artifact types, and apparent raw material preferences, and those recovered from Meadowcroft (Boldurian, 1985; Carlisle and Adovasio, 1982; Stanford and Bradley, 2012:93) and Miles Point (Lowery et al., 2010; Stanford and Bradley, 2012:98), which may have had an ancestral

Susquehanna River Watershed focus. Essentially, we may already be recognizing macro-band divisions in the pre-Clovis age occupations in the Middle Atlantic Region. Alternatively, the differences could have been due to temporal rather than spatial factors.

Site 44PY152 and several other sites are located at the east end of a notch in Smith Mountain, through which the Roanoke River flows (Blanton et al., 1996; Childress, 1993). This river is on a natural riverine corridor between the interior Tennessee and New River Drainages (Ohio River Watershed) and the Middle Atlantic Piedmont and Coastal Plain (Slide 36).

9.3 Saltville

A potential Pre-Clovis age association with extinct fauna at the SV-2 site in the Saltville Valley, along with the notable presence of fluted points from the same valley (McDonald, 2000: 5), fits within this hypothesized pattern. Site SV-2 is located in the headwaters of the Tennessee River Valley in southwestern Virginia (see Slide 35). The faunal concentration in the Saltville Valley could have attracted early groups from east of SMG and/or been part of the western range of pre-Clovis and Clovis age bands occupying territories in the Tennessee Valley.

9.4 "Across Atlantic Ice" implications

However, a contribution to the entry hypotheses (Slide 37), particularly the Solutrean hypothesis (Stanford and Bradley, 2012) might prove problematic. If the deep activity surface in Blueberry Hill is of pre-Clovis age and related to the other Cactus Hill pre-Clovis age expressions, then there appears to be a problem with the tool-like artifacts and the Solutrean hypothesis. They are not similar to those from Miles Point or Clovis. They have some similarities, such as burins, bifaces, points, macro-blades and cores but the rest of the Cactus Hill (and Blueberry Hill) tool kit is generally expedient. It is unlike the Clovis age tool kit at the end of the hypothesized trajectory. This is one of the more important lines of evidence that separates Clovis Age artifacts recovered at Cactus Hill from the site's pre-Clovis age artifacts.

The huge question this raises is, how do we get from the sophisticated technology, evident at the much older Miles Point through a less formal, 5,000 year later, technology like that found at Cactus Hill and Blueberry Hill and then back to the highly formal Clovis age technology that existed all across un-glaciated North and Central America by 13,000 BP (another 5,000 years later)? Bridging these gaps is one of the big questions acknowledged by Stanford and Bradley, (2012) for the Solutrean hypothesis.

Although, with such a paucity of sites and artifacts to work from, one can speculate that the vast time period between Miles Point and Cinmar at one end, and Clovis at the other end was very complex with macro-bands naturally forming and becoming isolated. The apparent divergence is tantalizingly evident between the Chesapeake Bay (ancestral Susquehanna Watershed) and the ancestral Roanoke Watersheds. The isolation may and probably should have led to divergent technologies.

For example, the preliminary examples of 23,000 year old technology in the ancestral Susquehanna looks different from the 18,000 year old technology emerging from ancestral Roanoke, which looks different than the 15,000 year old technology emerging from Buttermilk Creek (Gault and Friedkin sites) in Texas (Collins and Bradley, 2008; Waters et al., 2011). If this is the case then it would support diffusion of a powerful Clovis culture, based on a sophisticate technology, over pre-existing but culturally distinct groups across North America.

This would bring us back to the ultimate question about Clovis: Where did it come from? The results from Cactus Hill would argue that it did not go directly from Solutrean through Cactus Hill to Clovis. If this were not the path then where should archaeologists look?

If Stanford and Bradley (2012) are correct that Solutrean was a maritime, coastally adapted culture and technology, is it possible that a residual Solutrean-like culture and technology remained viable in North America and therefore continued for thousands of years along the exposed Continental Shelf? Eventually could it have spawned Clovis, which spread from east to west? The coastal ecotone would have remained a viable core area throughout the Late Pleistocene climatic fluctuations. Why would a culture, successfully adapted to such a relatively stable environment, have to make dramatic changes in its technology? Technologies possibly would have diverged as spin-off bands moved into new interior environments.

Clovis, to have spread as rapidly as it did across such diverse environments and cultures, could have had more than mere technology as a driving force. The obvious analogue would be the emergence of the horse culture on the Great Plains of central North America. It dramatically changed many linguistic and culturally different groups and even drew groups from outside the Great Plains to it. Archaeologically, the resulting differences are imperceptible to an untrained eye.

However, there are subtle but recognizable technological differences in Clovis age technologies from region to region. They may reflect the regional differences that existed before Clovis.

If a Clovis technology/culture diffusion model is valid, then how then did it spread across a whole continent? The answer may be present in selected, prominent water gaps like SMG. It also may be present in other water gaps such as Thoroughfare, Potomac, and Delaware Water Gaps to the north, and similar prominent landform east of the divides separating the French Broad River Basin from the Santee and Savannah River valleys to the south. The French Broad River provides a natural corridor from the Tennessee River Valley, through the almost impenetrable Smoky Mountains, to the southeastern Atlantic Seaboard. Its eastern divide is an ideal place to look.

A research design has been submitted and plans are in place to test the complex of Early Archaic, Clovis and possibly pre-Clovis age sites in the upper Leesville Lake Reservoir, immediately below SMG in southwest-central Virginia. A similar testing program is also underway downstream from Thoroughfare Gap in north-central Virginia (Slide 36).

Water gaps are being offered as inter-regional transportation funnels and/or macro-band coalescence (rendezvous) locations. Rivers, with their more gentle gradients and direct paths, are the simplest paths across diverse landscapes. They also often provide the most direct paths from one watershed to wind gaps across divides that separate major watersheds. Water gaps are natural transportation choke point and are often highly conspicuous features on the landscape. If groups were interacting between regions, then water gaps should have been natural funnels and meeting/stopping places.

REFERENCES CITED

Anderson, D.G. 1996. Models of Paleoindian and Early Archaic settlement in the Southeast: A historical perspective, in: D.G. Anderson and K.E. Sassaman (Eds.), The Paleoindian and Early Archaic Southeast. University of Alabama Press, Tuscaloosa, pp. 29-57.

Anderson, D.G., and Hanson, B.T. 1988. Early Archaic settlement in the Southeastern United States: a case study from the Savannah River Valley. American Antiquity 53: 262-286.

Blanton, D.B., Childress, W.A., Danz, J., Mitchell, L., Schuldenrein, J., and Zinn, J. 1996. Archeological Assessment of sites 44PY7, 44PY43, and 44PY152 at Leesville Lake, Piittsylvania County, Virginia. Virginia Department of Historic Resources Research Report Series No. 7. William and Mary Center for Archeological Research, Williamsburg.

Boldurian, A.T. 1985. Variability in Flintworking Technology at the Krajacic Site: Possible relationship to the Pre-Clovis Occupation of the cross Creek Drainage in Southeastern Pennsylvania. Unpublished Ph.D. dissertation, Department of Anthropology, University of Pittsburgh.

Bond, W.J. 1986. Illuvial band formation in a laboratory column of sand. Soil Society of America Journal 50: 265-267.

Carlisle, R.C., and Adovasio, J.M. (eds.) 1982. Meadowcroft: Collected Papers on the Archeology of Meadowcroft Rockshelter and the Cross Creek Drainage. Edited volume prepared for the Meadowcroft Rockshelter Rolling Thunder Review: The Last Act Symposium, Forty Seventh Meeting of the Society of American Archeology, Minneapolis.

Childress, W.A. 1993. The Smith Mountain Site: a buried Paleoindian occupation in the southwestern Piedmont of Virginia, Current Research in the Pleistocene 10: 7-9.

Coe, J.L. 1964. Formative cultures of the Carolina Piedmont. Transactions of the American Philosophical Society 54 (5): 1-30.

Collins, M.B. and Bradley, B.A. 2008. Evidence for Pre-Clovis occupation at the Gault Site (41BL323), Central Texas. Current Research in the Pleistocene 25: 70-72.

Daniel, I. R., Jr. 1998. Early Archaic Settlement in the Southeast: Hardaway Revisited. University of Alabama, Tuscaloosa.

Feathers, J.K., Rhodes, E.J., Hout, S., and McAvoy, J.M. 2006. Luminescence dating of sand deposits related to late Pleistocene human occupation of the Cactus Hill site, Virginia, US. Quaternary Geoarcheology 1: 167-187.

Fiedel, S. J. 2012. Is that all there is? The weak case for pre-Clovis occupation of Eastern North America, in: Gingrich, A.M. (Ed.), *The Eastern Fluted Point Tradition*, University of Utah (in press), Salt Lake City.

Johnson, M.F. 1989. The lithic technology and material culture of the first Virginians: An Eastern Clovis Perspective, in: Wittkofski, J.M. and Reinhart, T.R., Paleoindian Research in Virginia: A Synthesis. Special Publication No. 19. The Archeological Society of Virginia, Cortland, pp. 95-138.

Johnson, M.F. 1996. Paleoindians near the edge: a Virginia perspective, in: Anderson, D.G. and Sassaman, K.E. (Eds.), The Paleoindian and Early Archaic Southeast. University of Alabama, Tuscaloosa. pp. 197-212.

Johnson, M.F. 1997. Additional research at Cactus Hill: preliminary description of Northern Virginia Chapter-ASV's 1993 and 1995 excavations, in: McAvoy, J.M. and McAvoy, L.D., Archeological investigations of site 44SX202, Cactus Hill, Sussex County, Virginia, Appendix G. Research Report Series No. 8. Virginia Department of Historic Resources, Richmond.

Johnson, M.F. 2013. Cactus Hill, Rubis-Pearsall and Blueberry Hill: One Is an Accident; Two Is a Coincidence; Three is a Pattern – Predicting "Old Dirt" in the Nottoway River Valley of Southeastern Virginia, U.S.A. Unpublished PhD Thesis, University of Exeter (U.K.).

Johnson, M.F., and Anthony N.H. 2002. Archeological Investigation: Lorton Town Center 1 (LTC-2) (44FX2077). Fairfax County Park Authority, Falls Church, Virginia.

Jones, K.B., and Johnson, G.H. 1997. Geology of the /Cactus Hill archeological site (44SX202) Sussex County, Virginia, in: McAvoy, J.M. and McAvoy, L.D., Archeological investigations of site 44sx202, Cactus Hill, Sussex County, Virginia, Appendix C. Research Report Series No. 8. Virginia Department of Historic Resources, Richmond.

Lowery, D.L., O'Neal, M.A., Wah, J.S., Wagner, D.P., and Stanford, D.J. 2010. Late Pleistocene upland stratigraphy of the western Delmarva Peninsula, USA. Quaternary Science Reviews xxx: 1-9.

Lund, C.C. 1997. Late Pleistocene and Holocene Dunes along the Nottoway River, Sussex County, Virginia. Unpublished research paper, Department of Geology, The College of William and Mary.

McAvoy, J.M. 1992. Nottoway river survey – Part I - Clovis settlement patterns. Archeological Society of Virginia Special Publication Number 28, Richmond.

McAvoy, J.M., Baker, J.C., Feathers, J.K., Hodges, R.L., McWeeney, L., and Whyte, T.R. 2000. Summary of Research at the Cactus Hill Archeological site, 44SX202, Sussex County, Virginia. Unpublished grant report #6345-98 for the National Geographic Society.

McAvoy, J.M., and McAvoy, L.D. (Eds.) 1997. Archeological investigations of site 44SX202, Cactus Hill, Sussex County, Virginia. Virginia Department of Historic Resources Research Report Series No. 8, Richmond.

McCary, B.C. 1951. A workshop site of early man in Dinwiddie County, Virginia. American Antiquity 17: 9-17.

McCary, B.C. 1996. Survey of Virginia fluted points. Archeological Society of Virginia Special Publication Number 12. Richmond.

McDonald, J.N. 2000. An outline of the pre-Clovis archeology of SV-2, Saltville, Virginia, with special attention to a bone tool dated 14,510yr BP. Jeffersoniana 9: 1-59.

MacPhail, R. I. and McAvoy, J. M. 2008. A micromorphological analysis of stratigraphic integrity and site formation at Cactus Hill, an early Paleoindian and hypothesized pre-Clovis occupation in south central Virginia, USA. Geoarcheology: An International Journal 23: 675-694.

Pavich, Milan J., Markewich, H. W., Litwin, R.L., and Brook, G.A. 2008. Measurement of Galaceosostatic Adjustments in the Mid-Atlantic Region for the Last Two Glacial Cycles. Unpublished paper presented at the Quaternary Ice Sheet - Ocean Interactions and Landscape Resources session, Twentieth AMQUA Biennial Meeting, Pennsylvania State University.

Perron, J.T. 1999. Micromorphology of the Cactus Hill Site (44SX202), Sussex County, Virginia. Unpublished BA honors thesis, Department of Anthropology and Earth Sciences, Harvard University.

Rink, W.J., Dunbar, J.S., Tschinkel, W.R., Kwapich, C., Repp, A., Stanton, W., Thulman, D.K. 2012. Subterranean transport and deposition of quartz by ants in sandy sites relevant to age overestimation in optical luminescence dating. *Journal of Archaeological Science*. http://www.sciencedirect.com/science/article/pii/S0305440312004931

STANFORD, D.J. and BRADLEY, B.A. 2012: Across Atlantic Ice: the Origin of America's Clovis Culture. University of California Press, Berkley.

Thulman, David K. 2012. Bioturbation and the Wakulla Springs Lodge site artifact distribution. The Florida Anthropologist 65: 25-34.

Wagner, DP., and McAvoy, J.M. 2004. Pedoarcheology of Cactus Hill, a sandy Paleoindian site in southeastern Virginia, U.S.A. Geoarcheology: An International Journal 19: 297-322.

Waters, M.R., Forman, S.L., Jennings, T.A., Nordt, L.C., Driess, S.G., Feinberg, J.M., Keene, J.L., Halligan, J., Lindquist, A., Pierson, J., Hallmark, C.T., Collins, M.B., and Wiederhold, J.E. 2011 The Buttermilk Creek Complex and the origins of Clovis at the Debra L. Friedkin site, Texas. Science 331: 1599-1603.

Modeling Cactus Hill (44SX202)

Rubis-Pearsall (44SX360)

Blueberry Hill (44SX327)

By
Michael F. Johnson

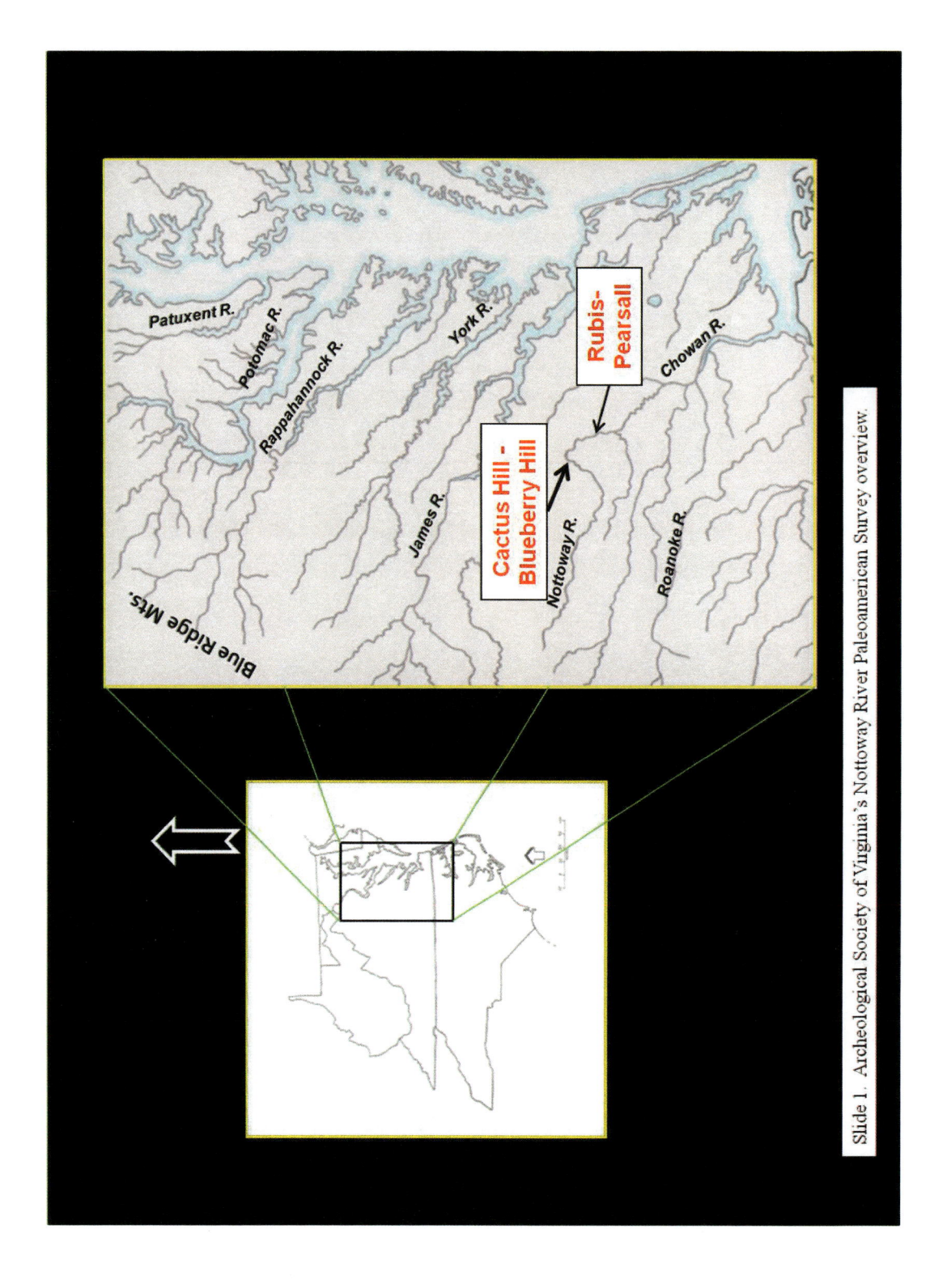

Slide 1. Archeological Society of Virginia's Nottoway River Paleoamerican Survey overview.

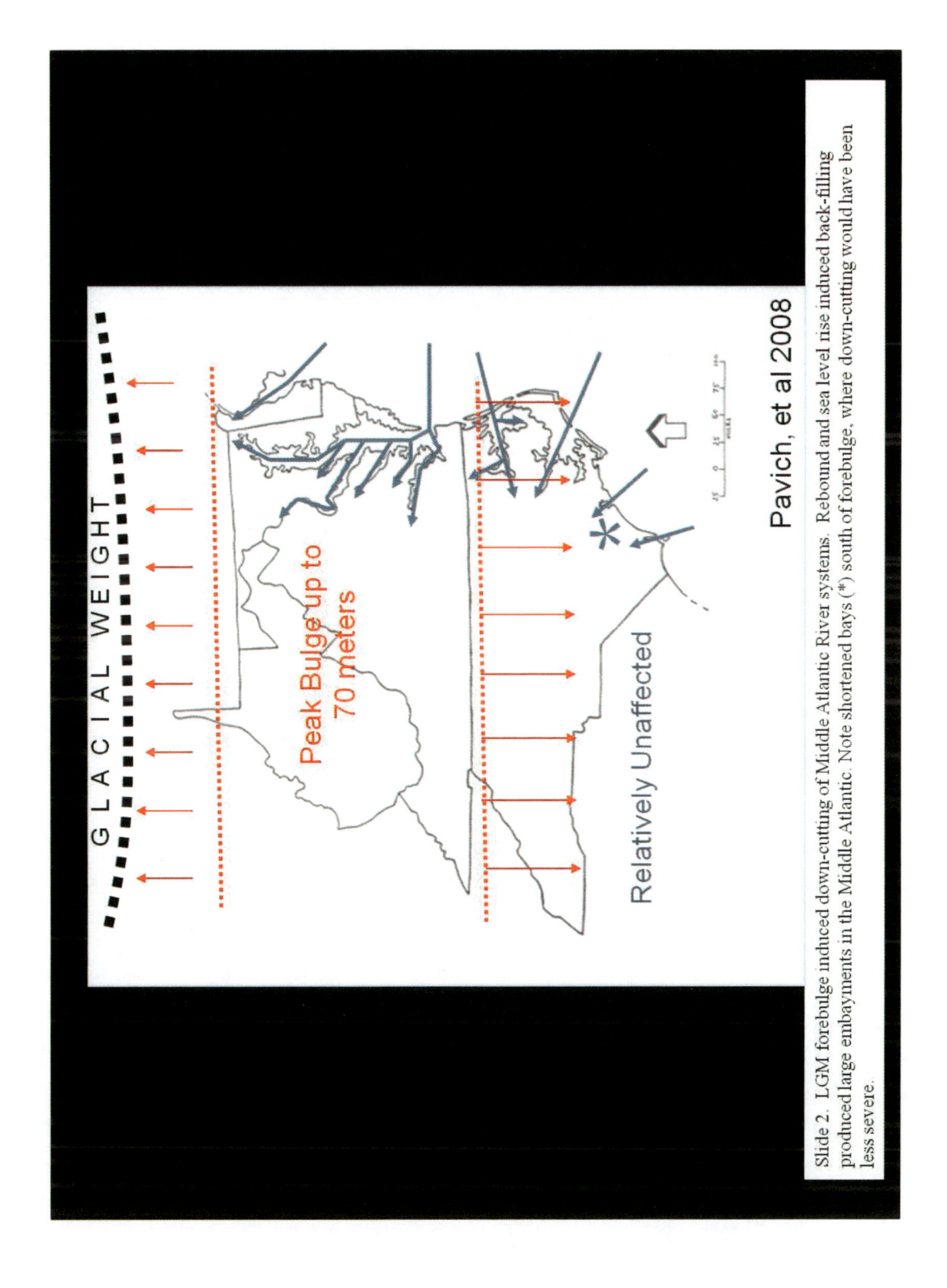

Slide 2. LGM forebulge induced down-cutting of Middle Atlantic River systems. Rebound and sea level rise induced back-filling produced large embayments in the Middle Atlantic. Note shortened bays (*) south of forebulge, where down-cutting would have been less severe.

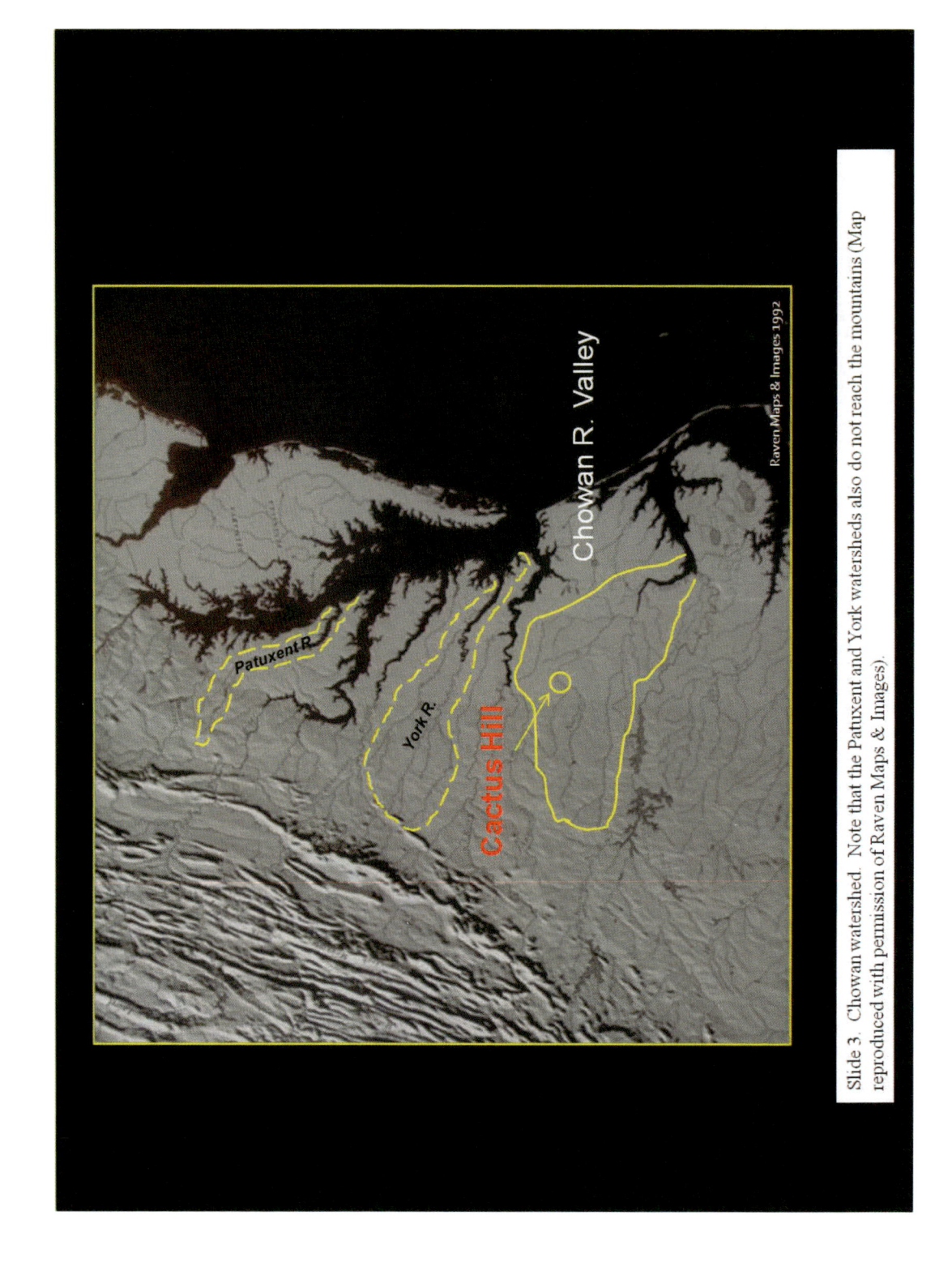

Slide 3. Chowan watershed. Note that the Patuxent and York watersheds also do not reach the mountains (Map reproduced with permission of Raven Maps & Images).

Slide 4. Local context for Cactus Hill, showing diagnostic pre-Clovis age points from Areas A/B and B (lower inset), and lithic feature A-14 (upper right inset) from Area A. (Photos by author)

Slide 5. Barr site local context.

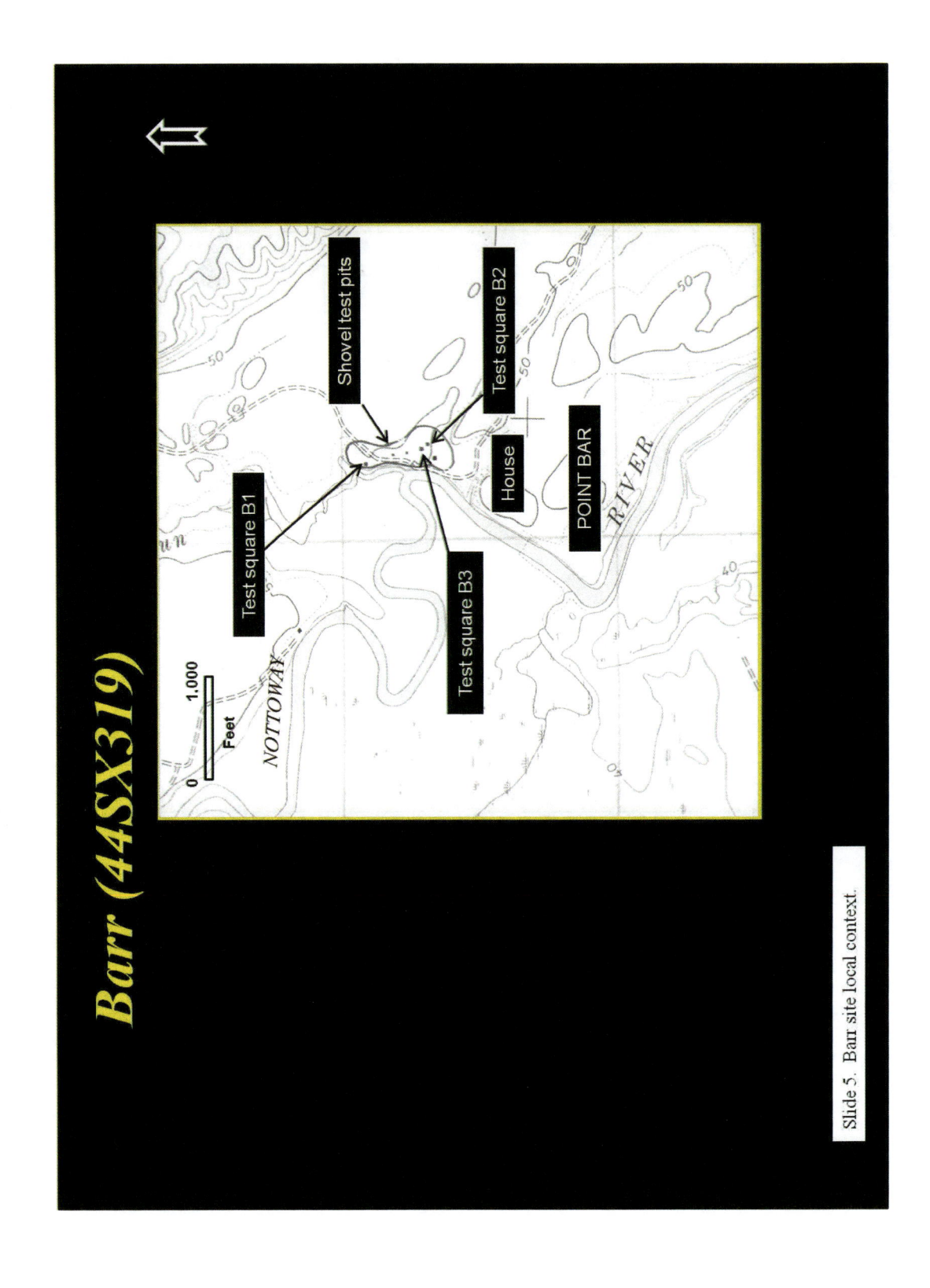

Slide 6. Diagnostic artifact and feature results from Barr testing. (Photos by author)

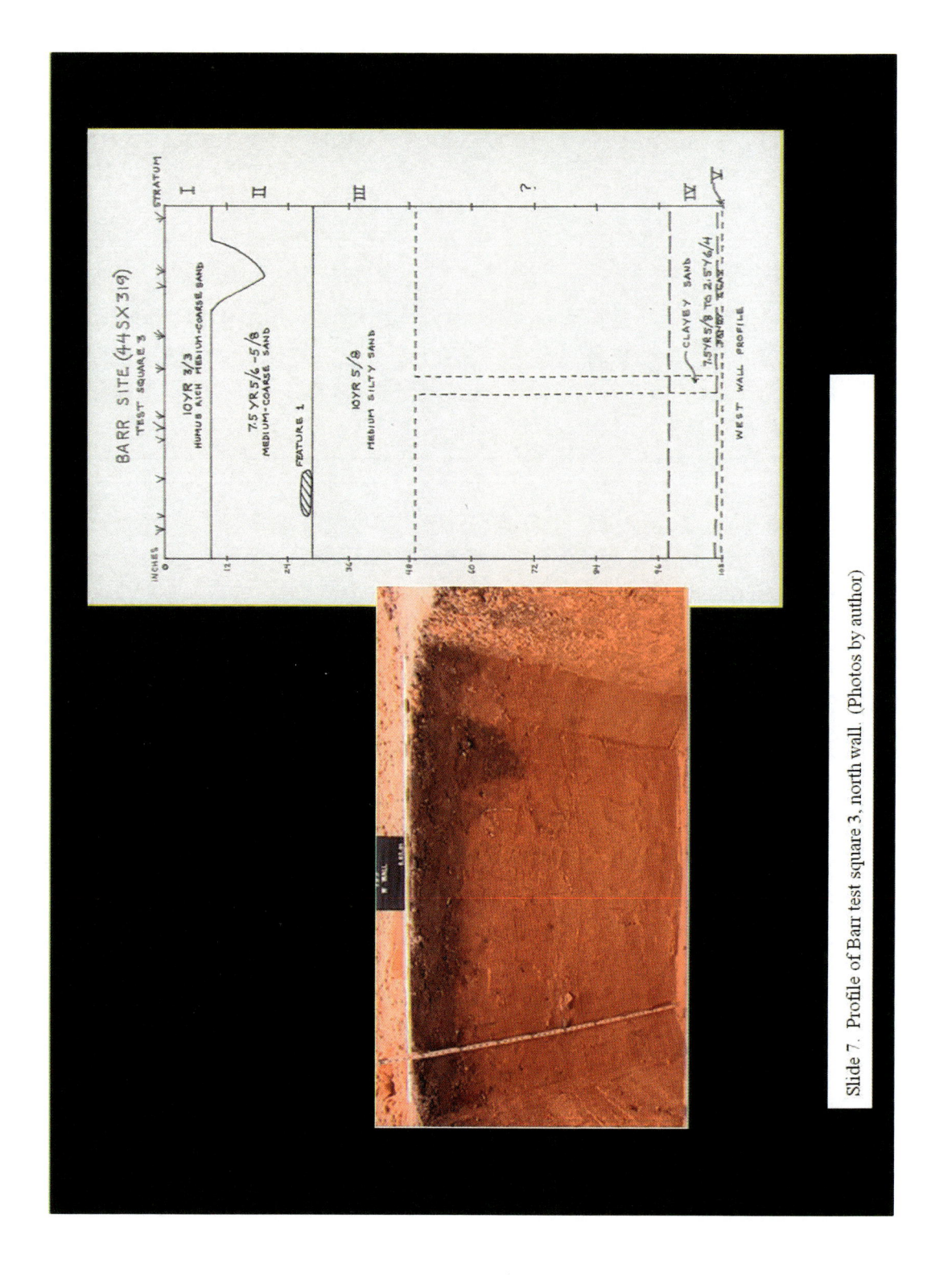

Slide 7. Profile of Barr test square 3, north wall. (Photos by author)

Warm Weather Model

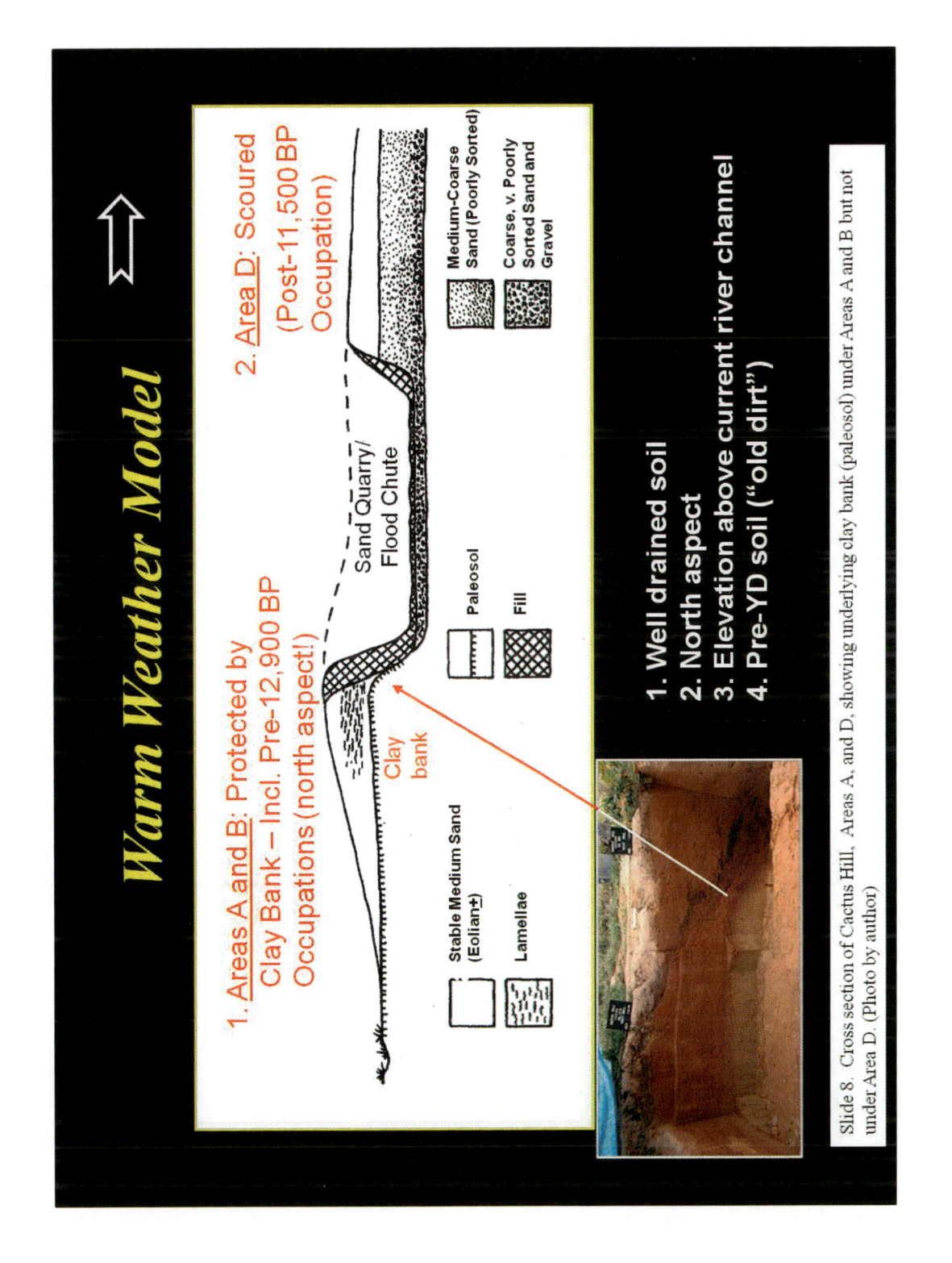

1. **Areas A and B**: Protected by Clay Bank – Incl. Pre-12,900 BP Occupations (north aspect!)

2. **Area D**: Scoured (Post-11,500 BP Occupation)

Sand Quarry/Flood Chute

Clay bank

Stable Medium Sand (Eolian±)
Lamellae
Paleosol
Fill
Medium-Coarse Sand (Poorly Sorted)
Coarse, v. Poorly Sorted Sand and Gravel

1. Well drained soil
2. North aspect
3. Elevation above current river channel
4. Pre-YD soil ("old dirt")

Slide 8. Cross section of Cactus Hill, Areas A, and D, showing underlying clay bank (paleosol) under Areas A and B but not under Area D. (Photo by author)

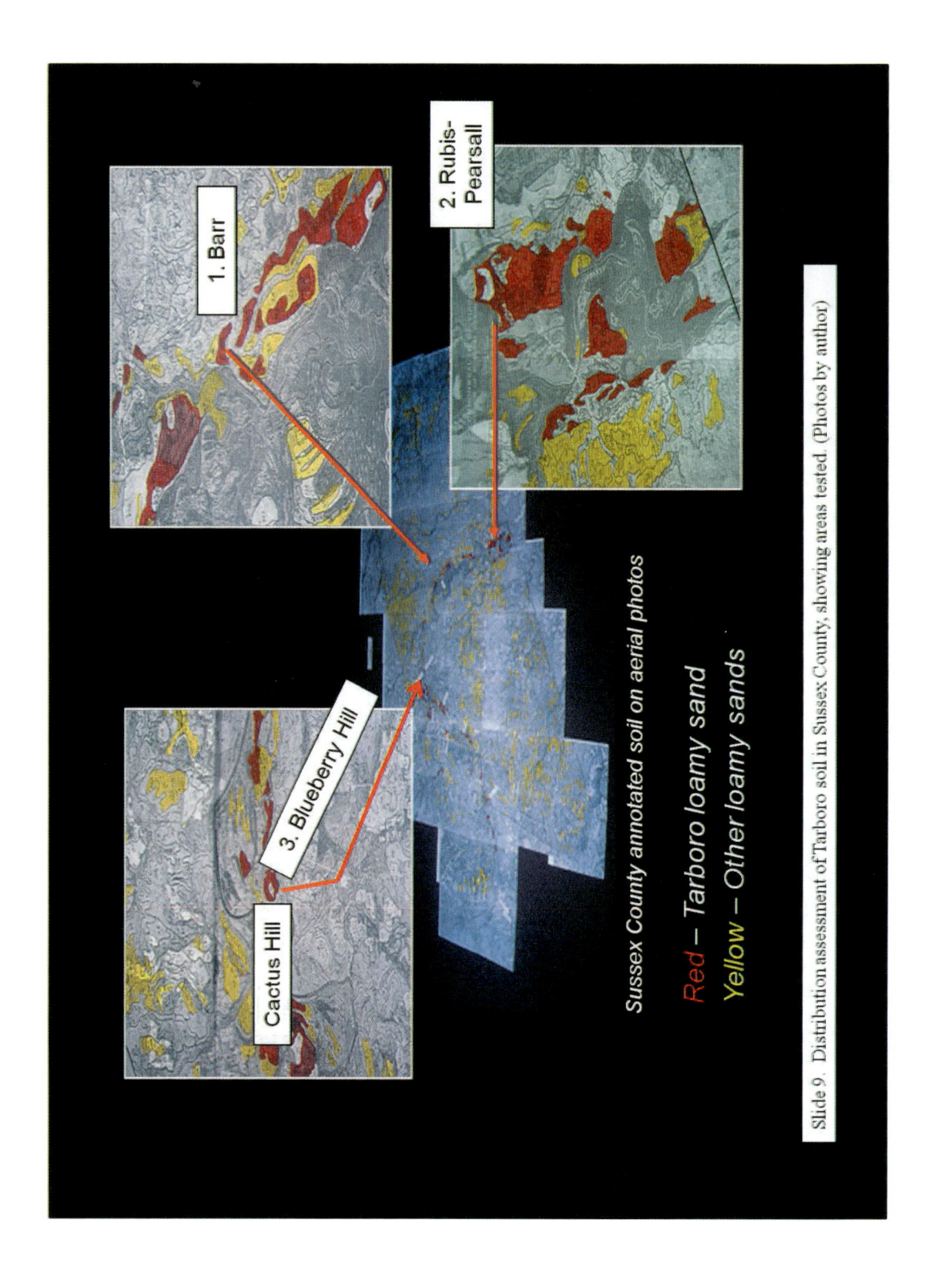

Slide 9. Distribution assessment of Tarboro soil in Sussex County, showing areas tested. (Photos by author)

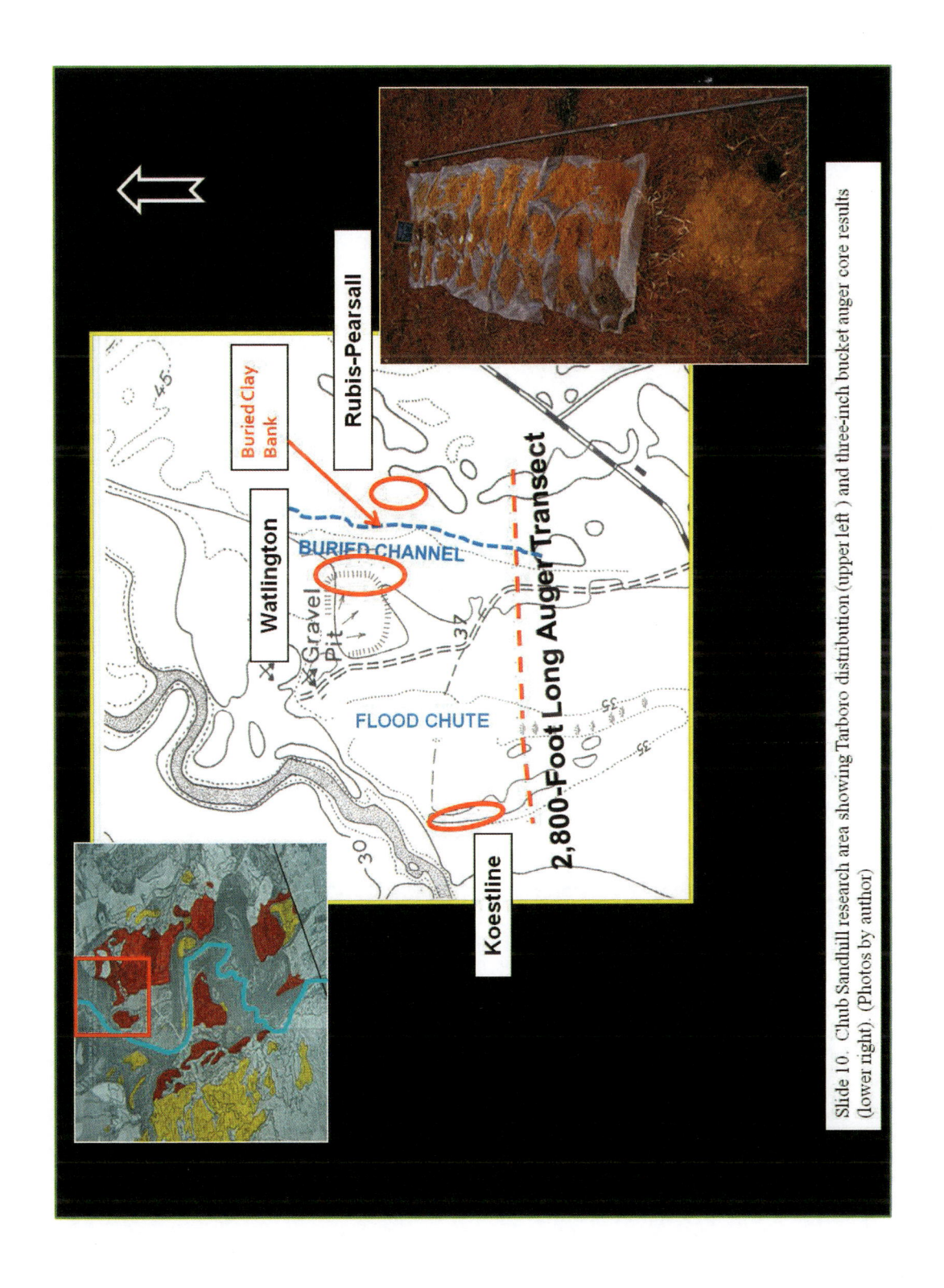

Slide 10. Club Sandhill research area showing Tarboro distribution (upper left) and three-inch bucket auger core results (lower right). (Photos by author)

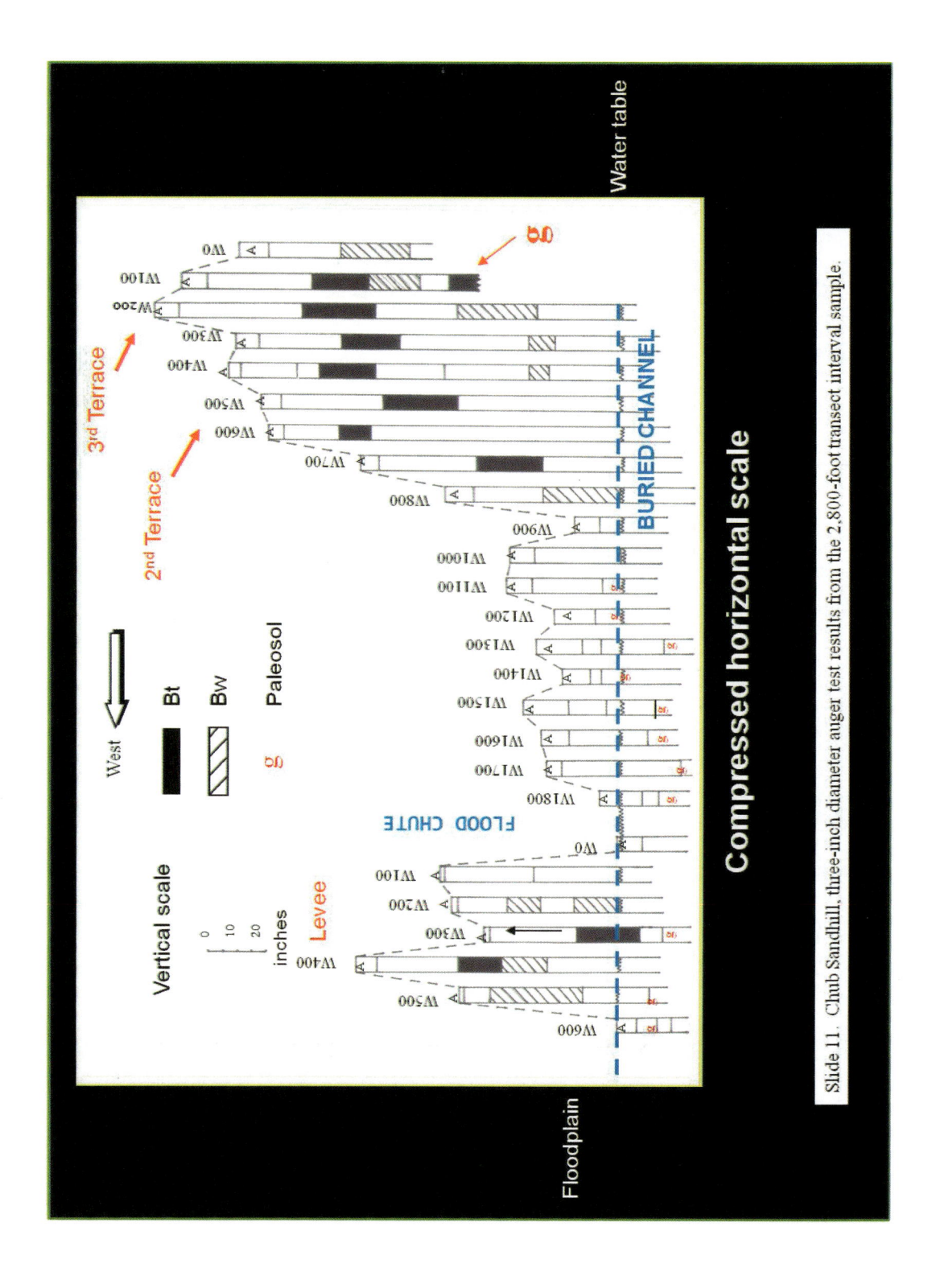

Slide 11. Chub Sandhill, three-inch diameter auger test results from the 2,800-foot transect interval sample.

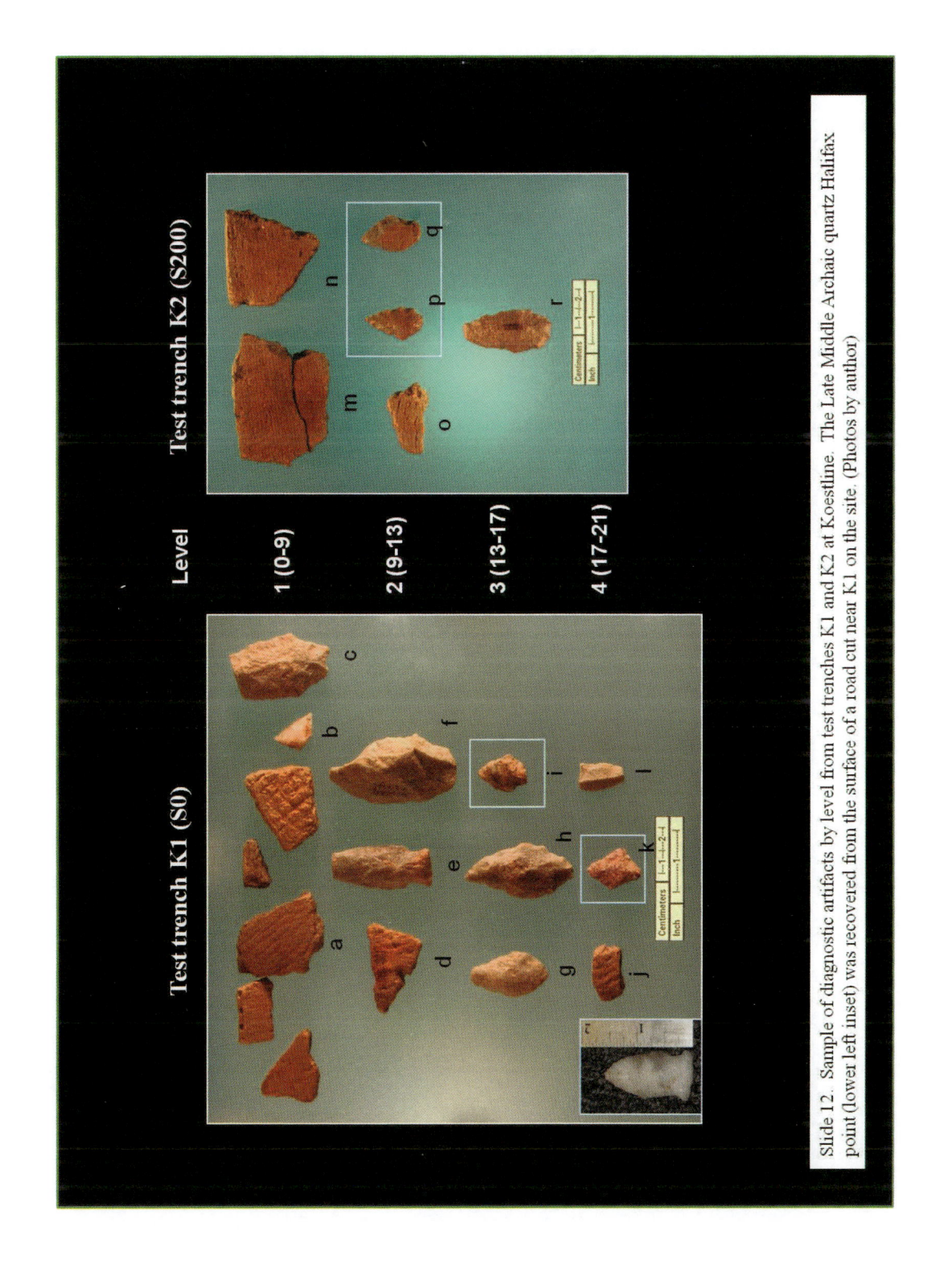

Slide 12. Sample of diagnostic artifacts by level from test trenches K1 and K2 at Koestline. The Late Middle Archaic quartz Halifax point (lower left inset) was recovered from the surface of a road cut near K1 on the site. (Photos by author)

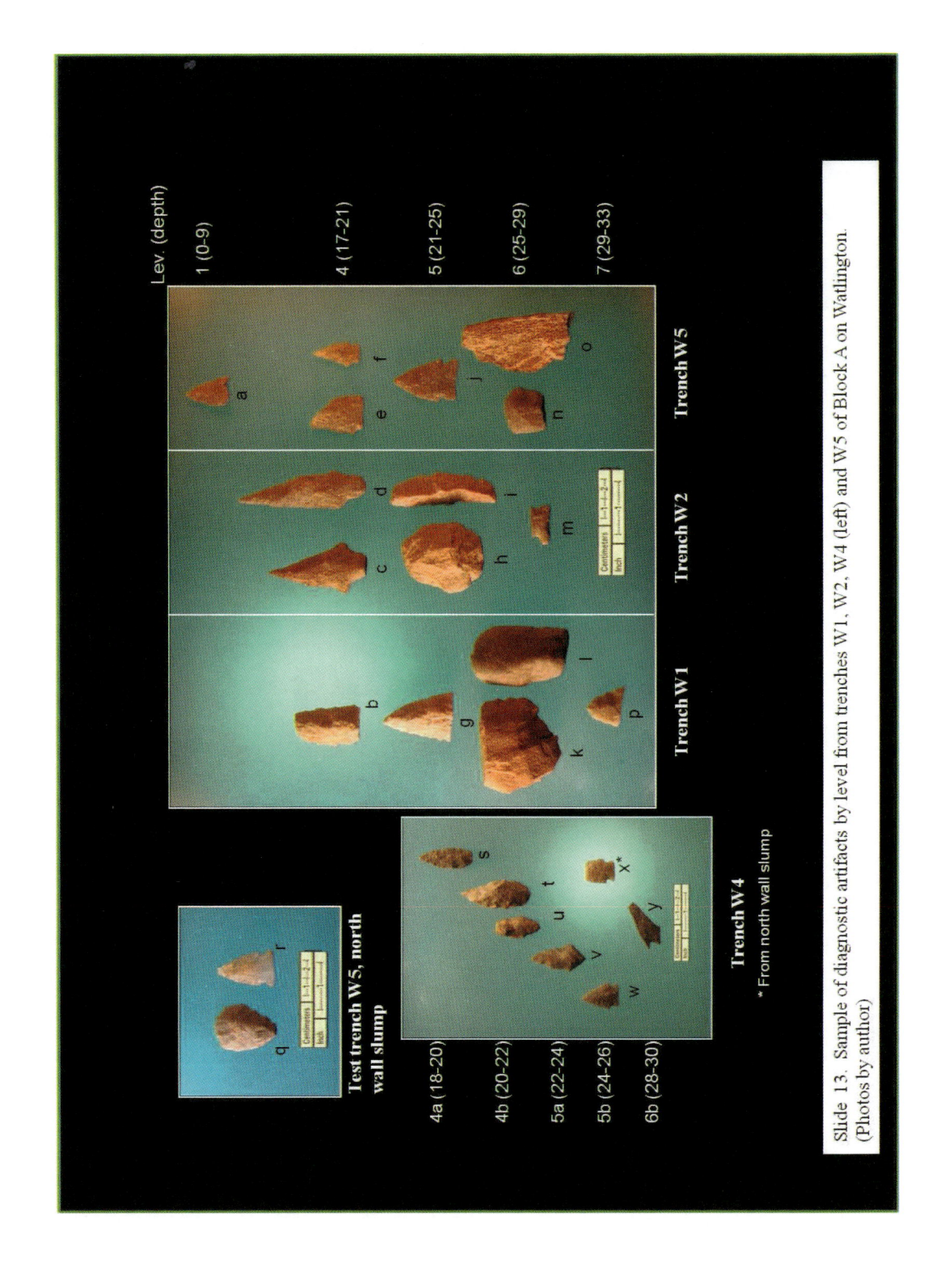

Slide 13. Sample of diagnostic artifacts by level from trenches W1, W2, W4 (left) and W5 of Block A on Watlington. (Photos by author)

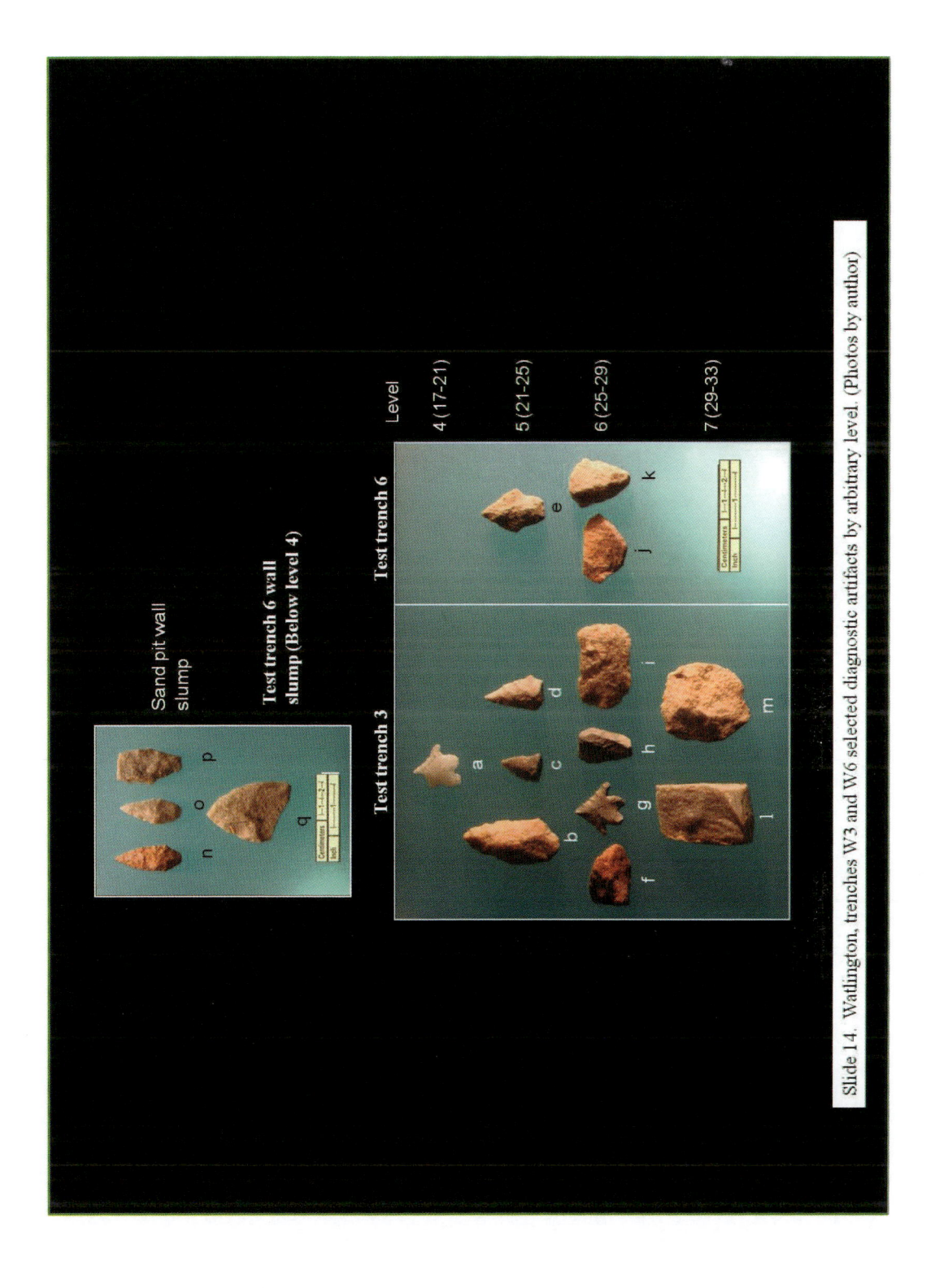

Slide 14. Watlington, trenches W3 and W6 selected diagnostic artifacts by arbitrary level. (Photos by author)

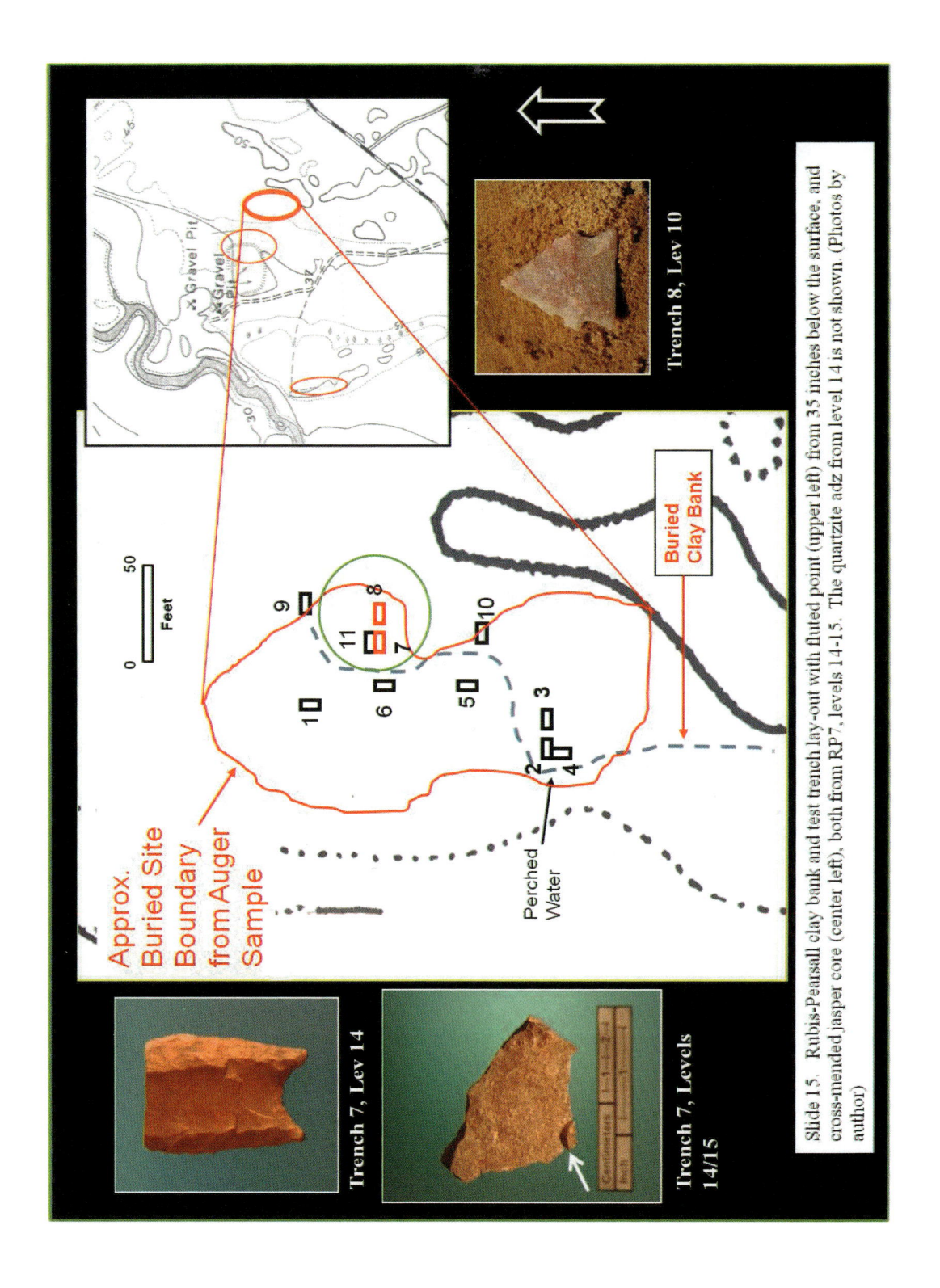

Slide 15. Rubis-Pearsall clay bank and test trench lay-out with fluted point (upper left) from 35 inches below the surface, and cross-mended jasper core (center left), both from RP7, levels 14-15. The quartzite adz from level 14 is not shown. (Photos by author)

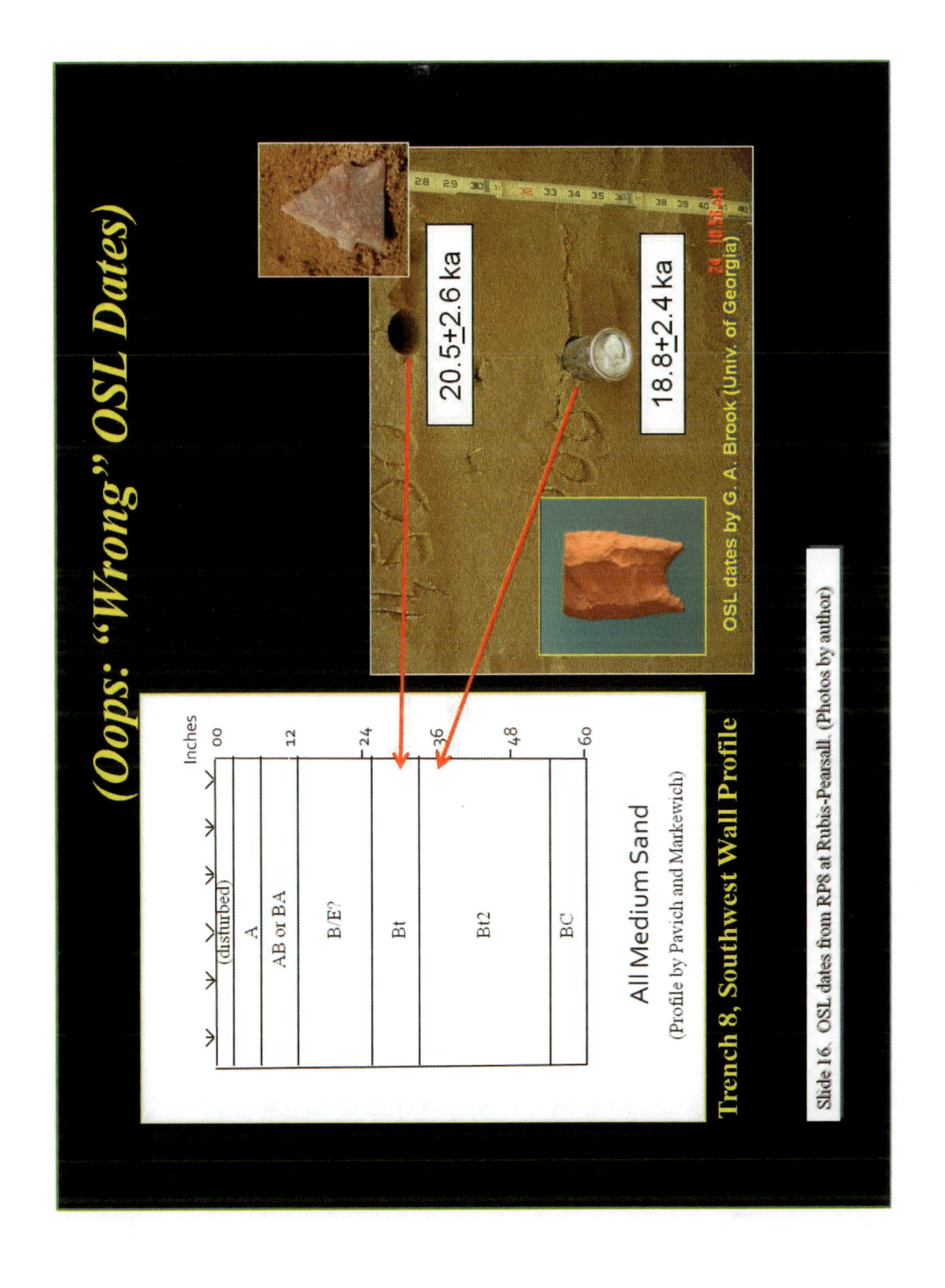

Slide 16. OSL dates from RP8 at Rubis-Pearsall. (Photos by author)

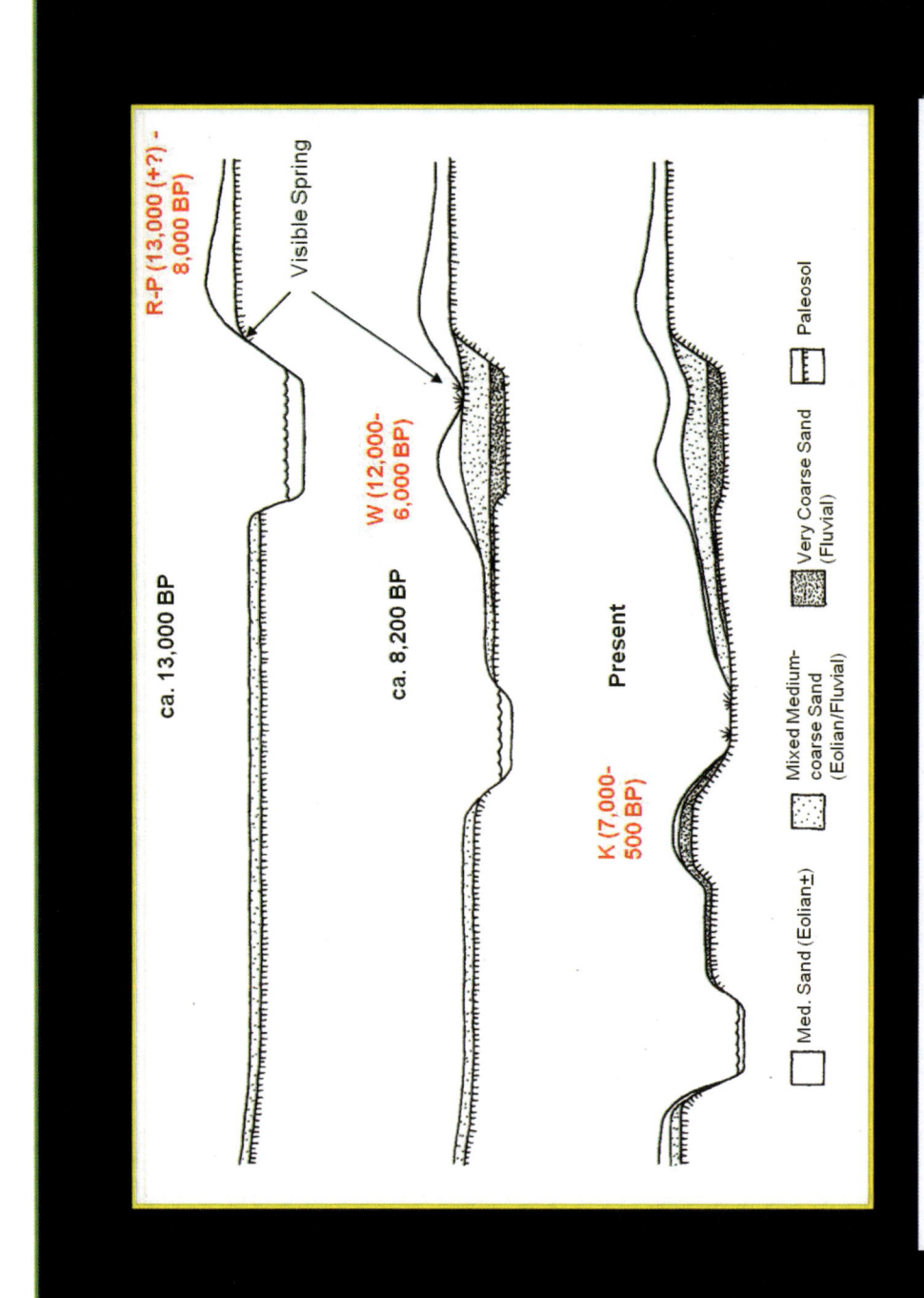

Slide 17. Idealized cross sections of the Nottoway River's migration from east to west across the Chub Sandhill point bar, and the apparent cultural abandonment of older terraces as indicated by site chronologies and soils/sedimentation.

Slide 18. Preliminary testing between Cactus Hill and Blueberry Hill prior to the discovery of Blueberry Hill. (Photo by author)

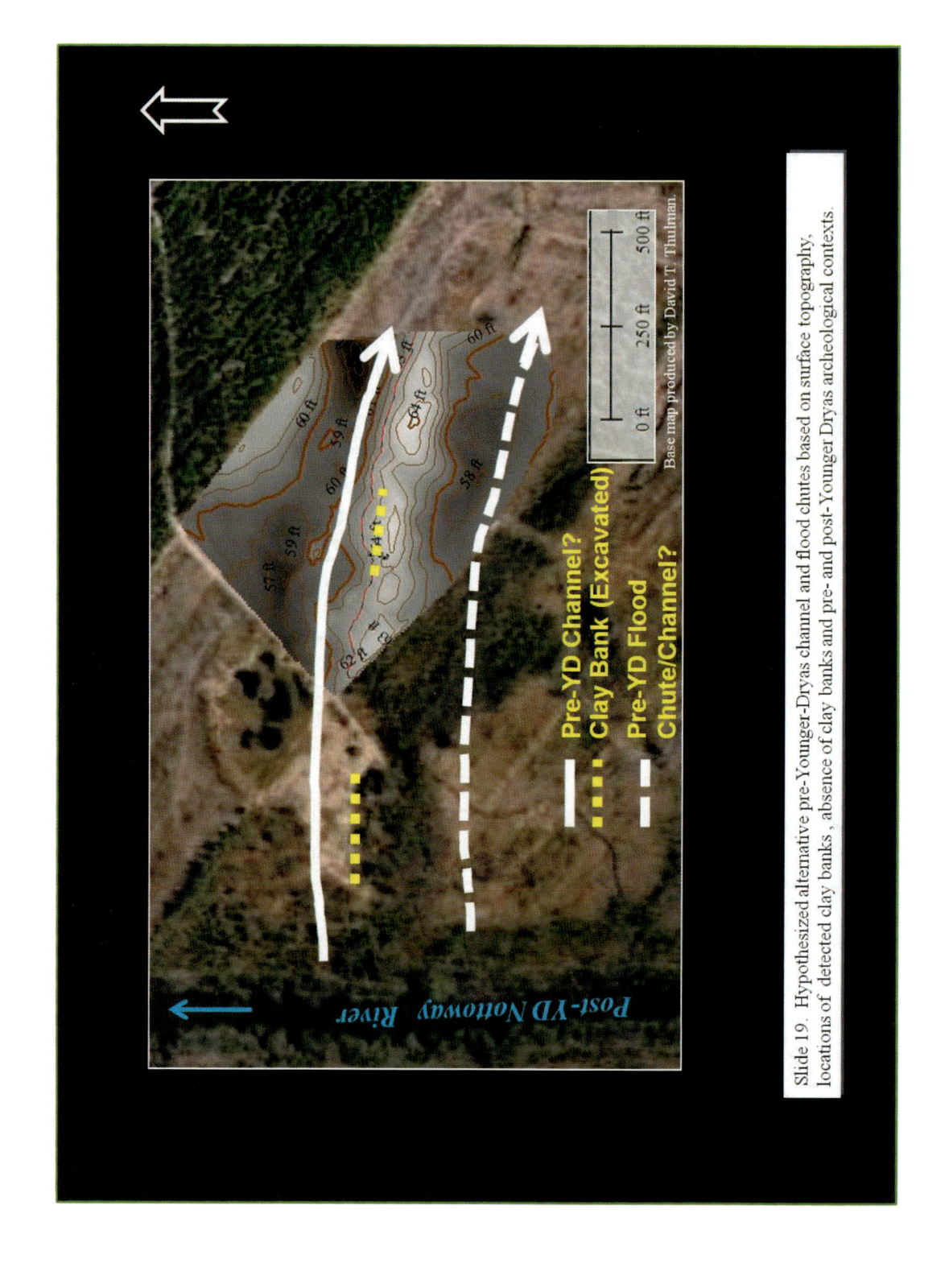

Slide 19. Hypothesized alternative pre-Younger-Dryas channel and flood chutes based on surface topography, locations of detected clay banks, absence of clay banks and pre- and post-Younger Dryas archeological contexts.

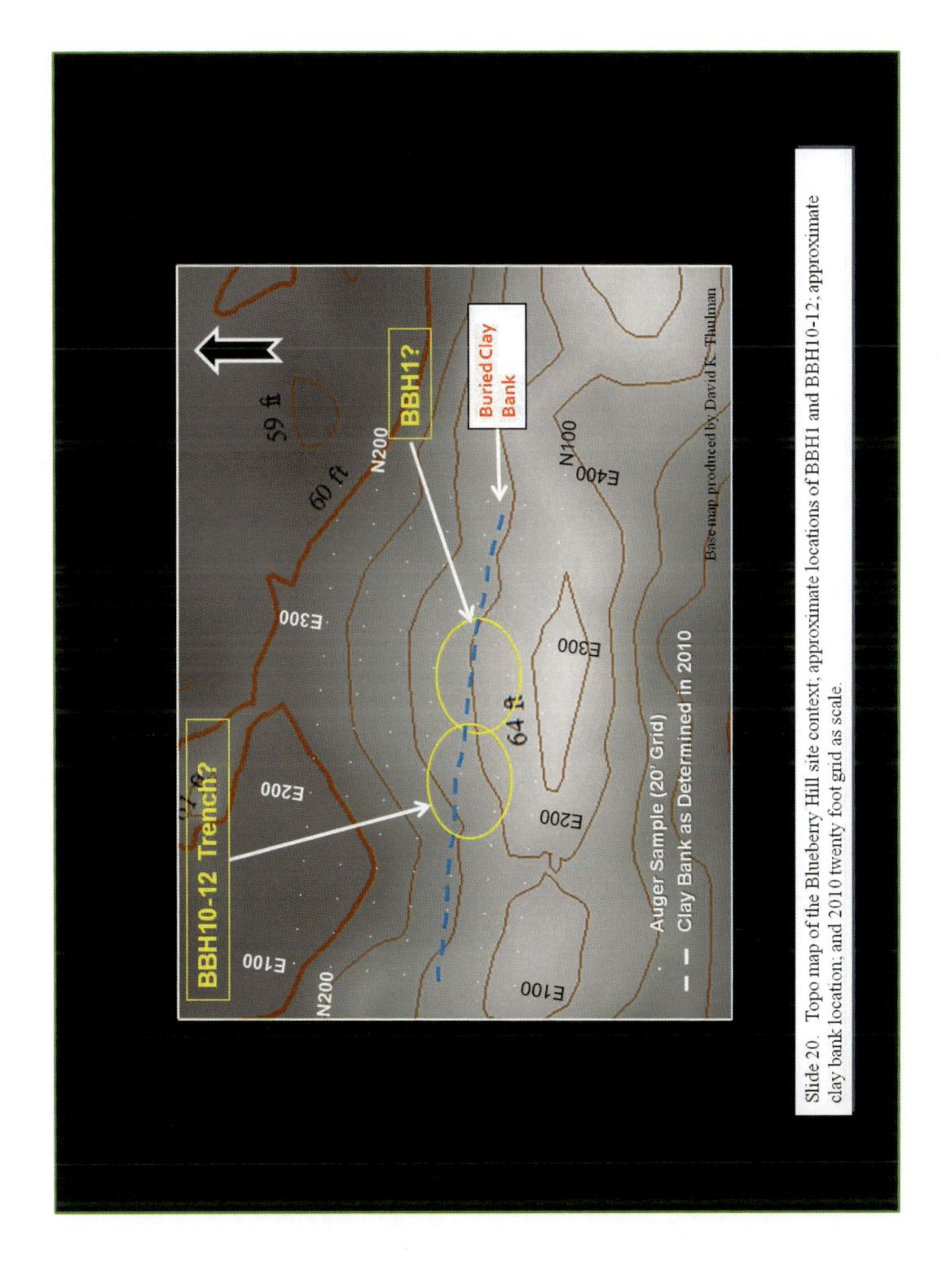

Slide 20. Topo map of the Blueberry Hill site context; approximate locations of BBH1 and BBH10-12; approximate clay bank location; and 2010 twenty foot grid as scale.

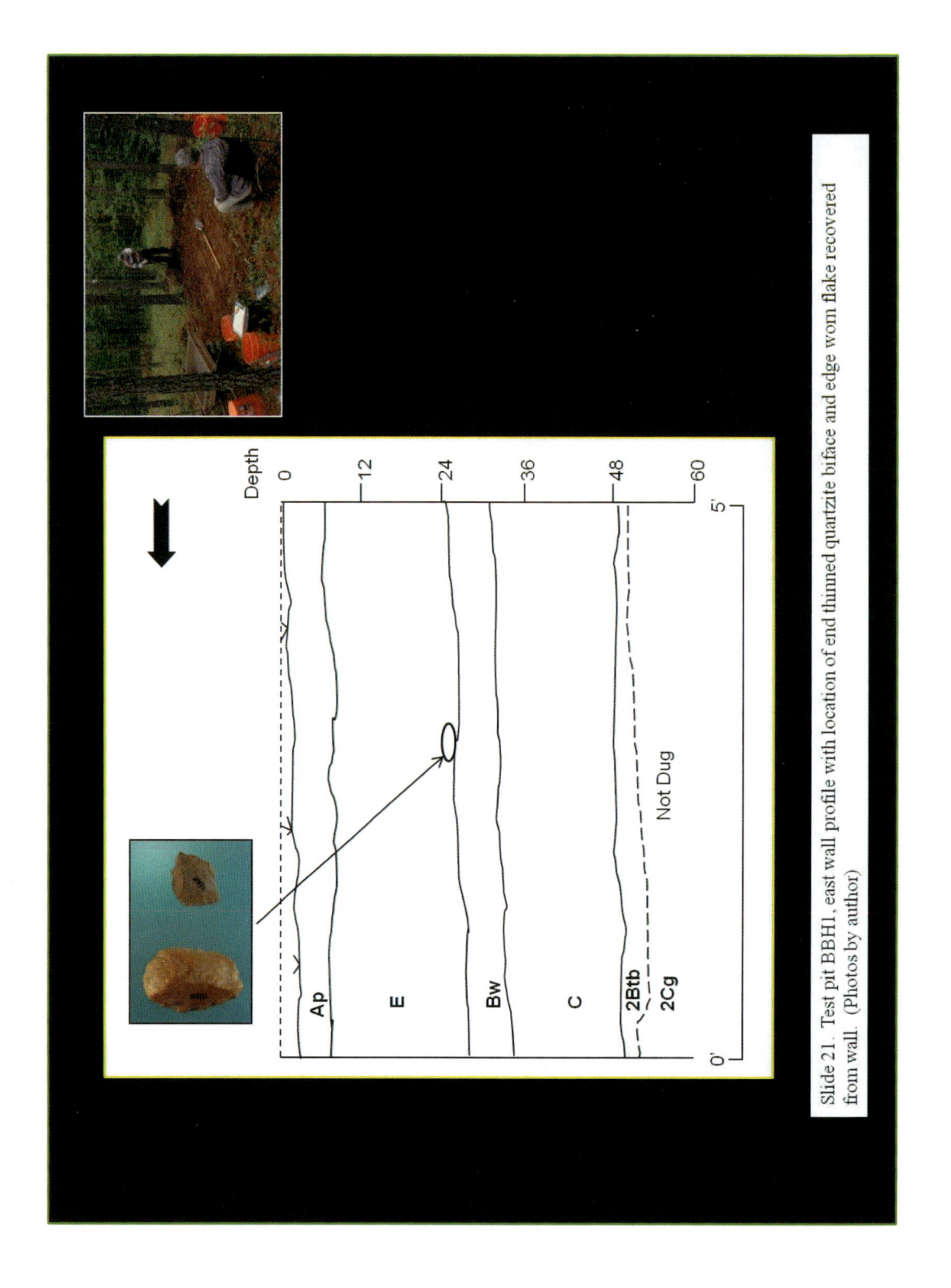

Slide 21. Test pit BBH1, east wall profile with location of end thinned quartzite biface and edge worn flake recovered from wall. (Photos by author)

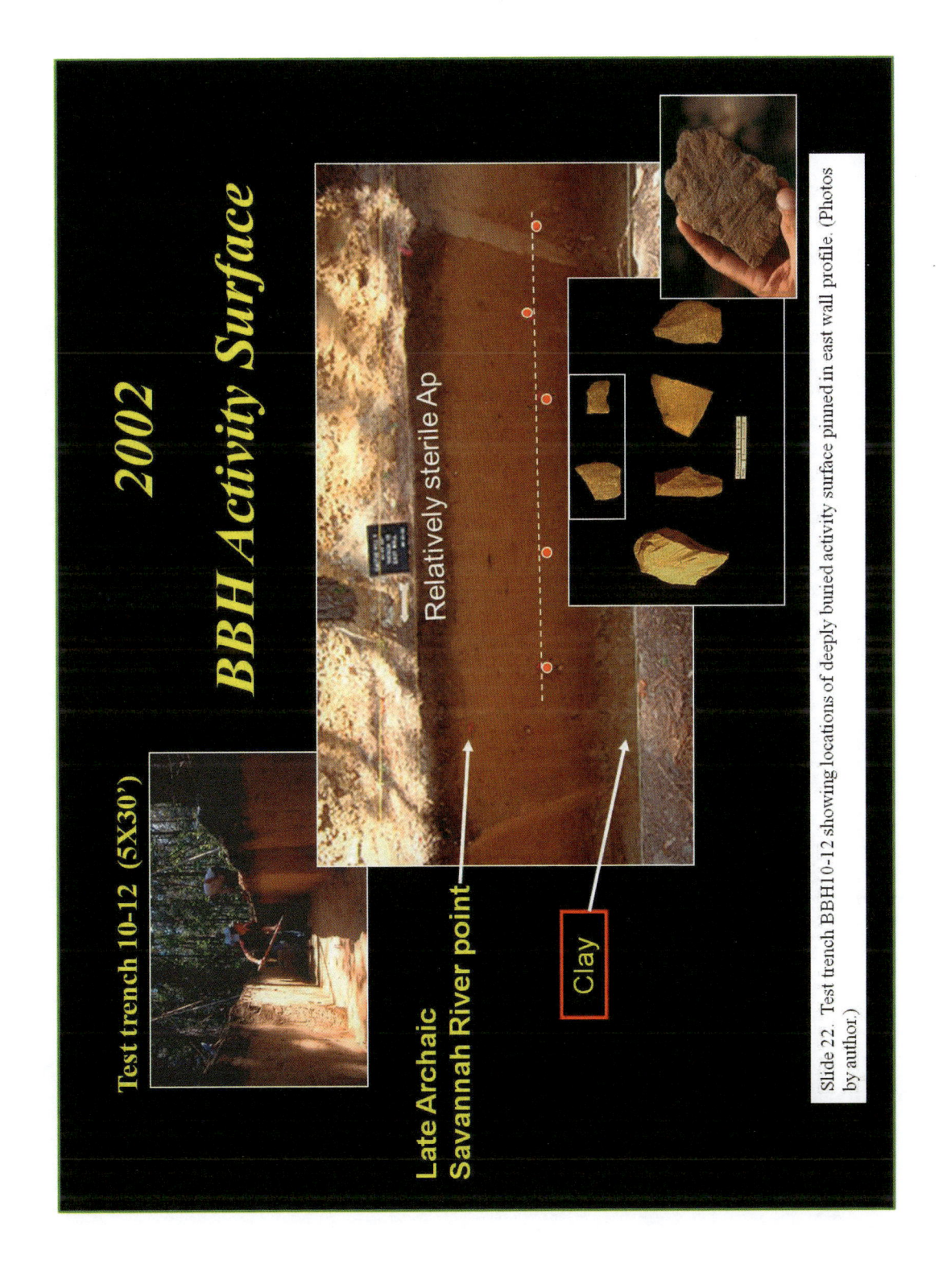

Slide 22. Test trench BBH10-12 showing locations of deeply buried activity surface pinned in east wall profile. (Photos by author.)

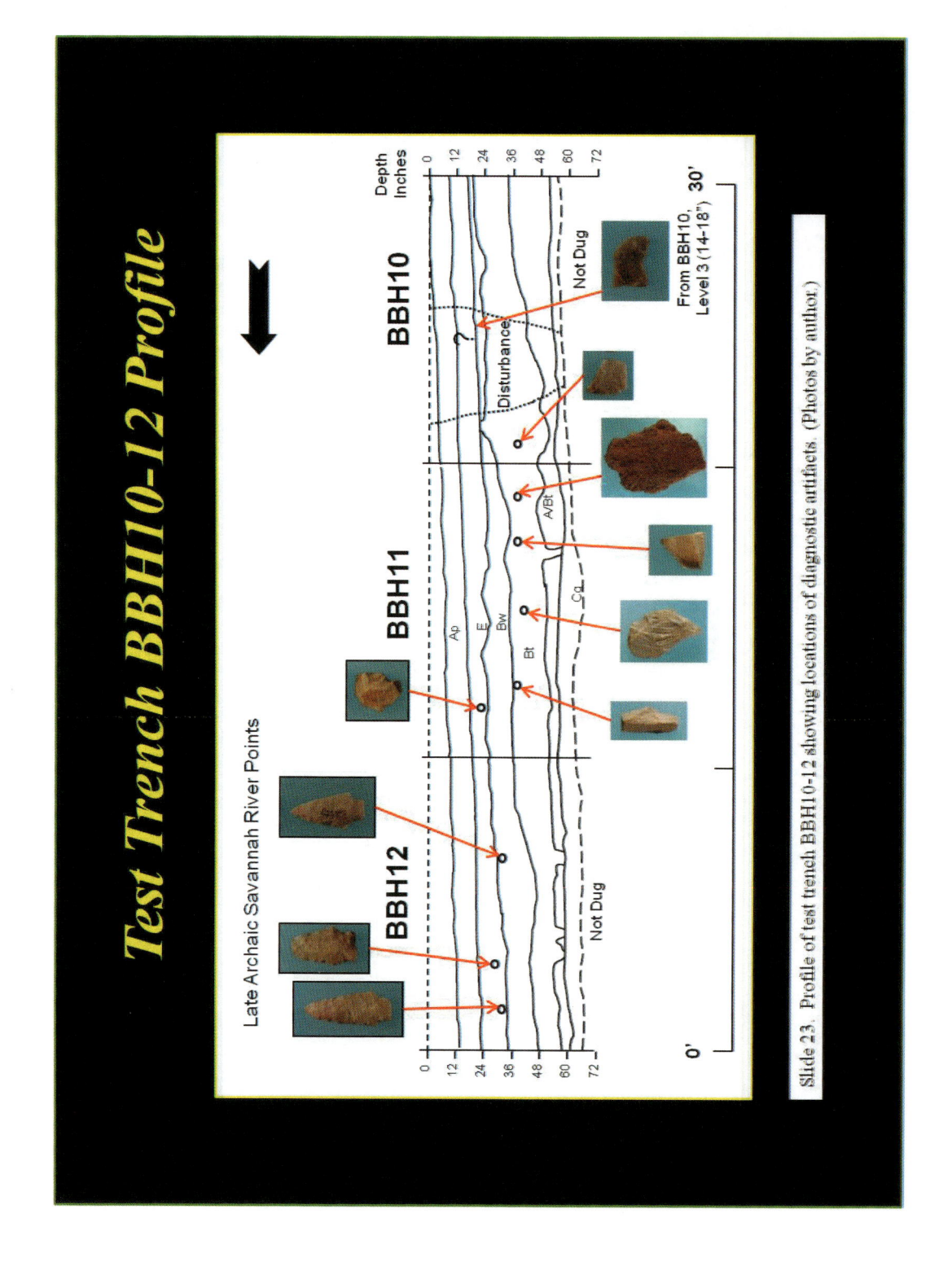

Slide 23. Profile of test trench BBH10-12 showing locations of diagnostic artifacts. (Photos by author.)

Slide 24. Quartzite Cactus Hill-like (Hardaway Blade-like?) point base (left) and quartzite fluted point base from BBH10. (Photo by author)

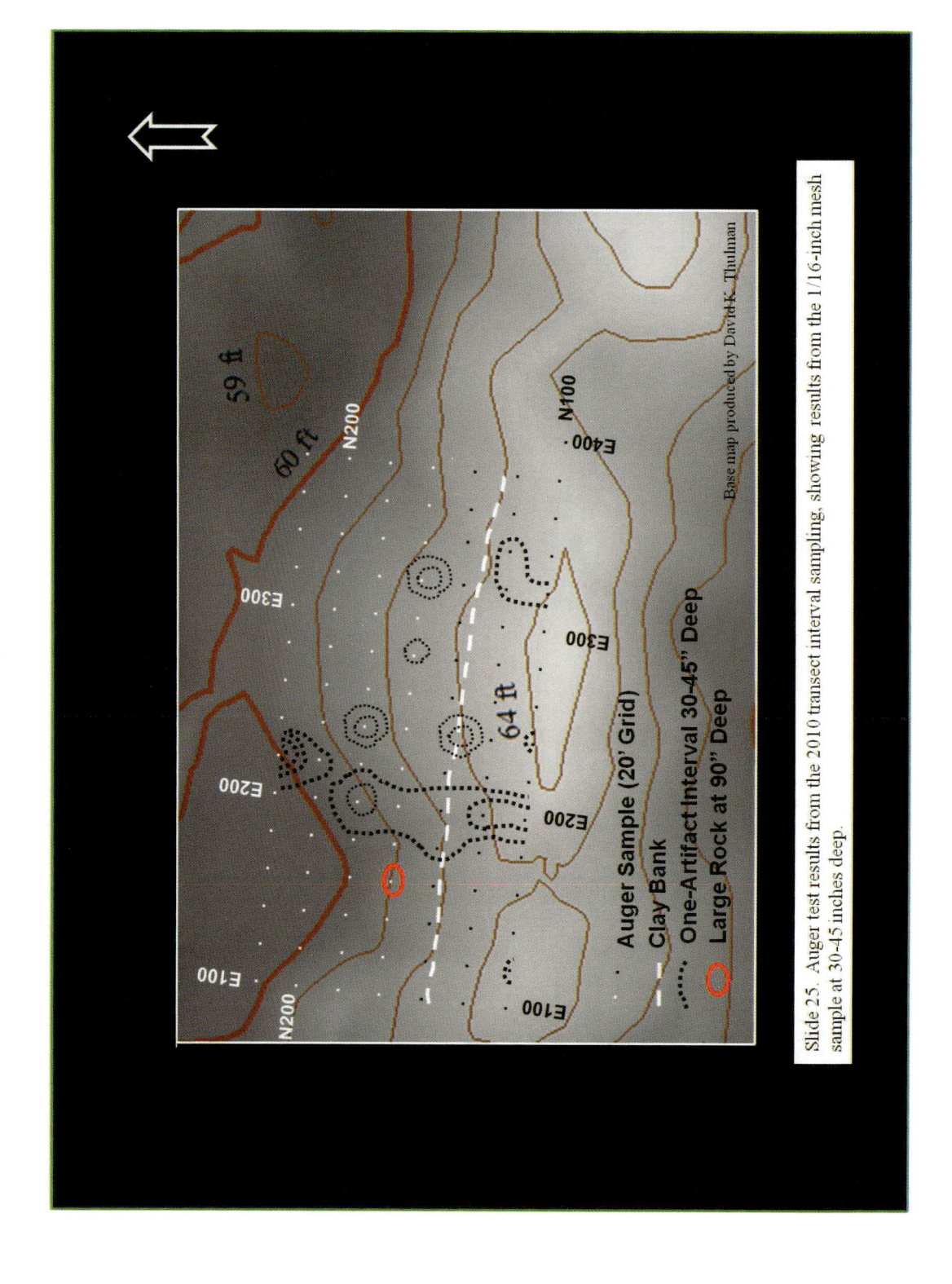

Slide 25. Auger test results from the 2010 transect interval sampling, showing results from the 1/16-inch mesh sample at 30-45 inches deep.

Slide 26. Test trench locations from the 2002 and 2010 excavations.

Slide 27. Selected artifacts from test trench BBH2B by level and depth in inches, showing quartzite biface cross-mend (b-c). (Photo by author)

BBH3A

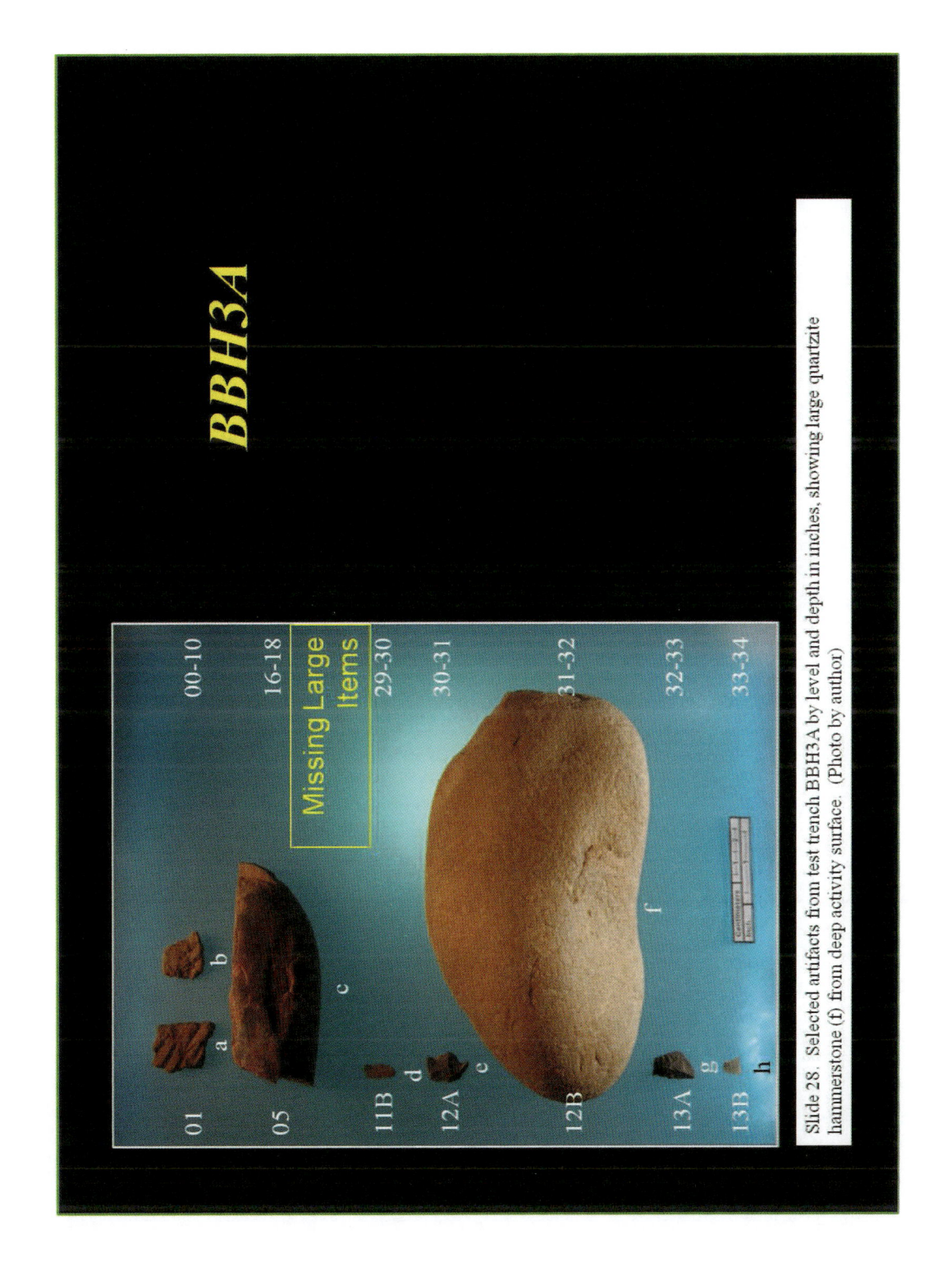

01	00-10
05	16-18
	Missing Large Items
11B	29-30
12A	30-31
12B	31-32
13A	32-33
13B	33-34

Slide 28. Selected artifacts from test trench BBH3A by level and depth in inches, showing large quartzite hammerstone (f) from deep activity surface. (Photo by author)

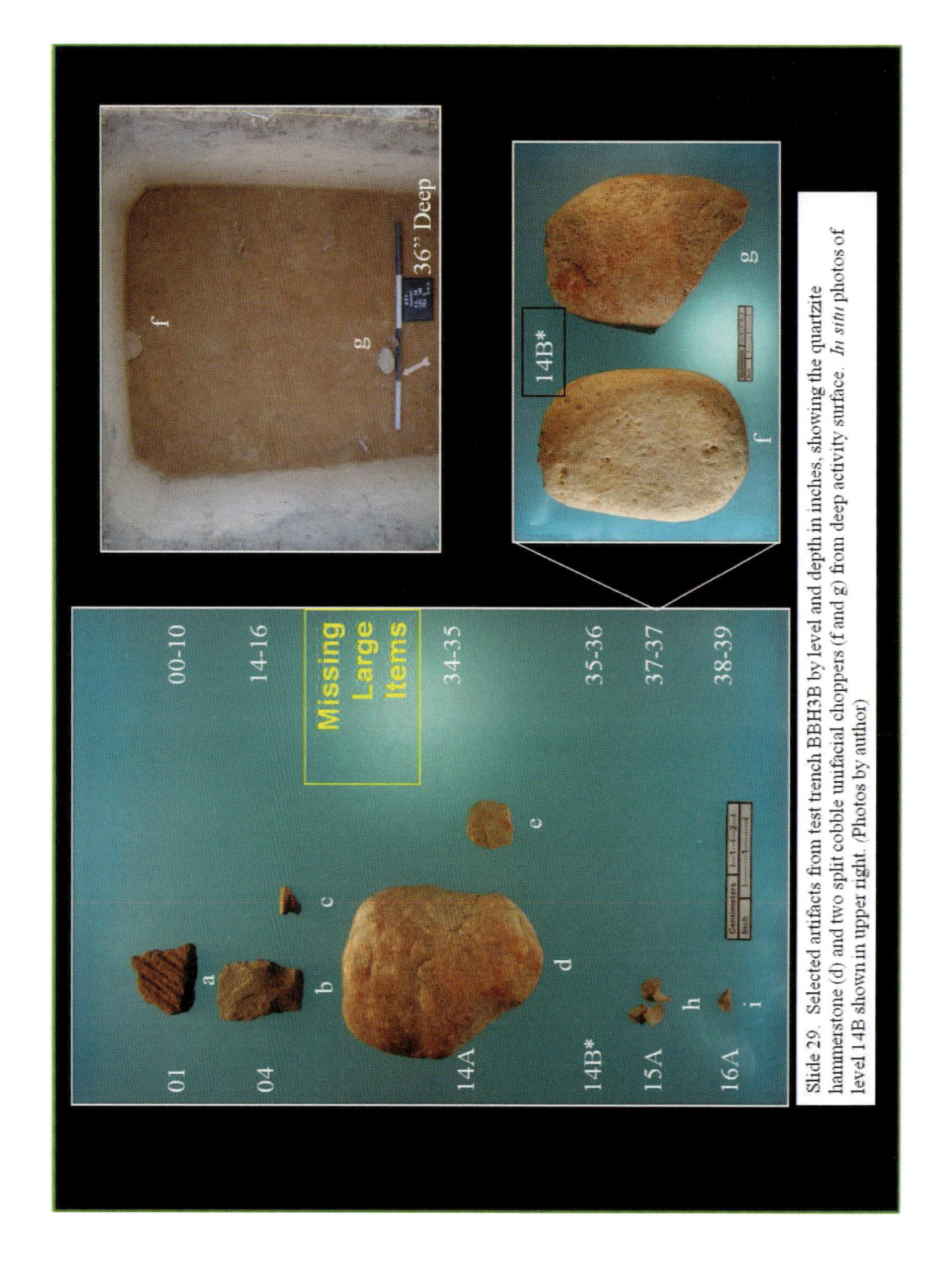

Slide 29. Selected artifacts from test trench BBH3B by level and depth in inches, showing the quartzite hammerstone (d) and two split cobble unifacial choppers (f and g) from deep activity surface. *In situ* photos of level 14B shown in upper right. (Photos by author)

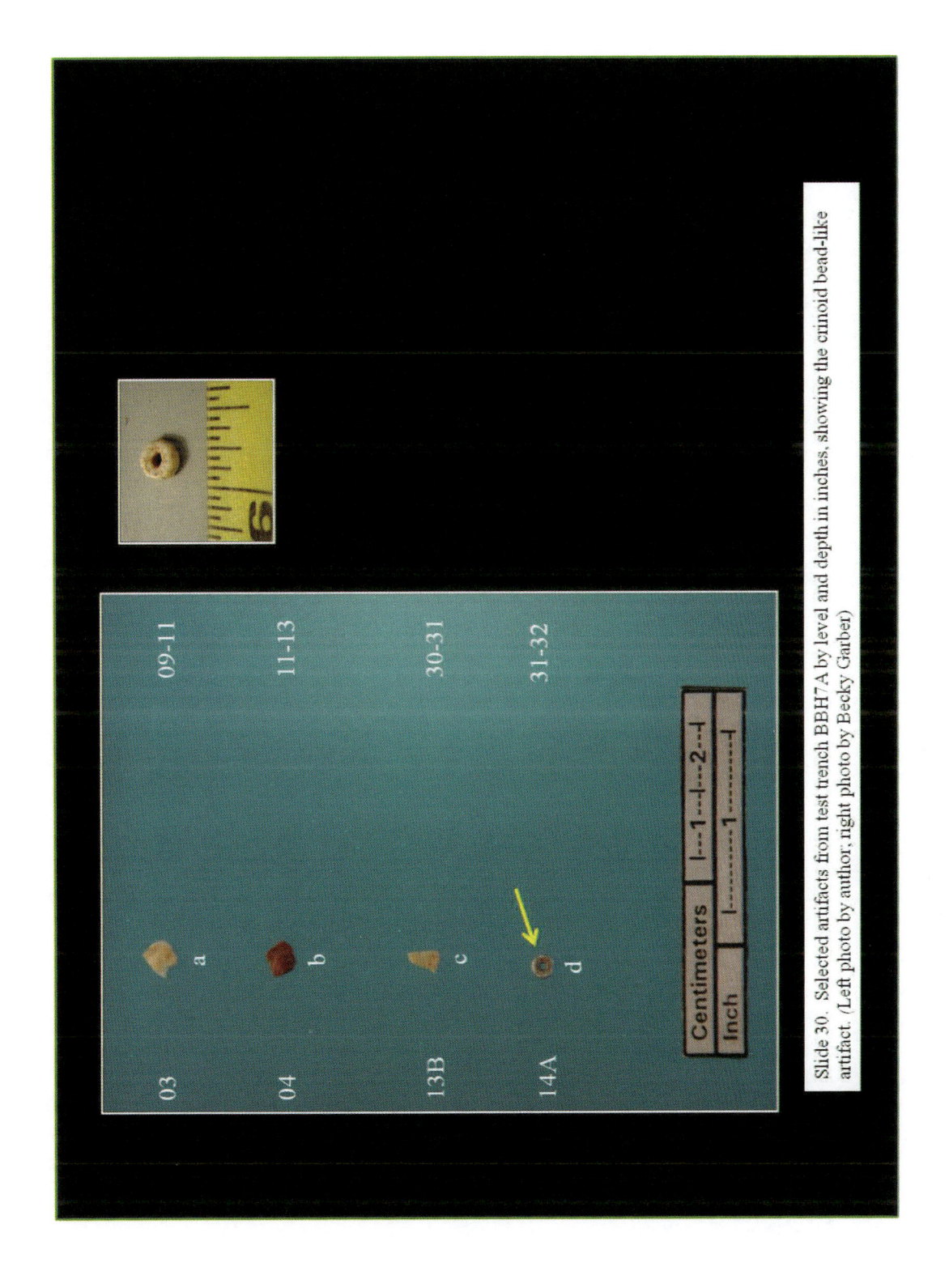

Slide 30. Selected artifacts from test trench BBH7A by level and depth in inches, showing the crinoid bead-like artifact. (Left photo by author; right photo by Becky Garber)

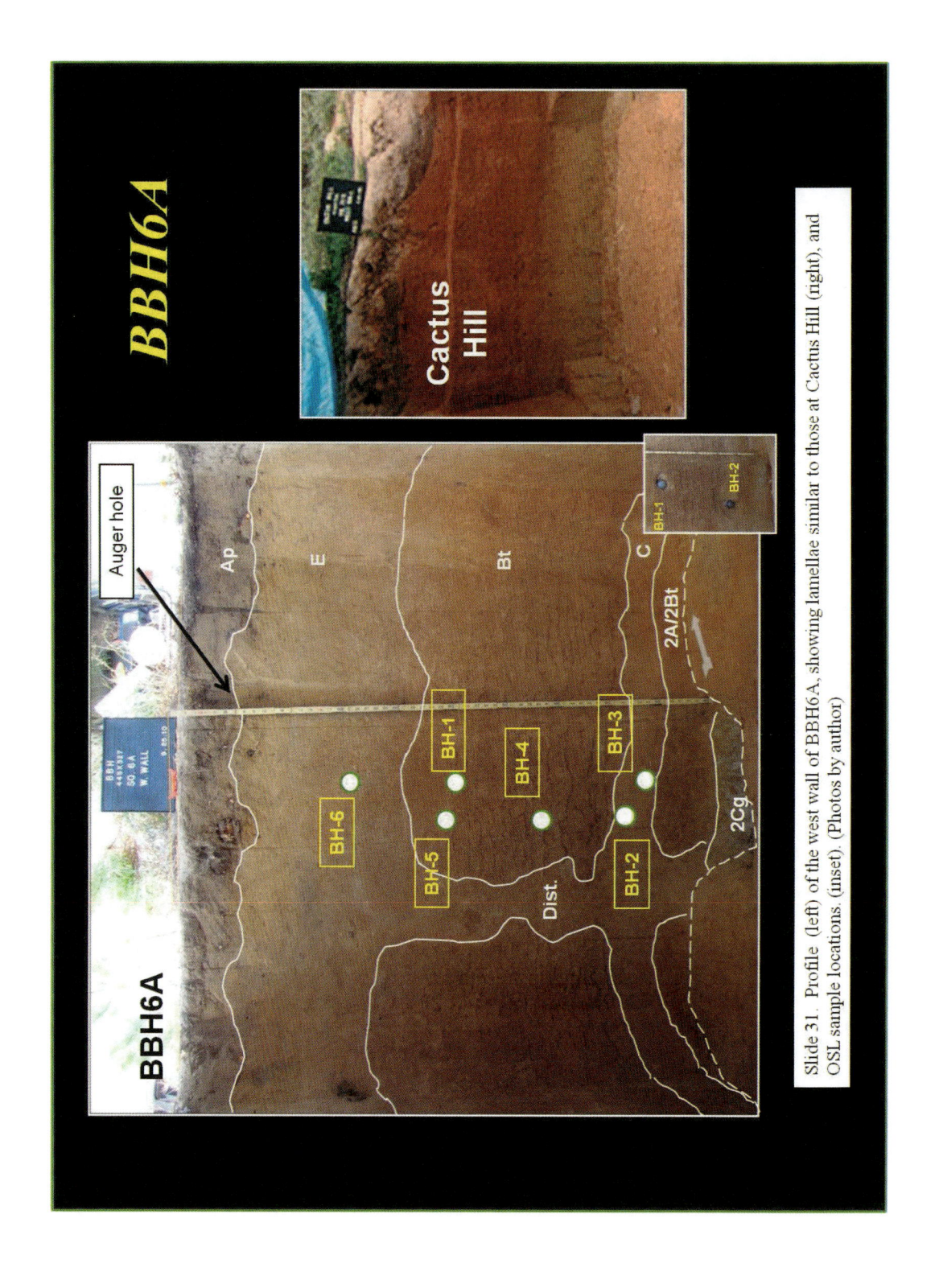

Slide 31. Profile (left) of the west wall of BBH6A, showing lamellae similar to those at Cactus Hill (right), and OSL sample locations. (inset). (Photos by author)

Sample	Depth	Mask size	Central age (ka) Cubic fit	Central age (ka) Linear fit	Minimal Age (ka) Cubic fit	Minimal age (ka) Linear fit
BH-6	50.8 cm (20")	1.0 mm	32.0±4.7	29.3±4.0	**13.0±1.8**	17.5±2.5
BH-5	76.2 cm (30")	1.0 mm	33.1±2.7	32.6±3.5	29.4±2.7	30.2±2.6
BH-1	78.74 cm (31")	1.0 mm	**15.8±1.7**	22.2±2.3	8.8±4.5	11.9±1.8
BH-4	101.6 cm (40")	1.0 mm	32.6±2.7	30.0±2.6	23.9±2.3	**25.4±1.9**
BH-2	129.54 cm (51")	1.0 mm	30.0±1.9	**30.6±1.1**	23.8±1.3	30.6±2.7
BH-3	132.8 cm (52")	1.0 mm	32.5±3.4	29.7±2.5	19.0±2.2	29.8±2.5

13.0±1.8 14-18"

15.8±1.7

25.4±1.9 34-36"

30.6±1.1

Slide 32. OSL results from BBH6A with highest quality dates highlighted. (Dates run by the USGS in Denver and analyzed by Kevin Burdett of the Florida Geological survey. (Photos by author)

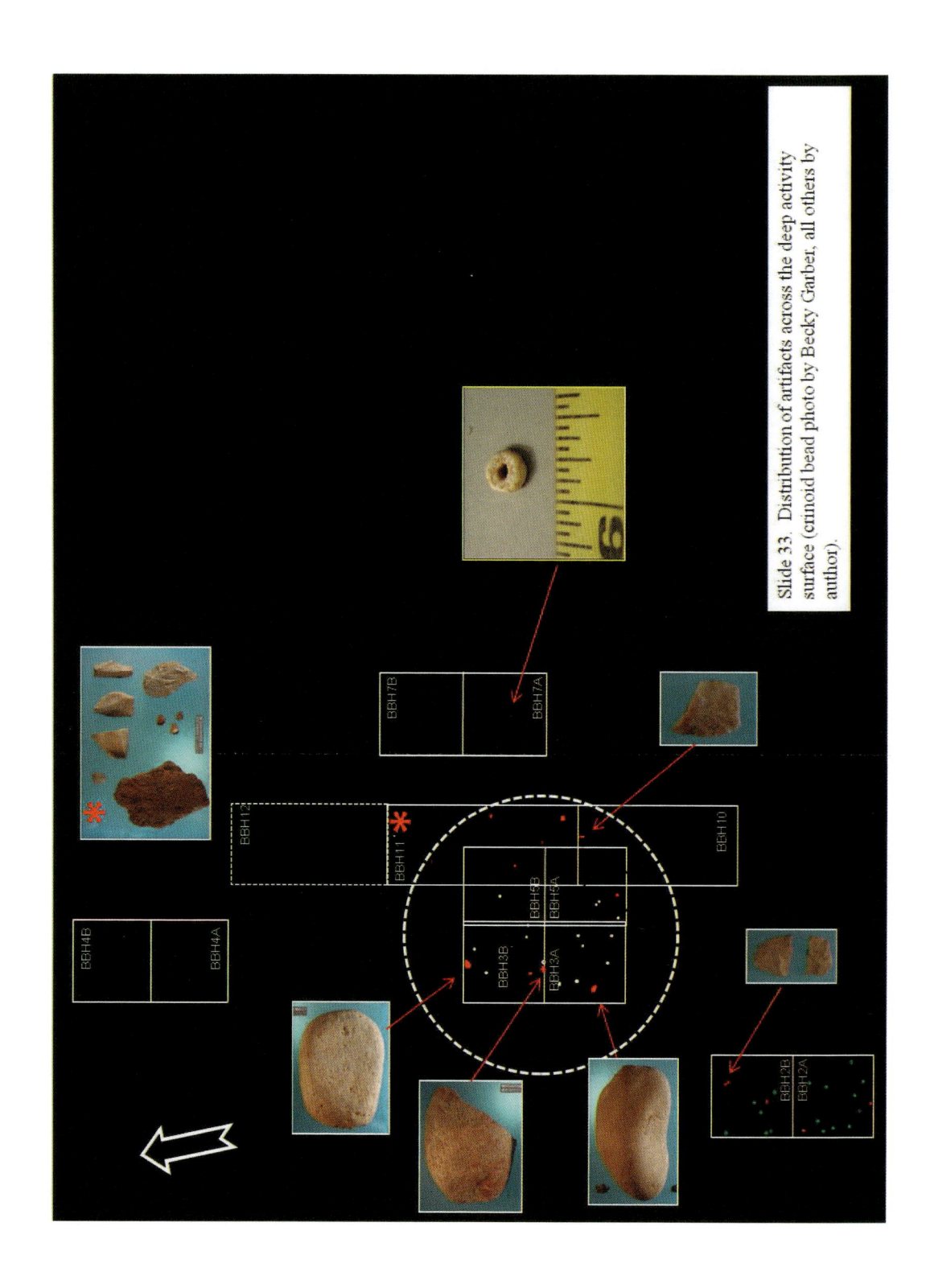

Slide 33. Distribution of artifacts across the deep activity surface (crinoid bead photo by Becky Garber, all others by author).

148

Slide 34. Hypothesized pre-Clovis age points from Cactus Hill, (Upper left photo by Author), 44PY152., (Center photo by William A. Childress), and Blueberry Hill, (lower right photo by author). Paoli, Kentucky chert Clovis age artifacts from apparent activity area are in upper right photo by Childress. Lower left photo of site is by the author.

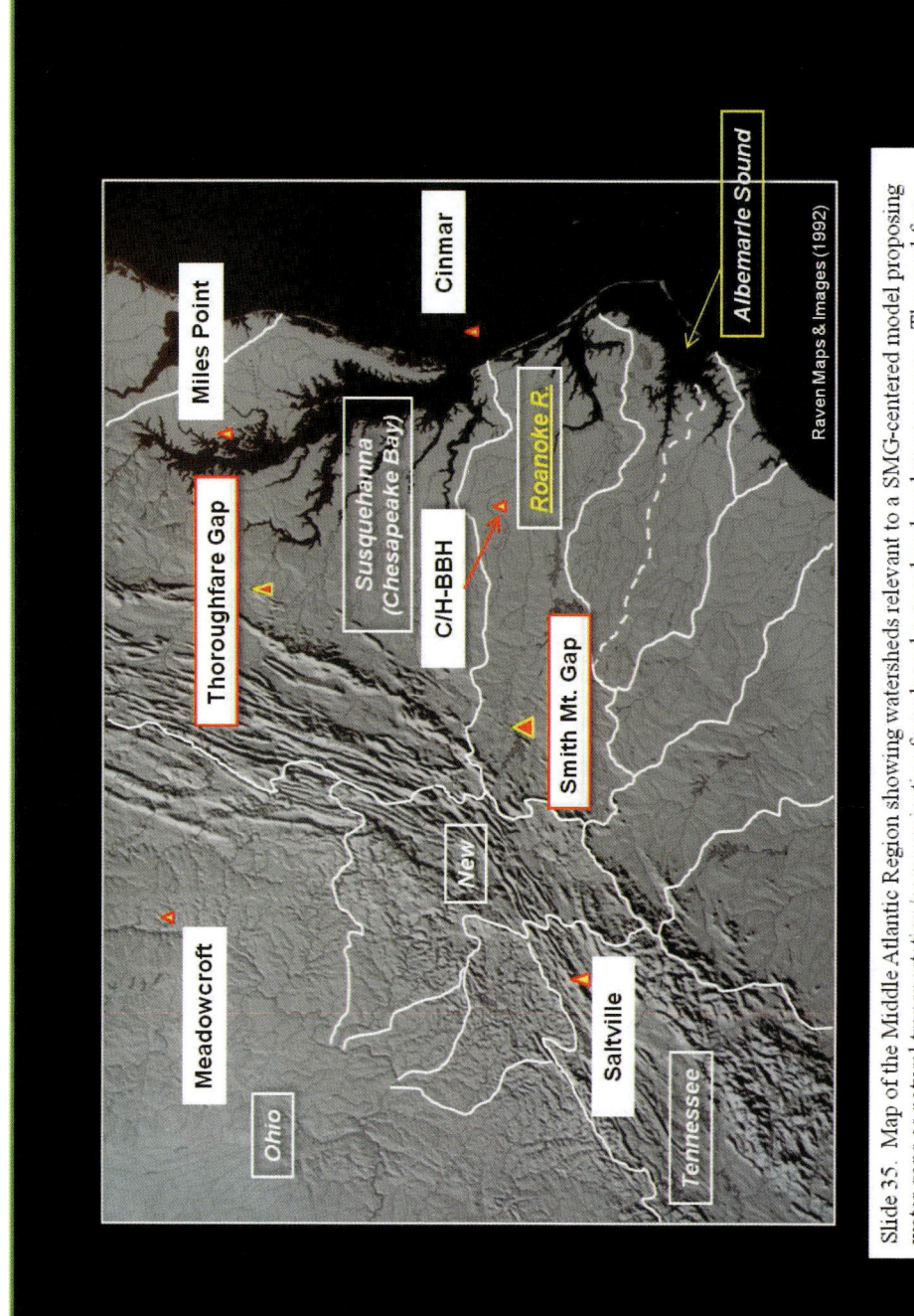

Slide 35. Map of the Middle Atlantic Region showing watersheds relevant to a SMG-centered model proposing water gaps as natural transportation/communication funnels and macro-band coalescence areas. Thoroughfare Gap, farther north, is also being tested for a connection between the Shenandoah Valley (Thunderbird site) and the Atlantic Coastal Plain/ancestral Susquehanna River Valley. Recorded pre-Clovis age sites are also shown (Background map reproduced with permission of Raven Maps & Images)

Slide 36. communication/transportation lines indicated by artifacts from SMG. Recorded pre-Clovis age sites and their watersheds, where east-west interaction at water gaps are proposed, also are included. (Background map reproduced with permission of Raven Maps & Images)

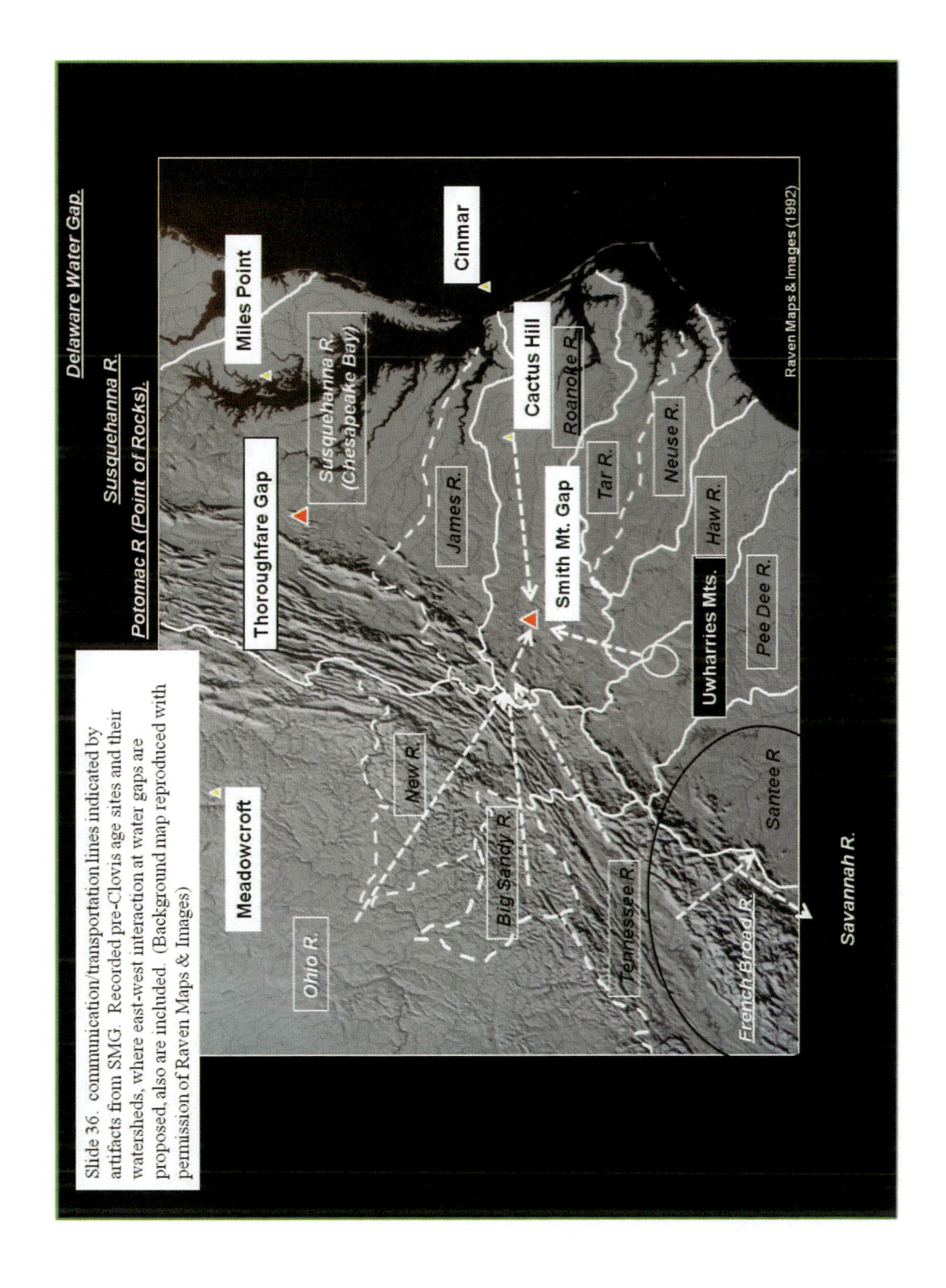

Figure 37. Hypothesized initial entry routes into the North America. (Photo of old Smithsonian Natural History Museum exhibit by author)

Pre-Fishtail settlement in the Southern Cone ca. 15,000-13,100 yr cal. BP: synthesis, evaluation, and discussion of the evidence

Rafael Suárez

Departamento de Arqueología, Universidad de la República and Sistema Nacional de Investigadores SNI- Agencia Nacional de Investigación e Innovación ANII. Magallanes 1577- Montevideo, Uruguay

"For decades, North American archaeologists have been discounting the South American evidence of a Pleistocene occupation of the New World because it does not correspond with their expectations or fit their models" Ruth Gruhn (2004:31).

The models discussed concerning the peopling of America have historically been generated by North American researchers, regardless of the variability and diversity of human occupation during the late Pleistocene of South America. Ruth Gruhn's observation sums up the problem that North American archeology has had to recognize: evidence of human occupation in South America, was as or more important than Clovis. While in the 1970s and 1980s the paradigm of the Clovis-First model expanded, there were some researchers who dared to oppose it in North America: James Adovasio and colleagues (1978) and K. R. Fladmark (1979) are only two examples, the first with very solid evidence from the Meadowcroft Rockshelter site and the second suggesting a new route model for the peopling of America which featured a circum-Pacific route, an idea taken up by others (see Gruhn 1994; Erlandson 2002; Goebel et al. 2008). As well, one can also recognize Alan Bryan and Ruth Gruhn as pioneers who sustained the evidence for early human occupation before Clovis both in South America and in North America (Bryan 1969, 1973, 1979, 1986; Bryan and Gruhn 2003; Gruhn 1991).

Archaeological evidence currently available suggests that the peopling of South America was from the North to the South, because during the late Pleistocene apparently the South Pacific, South Atlantic, the Drake Passage, and the Antarctic continent allegedly acted as barriers to effective human entry to the main land (Borrero 1999:322). Contrary to what happens in North America, where models of peopling of America are generated, and constantly reformulated, in South America original models proposed by South American researchers and also discussed in academic circles are scarce. An exception is the model published by Argentine archaeologist Laura Miotti (2006) which proposes a settlement model for South America by multiple routes. Settlement models from South America generally address local or regional developments, integrating data of different kinds to explain the arrival of the first humans in different environments.

The discussion regarding models of the peopling of the Americas in recent decades has led to the collapse of the Clovis-First model and various new attempts to reorganize the scenario of the origins of the first Americans have emerged (e.g., Goebel et al. 2008; Gruhn 2005; Stanford and Bradley 2012). Several South American sites have challenged the Clovis-First model since its inception, with conclusive arguments suggesting that human occupation of the Americas exceeds the 14,000 yr cal BP[1] (Bryan 1973, 1979); other researchers have suggested ages exceeding 20,000 yr BP (Collins 2012; Gruhn 1991, 2004, 2005, 2007; Miotti 2006; Stanford and Bradley 2012) for the entry of the first humans to America.

Perhaps the early site of South America best known in the archaeological literature is Monte Verde, in central Chile (Dillehay 1997). As well, however, there are several archaeological sites not as well known, located on Pampa-Patagonia in Argentina, south Patagonia in Chile, the Plains of Uruguay, and middle Uruguay River (Southern Brazil); and these sites have ages between 14,600 to 13,000 cal yr BP. In this paper I present an evaluation and discussion of known sites of the Southern Cone (below 28° S. Lat) that exceed 13,100 yr cal. BP (11,100 yr ^{14}C BP).

The recent redefinition of the age of Clovis (Waters and Stafford 2007) has reduced the chronological distance between the Clovis horizon of North America and the Fishtail complex of South America. Now we know that Clovis is at most only 100 years older than the Fishtail complex; and indeed both cultures are mostly contemporary. The synchrony in time and the great spatial distance between both complexes excludes any direct connection, in the sense that Clovis is the ancestor of the Fishtail complex, as has been proposed by several researchers. Other researchers have proposed an independent invention for Clovis and Fishtail points and technologies. Here I propose that both technologies, Clovis and Fishtail, have an older common ancestor; and both evolved in different ways in North and South America.

I use the term pre-Fishtail to describe the set of sites with dates exceeding 13,500 cal yr BP (11,500 yr ^{14}C BP) in the Southern Cone. This approach is justified for three main reasons; first, because in this part of the continent it is very clear that a Fishtail occupation

---

[1] Ages are presented in calendar years BP (yr cal BP) calibrated with Oxcal 4.2 program. In parentheses are ages without calibration in radiocarbon years BP (yr ^{14}C BP).

began to register just before 13,000 cal yr BP (11,000 yr ^{14}C BP). Second, it is incorrect to use the term pre-Clovis to refer to sites in South America, because there are no clear Clovis contexts excavated on this continent; and thus it is a mistake to use the term pre-Clovis to refer to early South American archaeological contexts. The term pre-Clovis, widely used by American archaeologists, should be restricted to North and Central America; and never used to refer to the earlier-dated sites in South America.

I choose the term pre-Fishtail because the Fishtail cultural complex is one of the oldest and best known related to the settlement of the Southern Cone. The pre-Fishtail concept does not imply cultural homogeneity for the earlier sites, but is used to characterize a chronological period in which there were certainly different human occupations, with diverse technologies and economies in various regions of the Southern Cone. The term pre-Fishtail is thus a way of referring to a set of human occupations representing one of the basal cultural levels of this continent.

This paper presents the evidence of the archaeological, stratigraphic, and chronological data for early human occupation of the southern cone of South America, initiated shortly after 15,000 yr cal BP. The paper presents the main archaeological sites with good stratigraphic contexts and precise chronological control (^{14}C dating). Several lines of evidence suggest that the human presence in this area of the continent was consolidated before 15,000 yr cal BP (Table 1). The paper also presents different examples of flaked stone collections registered in Uruguay by way of comparison with findings from other parts of the continent. This review compares the pre-Fishtail technology in regional and extra-regional context.

CHRONOLOGY AND MAJOR SOUTHERN CONE STRATIFIED SITES

Monte Verde site (central Chile)

The Monte Verde site is one of the best known early sites in the archaeological literature. It is located 960 km south of Santiago, Chile, on the banks of Chinchihuapi creek (Figure 1.1). It is an open site excavated by T. Dillehay and a large group of collaborators, indicating human occupation by 14,600 yr cal BP (Monte Verde II) (Dillehay et al. 2008). Evidence of another previous occupation (Monte Verde I) dated ~ 33,000 yr ^{14}C BP is presented with reservation and caution by Dillehay (1997:774). Monte Verde is considered by some researchers as the archaeological site that collapsed the Clovis-first model barrier (Bonnichsen and Lepper 2005:15; Meltzer 2009:123); its importance is comparable to the

Folsom site that showed that human occupation existed in the Americas towards the end of the Pleistocene (Meltzer 2009).

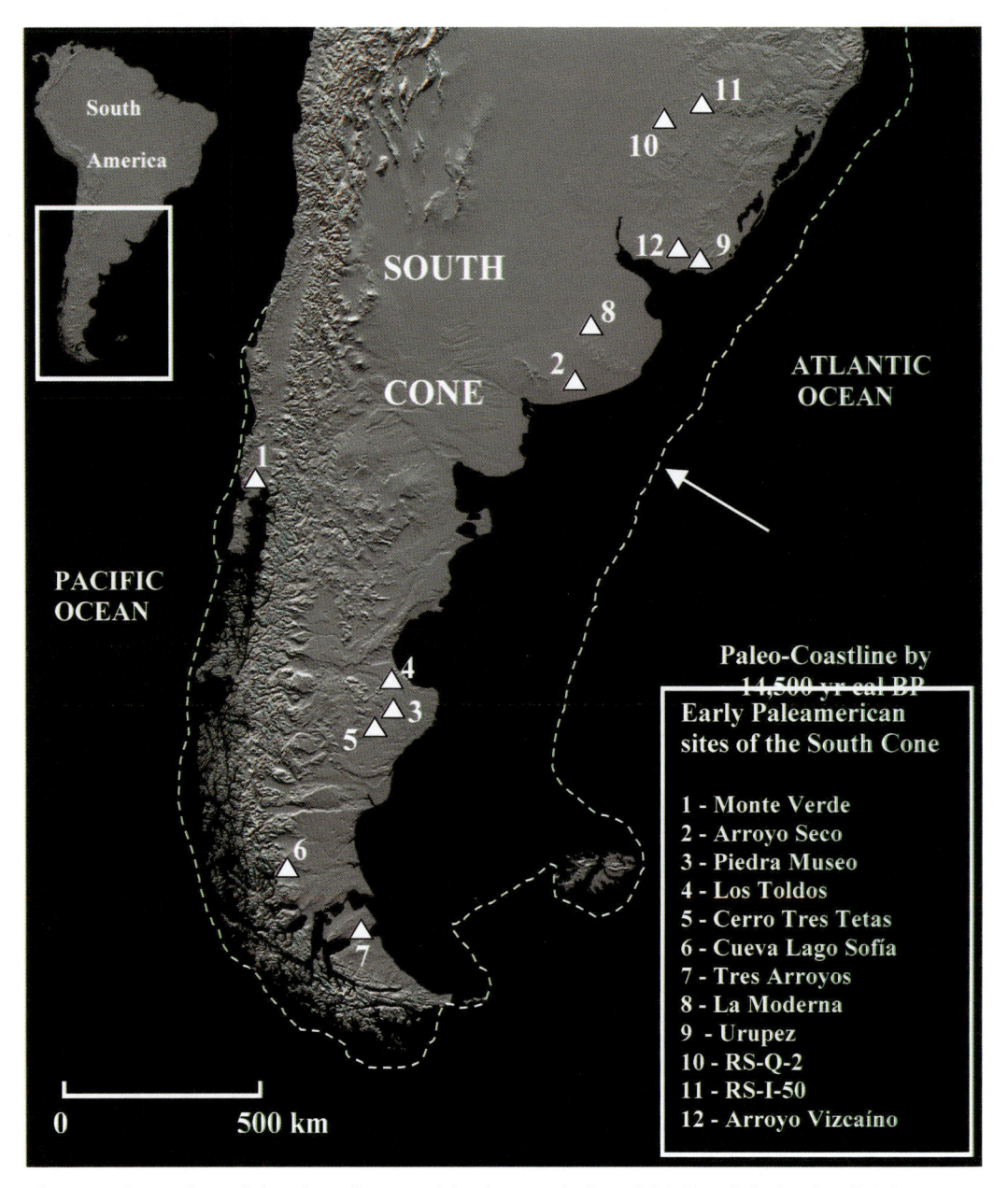

Figure 1. Location of the sites discussed in the text below 28 ° South latitude. 1) Monte Verde, 2) Arroyo Seco 2, 3) Piedra Museo, 4) Los Toldos, 5) Cerro Tres Tetas, 6) Cave Lake Sofia 1, 7) Tres Arroyos, 8) La Moderna, 9) Urupez, 10) RS-Q-2, 11) RS-I-50, 12) Arroyo Vizcaíno.

The evidence recovered at Monte Verde is contrary to the stereotype of the nature of the late Pleistocene archeology and the proposed models of the initial peopling from North America (Gruhn 2004:31, 2005:202).The very good preservation is quite unusual for an archeological site from the late Pleistocene. Below a peat deposit were recovered perishable objects of wood and bone, chewed and burned medicinal herbs, and remains of a number of species of edible land plants, and algae (*Gigartina* sp., *Mazzaella* sp.). The latter were collected on the distant seacoast, and dated at 14,625 yr cal BP (12,290 ± 60 yr ^{14}C BP) and 14,600 yr cal BP (12,310 ± 40 yr ^{14}C BP) (Dillehay et al. 2008:785). Also on the site were several wooden structures that correspond to habitations of daily use and an isolated ceremonial structure (Dillehay 1997; Dillehay et al. 2008).

The stone artifact assemblage of the Monte Verde II level dated 14,600 yr cal BP contains 692 pieces. Mike Collins performed the analysis and description of this material. Of these, certainly 24 have cultural modification, featuring four bifaces including two projectile points, three notches, two choppers, one polyhedral core and 14 flakes (Collins 1997:404 Table 14.3). Other artifacts found are one perforator, two grooved stones, 6 edge-battered, 45 matte-finished *bolas*, 34 *manos* and 11 hammerstones (Collins 1997:404). Another 454 pieces exhibited no clear cultural modification.

One biface is 143 mm long, 62 mm wide and 40 mm thick, made in quartzite from outcrops in the highlands about 80 km northeast of Monte Verde (Collins 1997:404). This artifact is very important, because it does not have overshot type flakes scars characteristics of the Clovis flaking technique (Collins 1997; Bradley et al. 2010).

Lanceolate projectile points are in direct contextual and stratigraphic association with mastodon bones. Other Pleistocene faunal bones with cultural modification are associated with wood artifacts recovered near hearths. According to Dillehay (1997), the Monte Verde II component is interpreted as a residential camp used for a period of approximately one year.

Monte Verde, like the Folsom and Calico sites, presents a curious case where in a "panel of experts" made an inspection to validate the site and its age (see Meltzer et al. 1997; Meltzer 2009:125-129). This 1997 event is a modern expression of what happened with the validation of the Folsom site in 1927, when a group of researchers composed of paleontologist B. Brown, and archaeologists F. Roberts (younger colleague of A. Hrdlička at the Smithsonian) and A.V. Kidder, visited the site while it was being excavated in order to corroborate and validate the association between extinct bison and Folsom points recovered by J. Figgins (Meltzer, 1993:53).This event put an end to the discussion of more of 50 years about the Pleistocene human antiquity in North America (Meltzer, 1993).

Figure 2. Points form Monte Verde II (photo courtesy of T. Dillehay).

Meltzer (2009:123), points out that the visit to Monte Verde was different from the Folsom site, because during the 1997 Monte Verde trip the site had already been excavated, there were no artifacts seen *in situ,* and much of the site area had been altered. However, the main argument to validate Monte Verde is the significantly detailed work done by T. Dillehay (1997, Dillehay et al. 2008) and his group of collaborators in analyzing a variety of site data.

Some researchers who participated in the evaluation of Monte Verde, such as C.V. Haynes, in subsequent publications continue with arguments and revisions of the Clovis-first model, which contradicts their validation of Monte Verde that they accepted in the 1997 expedition (see Haynes, 2005). Moreover, Haynes asserts that *"In the future the scientific investigation of all potential pre-Clovis sites must include on-site evaluation of the evidence by experienced and objective geologists and archaeologists within a reasonable time of its discovery"* (Haynes, 2005:114). This expression is clearly an attempt to justify and legitimize the notion that in the archaeology of the peopling of Americas there are the accused, police, judges and prosecutors (Politis 1999).

Table 1. Pre-Fishtail ^{14}C dates recovered in sites of the Southern Cone analyzed in this paper.

Site name	^{14}C yr B.P.	Lab. Number	Age calibrated years BP [a]	Reference
Monte Verde	> 33,000[b]	Beta 7825	38,420-37,034	Dillehay and Pino 1997
Monte Verde	33,370 ± 530[b]	Beta 6754	39,411- 36,710	Dillehay and Pino 1997
Monte Verde	13,565 ± 250[c]	TX-3208	17,199-15,424	Dillehay and Pino 1997
Monte Verde	12,780 ± 240[c]	Beta-59082	16,413-14,189	Dillehay and Pino 1997
Monte Verde	12,740 ±440[c]	TX-5375	16,715-13,894	Dillehay and Pino 1997
Monte Verde	12,650 ±130[c]	TX-4437	15,517-14,191	Dillehay and Pino 1997
Monte Verde	12,310 ±40[c]	Beta 239650	14,597-13,997	Dillehay et al. 2008
Monte Verde	12,290 ± 60[c]	Beta 238355	14,625-13,943	Dillehay and Pino 1997
Monte Verde	12,230 ± 140[c]	Beta-6755	14,910-13,795	Dillehay and Pino 1997
Monte Verde	12,000 ± 250[c]	OXA-105	14,889-13,358	Dillehay and Pino 1997
Monte Verde	11,990 ± 200[c]	TX-3760	14,600-13,371	Dillehay and Pino 1997
Monte Verde	11,790 ±200[c]	TX-5374	14,086-13,256	Dillehay and Pino 1997
Arroyo Seco 2	12,240 ± 110	OxA -4591	14,667-13,824	Politis 2008
Arroyo Seco 2	12,200 ± 170	CAMS58182	14,958-13,742	Politis 2008
Arroyo Seco 2	12,170± 55	OxA 15871	14,195-13,830	Steele and Politis 2009
Arroyo Seco 2	12,155 ± 70	OxA-10387	14,214-13,791	Politis 2008
Arroyo Seco 2	12,070 ± 140	OXA -9243	14,544-13,578	Politis 2008
Arroyo Seco 2	11,750 ± 70	CAMS-16389	13,776-13,415	Politis 2008
Arroyo Seco 2	11,770 ± 120	AA 62514	13,865-13,360	Politis 2008
Arroyo Seco 2	11,730± 70	OxA-9242	13,765-13,401	Politis 2008
Arroyo Seco 2	11,590 ± 90	AA-7965	13,685-13,265	Politis, 2008
Arroyo Seco 2	11,320 ± 110	AA-39365	13,413-12,932	Politis 2008
Arroyo Seco 2	11,250 ± 105	AA-7965	13,359-12,860	Politis 2008
Piedra Museo	12,890± 90	AA-20125	16,166-14,976	Miotti et al. 2003
Cerro Tres Tetas	11,560 ± 140	LP-525	13,736-13,165	Paunero 2003
Cueva Lago	11,570 ± 60	PITT-0684	13,601-13,277	Prieto 1991

Cueva Lago Sofía 1	11,570 ± 60	PITT-0684	13,601-13,277	Prieto 1991
Tres Arroyos	11,820 ± 250	Beta-20219	14,266-13,146	Massone 2004
RS-I-50	12,700 ± 220	SI-801	16,214-14,103	Miller 1987
RS-Q-2	12,690 ± 100	SI-2351	15,584-14,479	Miller 1987
Urupez	11,690 ± 80	Beta-211938	13,752-13,357	Meneghin 2006
La Moderna	12,350 ± 370	TO-1507	15,964-13,442	Politis 2008

Note: [a] Calibration with Oxcal 4.2 program; [b] Monte Verde I; [c] Monte Verde II.

Arroyo Seco 2 site (Pampa, Argentina)

Site 2 of the Arroyo Seco locality is located in Pampa Argentina adjacent to a stream and lagoon (38 ° 21 '38" S, 60 ° 14'39" W) (Figure 1.2). It is an open site with evidence of multiple occupations (i.e., multicomponent). The site was discovered and initially excavated by Mr. Mulazzi (an amateur archaeologist) in early 1940; since the 1970s, Gustavo Politis has carried out excavations at the site. From the oldest level there are eleven ^{14}C dates (see Table 1) between 14,667 yr cal BP (12,240 ± 110 yr ^{14}C BP) to 13,359 yr cal BP (11,250 ± 90 yr ^{14}C BP) (Politis 2008; Steele and Politis 2009).

Recovered at the site were bones of modern fauna such as *Lama guanicoe, Ozotoceros bezoarticus, Rhea Americana*; and also extinct fauna including *Megatherium americanum, Equus neogeus, Hippidion sp., Toxodon platensis, Glossotherium robustus and Paleolama wedelli* (Steele and Politis 2009).The site has been interpreted as a base camp for multiple activities of the early explorers of the Pampa (Politis, 2008).

A total of 47 artifacts were recovered in the S/Z stratigraphic unit containing the oldest archaeological material. Eight different rock types (quartzite, silicified tuff, basalt, chert, etc.) are represented (Leipus and Landini 2013). This region of the Pampa offers no outcrops of rocks that can be used to manufacture artifacts; stone resources are located distantly from the site. The quartzites were probably transported from Arroyo Diamante sources located 150 km from the site. The chert must have been transported from sources located 150 km from the site. The chert (ftanita) must have been transported from sources located between 185 and 65 km of the site. Tuff sources are located 50 km from the site (Leipus and Landini 2013).

The flaked stone artifacts are simply modified. The technology of the ancient component is characterized by unifacial artifacts, very little standardization in form, with marginal retouching. No diagnostic artifacts such as projectile points were recovered. The artifacts are small; and were manufactured by marginal retouch and micro retouch, with bipolar and primary flakes of basalt and quartzite serving as blanks for artifacts. There are no multiple-use tools, scrapers were manufactured from flakes derived from beach cobbles; and some *piece esquillée* were also recovered (Leipus and Landini 2013). The technology and artifact assemblage of the oldest component of Arroyo Seco 2 are consistent with the conclusions of Gruhn (2004:31) that the technology of human groups towards the end of the Pleistocene in South America was relatively simple.

Piedra Museo site (Central Plateau of Patagonia, Argentina)

The site Piedra Museo AEP-1 (47°53'42" S - 67°52'04" W) is a rockshelter located in the Central Plateau of Santa Cruz Province in Argentinean Patagonia (Figure 1.3).The site was excavated by Laura Miotti and associates (Miotti 1995, Miotti et al. 2003). It presents six stratigraphic units representing different events since the late Pleistocene (U6), the Pleistocene /Holocene transition (U5), and the early and middle Holocene (U4, U3, U2) as well as the recent Holocene (U1) (Miotti et al. 2003:100 Figure 1).The lowest unit, U6, features a sparse unifacial flaked stone assemblage and a bone point. In the overlying unit, U5, two Fishtail points were recovered, one with fluting on both sides, made on reddish jasper. The site has three Pleistocene fauna species, *Hippidion saldiasi* (horse), *Mylodon sp.* (ground sloth), and *Lama (Vicugna) gracilis*. There were also remains of *Lama guanicoe*, Rheidae (American ostrich), and medium sized birds which were exploited and consumed by humans during the early stages of occupation of the site: several bones exhibit cut-marks (see Miotti y Salemme 2005:211-216 Figure 6). The site has 11 ^{14}C ages, one of the dates obtained by AMS on charcoal from the bottom of Unit U6 yielded an age of 14,970 yr cal BP (12,890 ± 90 yr ^{14}C BP) (Miotti et al. 2003). Others ages of U6 are 13,080 yr cal BP and 12,970 yr cal BP and from Unit U5 there is two dates of 12,570 yr cal BP (Miotti et al. 2003). Distributional, taphonomic, and bone modification analysis indicate that in the late Pleistocene, Piedra Museo AEP-1 was a locus of confined activities related to butchering megamammals (Miotti and Salemme 2005:216).

Cerro Tres Tetas site

Cave 1 of the Cerro Tres Tetas locality (48° 08' 58''S - 68°56' W) is a multicomponent site located 75 km from the Piedra Museo locality in the Central Plateau of Santa Cruz province (Figure 1.5). The site was excavated by the archaeologist Rafael Paunero, and presents evidence of a long process of human occupation that began at the end of the Pleistocene and ended in historic times (Tehuelche Indians) (Paunero 2003). An area of 12.25 m^2 was excavated in the interior of the cave. A total of 12 radiocarbon ages date the site, and a sample recovered at the base of the U5 stratigraphic unit gave an age of 13,730 yr cal BP (11,560 ± 140 yr ^{14}C BP) (conventional method). A total of 523 stone artifacts were recovered from the lowest occupation level, including side-scrapers, retouched flakes, knives, one chopper, and one hammer stone, cores and 474 flakes. Paunero (2003) reported the presence of several hearths. A more detailed report on this site is pending.

Cueva Lago Sofía 1 site

This site is a cave located in the southern Chilean Patagonia province of Ultima Esperanza (51° 32 'S, 72° 32' W) (Figure 1.6). In the lower level of the cave a hearth with burned and fractured bones of Pleistocene fauna was discovered, yielding two dates of interest, one of 13,600 yr cal BP (11,570 ± 60 yr ^{14}C BP) and another of 16,955 yr cal BP (12,990 ± 490 yr ^{14}C BP) (Prieto, 1991). The age of 16,955 yr cal BP was obtained from a bone of Pleistocene fauna, and Jackson and Massone (2005) argue that the bone was from a natural paleontological deposit and not in a cultural context. However, stone artifacts were recovered in association with the date of 11,570 ± 60 yr^{14}C BP; including bifacial and unifacial flakes, as well as one bone retoucher and one bird bone awl. A more detailed report is needed.

Tres Arroyos

This site, a rockshelter located on the island of Tierra del Fuego (53° 23 'S, 68° 47' W), is the most southern early site of America, only 260 km from the southernmost point of South America (Figure 1.7). Tres Arroyos was excavated by M. Massone (2004) and collaborators between 1996-1999. The stratigraphy indicates a deposit of between 0.70 to 0.90 m deep. In the oldest level five hearths were identified, with five radiocarbon dates over 10,000 yr BP, the oldest being 14,266 yr cal BP (11,820 ± 250 yr ^{14}C BP by conventional method) (Massone 2002, 2004).

Different Pleistocene fauna species were recovered: *Hippidion sp.* (horse), *Mylodon darwini*) (giant sloth), *Vicugna vicugna* (vicuña), and *Lama sp.* (camelid) and *Panthera onca mesembrina* (feline). Massone (2002, 2004) describes in detail the hearths excavated, indicating that they have a variable diameter between 21 and 45 cm and variable depths of 5 to 12 cm. Hearth 1 is the largest, with burned bones recovered inside it. One of the bones was dated to 14,266 yr cal BP (11,820 yr ^{14}C BP) (Massone, 2004:66): other hearths gave dates of 12,690 yr cal BP. Stone artifacts recovered in the oldest level are scrapers, knives, cores, and flakes; some of which were bifacially retouched. Materials used to manufacture the artifacts are petrified wood, rhyolite, and chert (Jackson, 2002).

INCONCLUSIVE EARLY SITES

Urupez site

U. Meneghin (1977) a collector, enthusiast, and amateur archaeologist, presents important research findings and excavations at Cerro de los Burros in the department of Maldonado in Uruguay. Here two fractured Fishtail points were recovered on the surface. Recently Meneghin returned to research in this locality with a series of excavations at a new site called Urupez (Figure 1.9), where he has obtained interesting data. The Urupez site (34° 49'05 .8" S - 55° 18'56 .4" W) has little stratigraphic depth; only between 0.35 to 0.40 meters of sediment (Meneghin, 2005:9-10). Two outstanding AMS radiocarbon ages, one of 13,745 yr cal BP (11,690 ± 80 yr ^{14}C BP) and another of 12,730 yr cal BP (10,690 ± 60 yr ^{14}C BP) (Meneghin 2004, 2005, 2006), are associated with Fishtail points, according to Meneghin. However, the first date exceeds by ~ 700 ^{14}C years the reliably known age for Fishtail points regionally; and should therefore be treated with caution, until the site is published in detail with descriptions of artifact associations and stratigraphic studies which clearly demonstrate the archaeological and contextual association between artifacts and radiocarbon dates. As well, the oldest age should be confirmed with new dating. The second date -12,730 yr cal BP- is in the chronological range of Fishtail points in the region, so this age is more reliable than the first. Recently Meneghin (2013 personal communication) indicates the presence of another Fishtail point in stratigraphic context at the site. However, as indicated, the publications of Meneghin (2004, 2005, 2006) neglect the stratigraphic description of sedimentary sequences in which levels of archaeological materials were found; nor does he take into account the associated problem of working in low-productivity sites with wide-ranging radiocarbon chronologies. Despite these observations, the contribution of Meneghin's work over the last 30 years in the locality of Cerro del Burro, has the merit of having recovered the oldest known radiocarbon dates on the current shoreline of the Atlantic Ocean. The age of 13,745 yr cal BP for the Urupez site is not conclusive: only new dates that are in this range and with a clear association with cultural material will allow acceptance of the site's chronology.

Los Toldos site (Cave 3)

Los Toldos is an important late Pleistocene archaeological site in Southern Patagonia (Figure 1.4), known since the 1950s. The site, one of several rockshelters in the wall of a small canyon, was excavated by A. Cardich (et al. 1973). From the lowest cultural level, a date of ~15,000 yr cal BP (12,600 ± 600 yr ^{14}C BP) was recovered. According to Cardich, a distinctive unifacial flaked stone industry was recovered at this level, with standardized and well-made artifacts, characterized as large side-scarpers. Also recovered were remains of Pleistocene fauna, horse (*Onohippidium (Parahipparion) saldiasi*), one extinct camelid (*Lama (Vicugna) gracilis*), and guanaco. The problem of the dating of ~15,000 yr cal BP is that in the original publication the laboratory number of the sample was not presented, the sample was on scattered charcoal fragments, and the association with artifacts is unclear (Borrero 1999).

REJECTED EARLY SITES IN THE SOUTHERN CONE

This section discusses what are regarded as sites that have problems of context, associations, and other serious doubts as to be accepted as prior to 13,100 yr cal BP. The main problems are: 1) in some cases archaeological excavations were not conducted, 2) there area lack of detailed descriptions of the cultural context and the stratigraphy of the sites, and 3) the early radiocarbon dates are unclear or ambiguous.

RS-I-50 and RS-Q-2 Sites

These two are open air sites located in southern Brazil (Rio Grande do Sul state), investigated by Brazilian archaeologist Eurico Th. Miller in 1970. Miller (1987) defines the "Ibicuí phase" from two open sites, one located in the Ijuí River (RS-I-50) and another on the Quaraí River (RS-Q-2) (Figure 1.10-11). Miller (1987:41, Table 1) dated the RS-I-50 site at ~ 15,000 yr cal BP (12,700 ± 220 yr ^{14}C BP, sample SI-801) and the RS-Q- 2 site at ~ 15,000 yr cal BP (12,690 ± 100 yr ^{14}C BP, sample SI-2351).

The evidence of this "phase" is provided by the presence of 46 stone artifacts and two bones of Pleistocene species with thin and shallow parallel grooves, and a *Glossotherium robustum* skull; remains that were exposed in "discontinuous agglomerates" (Miller, 1987:48). The association between the artifacts and *Glossotherium robustum* in RS-I-50 was not demonstrated in an archaeological excavation, but deduced from the natural stratigraphic profile within the ravine, as shown in the photograph by Miller (1987:45 Figure 4).

Miller (1987:54) notes that the date of ~ 15,000 yr cal BP (12,690 ± 100 yr ^{14}C BP SI-2351) for the RS-Q-2 site *"is indirectly (only stratigraphically) associated, dating plant remains on the horizon IX for the site RS-Q-2B. Indirectly dates the same horizon of the site RS-Q-2. In RS-Q-2 does not include projectile points in the excavations. The lithics are similar to the Ibicuí Phase, reason this site and this indirect date are hypothetically attributed to Ibicuí Phase"*.

In Table 1 Miller (1987:41) lists the radiocarbon dates, the date of 12,690 ± 100 yr ^{14}C BP (SI-2351) is included twice, belonging to two different sites, RS-Q-2 and RS-Q-2B. The SI-2351 date was obtained from RS-Q-2B without cultural association (non-lithic artifacts); however, Miller extrapolates and correlates the SI-2351 date to the site RS-Q-2. Thus Miller (1987:41) at one time includes the SI-2351 date in the "*Ibicuí phase*" and then lists the same date with the sub-title "*without cultural association*".

Therefore, in relation to the "Ibicuí phase", there is some confusion when trying to correlate the date of the RS-Q-2B site, where no association with cultural material exists, with the site RS-Q-2.

Moreover, the absence or limited extent of archaeological excavations in Miller's sites causes problems of confidence in his report of stratigraphic and contextual relationships. Finally, the low number of artifacts recovered, and the low number of ^{14}C dates raise doubts regarding the existence of this cultural phase. In this sense the 46 stone artifacts recovered from the three sites of this phase indicate an average of only 15.3 artifacts per site. Some researches argue that in fact the 46 artifacts are geofacts (Diaz 2004). But, if we consider that there is a single ^{14}C date of the sample represented by SI-801 for the RS- I-50 site, it becomes difficult to characterize a "cultural phase" with the archaeological evidence presented by Miller (1987). Other Brazilian authors have discussed the sites. Milder (1994) agrees with the idea that the association between Pleistocene fauna and cultural material is questionable and not derived from an archaeological excavation.

Based on the arguments presented, the existence and age of the "Ibicuí phase" should be taken with care and caution. This caution does not mean that there cannot be human occupation and sites of ~ 15,000 yr cal BP (12,500 yr ^{14}C BP) in the region; I only indicate that the data presented by Miller (1987) are insufficient to justify solid and decisive evidence of occupations of around 15,000 yr cal BP in the middle Uruguay River. Nevertheless, we must recognize that the research of Miller (1987) in other river sites was very important, because with 18 radiocarbon dates it provided the first database for the region.

La Moderna site

La Moderna, an open air site in the Argentina Pampa (Figure 1.8), was the first site in which an association between Pleistocene fauna and humans was demonstrated, in the early 1970s (Politis et al. 2003). From the beginning, the chronology of the site has been problematic and controversial. The site is currently being investigated by Gustavo Politis, who has been very critical of the radiocarbon dates of the site. However, it is a very interesting site to discuss the role of humans in the extinction of Pleistocene fauna, as there is strong evidence that *Doedicurus clavicaudatus* (a glyptodon, like a giant armadillo) survived until ~ 7000 yr ^{14}C BP, in the early Holocene. The oldest archaeological component is characterized by the presence of expedient lithic artifacts manufactured on crystalline quartz, and other curated tools made on quartzite and chert.

The material used for dating the site by radiocarbon was primarily collagen from bones of *Glyptodon*, and about ten years ago there was also radiocarbon dating of organic sediment. Originally the collagen of bones of *Glyptodon* sp. was dated to 6550 ± 160 yr ^{14}C BP (Beta-7824, by conventional method) and 12,350 ± 370 yr ^{14}C BP (TO 1507 - by AMS method); thus one sample indicated a very old age and another sample a very young age (Politis et al. 2003:46). New samples of the same bone yielded dates of 7010 ± 100 yr ^{14}C BP (TO-1507-1) and 7510 ± 370 yr ^{14}C BP (TO 1507-2). Organic sediment samples yielded ages of 7500 and 8300 yr ^{14}C BP (Politis et al. 2003).

The old age of ~ 14,920 yr cal BP (12,300 yr ^{14}C BP) was rejected by the researchers of the site, due to contamination of the sample by secondary carbonates. A chronology of 8,000 yr cal BP (7,500 to 7,000 yr ^{14}C BP) is now suggested for the oldest component of the site (Politis and Gutierrez 1998:118, Politis et al. 2003).

Arroyo del Vízcaino site

The site contains three bone beds that appear when the Arroyo Vizcaino in Uruguay (34 ° 37'03 " S , 56 ° 02 ' 33" W) runs dry, exposing Pleistocene bones at the base of the arroyo channel. The site has yielded remains of 11 species of Pleistocene fauna, and Fariña et al. (2014) suggest that some bones have cut marks made by stone artifacts. Nine radiocarbon dates are between 32,305 to 35,495 cal yr BP (or 27.000 to 30.100 14C yr BP). There are several problems with the presentation of the dates. The methods of dating were not discussed; but to judge from the laboratories used and the range of error (±) given, six of the dates are with the standard method (Uruguay laboratory) and three probably by AMS (Beta and Oxford). No references are made to local or regional chronologies regarding early sites of Uruguay, Pampa, and southern Brazil .

The excavation was done with a paleontological perspective, unfortunately neglecting important archaeological aspects that must be considered when working on a Pleistocene site with any possibility of human presence. The main problems with the site are: First, severe taphonomic problems are not addressed. There is no stratigraphic profile presented; and although it is argued that there is no lag effect and that the accumulation of bones may be caused by human activity, the site is located in the bed of an active arroyo, and proof of the integrity of the bone bed is not presented.

Second , it is argued that the cut marks on the bones are products of human activity, simply by referring to sources in the literature. This is the weakest line of evidence for human presence in the report. The study of possible cut marks left on bones by lithic artifacts is in itself a specialization in archeology, and the issue cannot be treated lightly. Therefore, the authors of the report should have illustrated microphotography of the cut marks (rather than 3D reconstructions); and it is essential that the cut marks be examined by specialists in this field of archaeology, to dispel any shadow of doubt. It is necessary to eliminate trampling, stream abrasion, or carnivore activity as the main causes of the cut marks. From my admittedly subjective view (without being a specialist in the study of cut marks), a cut mark as argued by Fariña et al (2014) may be seen by an expert as a natural mark generated by the contact of bone with a sharp angular clast. As indicated in the report, angular clasts are present in the site matrix, and their presence has been used by the authors to justify an argument that the material has not been water-rolled and that the material is not a mixture derived from distant sources.

The paleontological description is excellent, with detailed calculations of body mass (a speciality of Fariña), so a critique of that aspect of the research is not called for in my discussion of the site. But there are other problematic issues. First, the diversity of species -- 11, including 17 individual giant sloth ((*Lestodon)* and large carnivores such as saber-toothed tiger—is remarkable for an archaeological site, and suggests more a natural accumulation than human involvement in the bone bed. It must be asked: What type of human site is it? a kill site?, a trap site ? The earliest archaeological sites in North America or South America generally have only one, or at most two and no more than three or four large Pleistocene faunal species. The kill sites that the authors cite, such as the classic Folsom site in New Mexico, generally present only one fauna species, as extinct bison in that case, associated with 25 projectile points (Meltzer et al. 2002). Given the known archaeological record in the Americas, a kill site with remains of 11 large species associated with only one pseudo archaeological lithic piece is very difficult to conceive. The main text of the report presented only one possible artifact of human origin, a specimen that can be described as microlithic (2.9 cm in length, 1.4 cm in width). The presence of a bulb of percussion as a single criterion is insufficient to argue that a lithic specimen is the product of human activity. The few other "artifacts" presented in Figure S14 (n = 6) appear to be excellent samples of geofactos, similar to specimens at the Calico site in California (Haynes, 1973, Jennings 1974, Meltzer 1993).

The question of whether or not lithic artifacts are present brings up another serious problem with the research. This can be seen in the following quotation: "*Although during the fieldwork no systematic effort was made to collect lithic material and only a part of this site has been exhaustively explored, a few lithic elements were found to have seemingly anthropogenic features*" (Fariña et al. 2014:6). This is a serious methodological error. The researchers should have mapped and collected all lithics –angular clasts, blocks, rounded stones, etc. --that might be associated with the bone bed. Here we see an all-too-common paleontologist bias, which prioritizes the recovery of bones, with no efforts made to recover all possible lithic artifacts.

In short, the work done to date at the site of Arroyo Vizcaino is a clear example of what not to do when working on a possible archaeological site of the late Pleistocene. Sadly, the weaknesses and flaws in the published and publicized report on this site can be used by skeptics and critics to discredit early Pleistocene sites in South America. The Arroyo El Vizcaino site, due to the scarcity of robust and reliable data, must be rejected as a place where early humans coexisted with fauna of the Pleistocene. It may be considered an important paleontological site, but it has too many serious methodological, technical, and interpretive problems to be considered an early site with human occupation.

EARLY FLAKED STONE TECHNOLOGIES IN URUGUAY

The area in which I carry out my research on early occupation is on the plains of Uruguay (30°-34° S. Lat., 58°-53° W. Long.). The presence of the Fishtail complex in

Uruguay has been confirmed by the high density of Fishtail points (Figure 3) (Bosch et al. 1974; Figueira 1892; Flegenheimer et al. 2003; Politis 1991; Suárez 2000, 2001, 2006).

Additionally, an oriented technology to produce specialized bifaces and bifacial knives (Figure 4), has been recognized in Uruguay (Suárez 2001, 2011a).

Figure 3. Fishtail points from Uruguay in different stages of life use.

Figure 4. Paleoamerican specialized bifaces from early sites of the Middle Uruguay River (A-B) and (C) Cuareim River, North of Uruguay.

However, radiocarbon dated stratified sites with Fishtail points are scarce. So far, no clear stratigraphic contexts that indicate a pre-Fishtail occupation have been found in Uruguay. On the other hand, analysis of the flaked stone material recovered in sand dunes and surface sites generated in nearly 140 years of collections (Ameghino 1877; Figueira 1892) may offer some results that help set the stage for an old colonization during the late Pleistocene in this region of the continent.

Surveys of archaeological collections from various locations in Uruguay have recovered material that has not been described and could be evidence of pre-Fishtail occupations in the plains of Uruguay. The flaked stone technology of the first human groups who entered South America in the late Pleistocene was relatively simple (Gruhn 2004:31), and the technology of the sites analyzed for the Southern Cone confirm this interpretation; for example, the Arroyo Seco 2 site includes artifacts made mostly on flakes, and an assemblage with low standardization in forms. The only early site in which diagnostic artifacts were recovered is Monte Verde II, in which two projectile points were found. The projectile points of Monte Verde II are lanceolate and strongly biconvex in cross section, and exceed 100 mm in length. *The flaking is even and well controlled, the flakes scars are roughly parallel and oriented perpendicular to the edges* (Collins 1997:423).

Two projectile points from surface contexts in Uruguay (Figure 5) are reminiscent of the "style" of the Monte Verde II points, even though the Monte Verde site is located 1,500 km from the site on which we recovered the points in Uruguay. The two points presented are fractured; one is 52 mm long, 22 mm wide, 10 mm thick and is manufactured in a black rhyolite (Figure 5B). The other point is 48 mm long, 18 mm wide and 7.4 mm thick manufactured in quartzite (Figure 5A). Both points should have been ~100 mm long when they were complete. The flanking technique of the points is well controlled; the flake scars are parallel and perpendicular to the edges. These points from Uruguay have the same technological features of the Monte Verde II points.

Figure 5. Possible pre-Fishtail points recovered in Uruguay (Museo del Indio and Megafauna, Maldonado), note design and "style" similar to the Monte Verde II points.

Furthermore, indented base points have been recovered in Uruguay. The example in Figure 6 is 83 mm long, 39 mm wide, and 9 mm thick: basal concavity depth 3 mm, base width 26 mm, manufactured in Sierra de la Ballena brown quartzite. The base was thinned on one face with a short and wide flake scar, and on the opposite face with staggered parallel scars. The edges are sinuous, and the flake scars are deep on both faces; some finish abruptly. This point has similarities with the points that are being recognized for the pre-Clovis epoch in Eastern North America (e.g., the Miller point and the Cactus Hill point) (Stanford and Bradley 2012). This point design has not been recognized in Holocene archaeological contexts (Iriarte 2006), nor for the late Pleistocene or early Holocene archaeological sites of Uruguay (Suárez 2011a).

Figure 6. Indented base point from Uruguay (Museo del Indio y la Megafauna, Maldonado).

POST-FISHTAIL TIMES

The research done in the North of Uruguay on the Uruguay middle river at Pay Paso 1 and K-87 site, suggests based on the stratigraphic, chronological, and techno-morphological observations of the artifacts tools, the existence of two designs of points during the Pleistocene-Holocene transition in Uruguay (Suárez 2011a). The name of each one of the designs defined refers directly to the site where the artifacts were first dated by ^{14}C.

Tigre points

The Tigre points (Figure 7) have wide stem, straight or slightly convex stem sides, very pronounced notched shoulders *ca.* 70-90°, short or long triangular blades, convex base, thinned by retouch, and complete bifacial thinning.

The Tigre points recovered was recovered buried at three stratified and dated sites (K87, Pay Paso 1 and Laguna de Canosa) are points that were reshaped and rejuvenated. Six radiocarbon ages dates the Tigre points between 12,500 to 11,100 yr cal BP (10,420 yr ^{14}C BP to 9,730 yr ^{14}C BP) (Suárez 2011a).

Figure 7. Tigre points ca. 12,500-11,100 yr cal BP. A, B, C) From Pay Paso 1 site. D) From Laguna Canosa Site.

Pay Paso points

The design and main tecno-mophological characteristics of Pay Paso points (Figure 8) include a short stem, profound concave stem base, divergent concave stem sides expanded toward the base, convex or straight blade sides, regular laminar retouch of the blade, and very careful basal thinning of the stem done by triangular and short flake scar.

Pay Paso points was recovered buried in K87 site and Pay Paso 1 sites. Twenty-two Pay Paso points were recovered in other archaeological sites on the middle Uruguay and Cuareim Rivers. These type of points also appear in the middle Negro River and the Tacuarembó Grande River in archaeological collections of Central Uruguay and South of Brazil (Corteletti 2008).

The Pay Paso points, like to the Fishtail and Tigre points, suffered an intensive process of maintenance and rejuvenation.

Ten AMS ages date the Pay Paso points during the early Holocene between 11,000-9,100 yr cal BP (Suárez 2011a).

Figure 8. Pay Paso points ca. 11.000-9.100 yr cal BP. A) Early stage of life use, from the Tacuarembó Grande River (silicified sandstone). B) Intermediate stage of life use, form component 3 of the Pay Paso 1 site. C) Final stage of life use, from Pay Paso 1 site.

DISCUSSION

The first Fishtail points were recovered in stratigraphic context by J. Bird (1938) at the classic Fell's Cave site, associated with Pleistocene fauna (*Equus* and *Paleolama*). Interestingly, these findings were made just ten years after the discovery of the Folsom site in North America. However, the first Southern Cone Fishtail points were originally published in Argentina by F. Ameghino (1880) and some years later in Uruguay by J.H. Figueira (1892), these authors not realizing that this projectile point form was one of the oldest on the continent.

Today we know that the Fishtail complex and Clovis complex emerge synchronously and practically simultaneously (Water and Stafford 2007, Steele and Politis 2009, Suárez 2011a), one in North America and the other in South America. There is an extensive literature on Clovis technology (Bradley et al. 2010; Collins 2002, Collins and Lohse, 2004; Frison and Bradley 1999) and Fishail technology (Bird 1969, Mayer-Oakes 1986, Nami 2003, Politis 1991 Suárez, 2001, 2006, 2011a) so I do not go into details here. The important thing to note is that some chronological and technological similarities are now beginning to appear. There are two interpretations of the origin of the Fishtail points. The first, based on the traditional Clovis-first model, argues that Clovis had given rise to Fishtail, because Clovis was a few hundred years older (Fiedel 1999; Morrow and Morrow 1999). Another interpretation suggests that Clovis and Fishtail had an independent origin (Borrero, 1983; Politis 1991). Now we know that earlier cultural complexes existed, occupying different environments, both in South America and in North America before the Clovis and Fishtail complexes. By 13.000 yr cal BP the cultural complexes Clovis and Fishtail had emerged on both continents, indicating a common ancestor that influenced both lithic technology and lifestyle in both cultural groups.

A number of sites in the Southern Cone indicate a pre-Fishtail occupation, although some are problematic at the moment. The analysis of the data and evidence from sites exceeding 13,000 yr cal BP in age indicates that a group of five sites -Cerro Tres Tetas, Cueva Lago Sofia 1, Tres Arroyos, Los Toldos, and Urupez- need new radiocarbon dates to confirm that they are older than 13,000 yr cal BP (11,000 yr ^{14}C BP). Except for the Tres Arroyos site, which has been published in detail, in terms of both stratigraphic context and flaked stone technology (see Massone 2004, Jackson 2002), the other sites need more general information presented in a publication in detail, on the archaeological context, stratigraphy, technology, and new ^{14}C dates, to provide more integrity to the sites and the evidence for human occupation previous to 13,000 yr cal BP.

Recently, ^{14}C dates were obtained on new samples from the sites of Arroyo Seco 2, Cerro Tres Tetas, Cueva Lago Sofia 1, Piedra Museo, and Tres Arroyos (see Steele and Politis 2009). For the Cerro Tres Tetas, Cueva Lago Sofia 1, and Tres Arroyos sites, the new dates obtained are relatively younger ages than those originally obtained (Steele and Politis2009). For Arroyo Seco 2, however, the dates on bones of Pleistocene fauna confirm the early ages previously obtained in past decades. The bones were selected for dating because they had cultural modifications, probably due to bone marrow extraction or bone quarrying (Steele and Politis 2009).The new bone samples duplicate dates obtained previously, for the following: ~ 14,214 yr cal BP (12,155 ± 70 yr ^{14}C BP) (*Megatherium*), ~ 14,544 yr cal BP (12,070 ± 140 yr ^{14}C BP) and ~ 13,765 yr cal BP (11,730 ± 70 yr ^{14}C BP) (megamammal, indeterminate), and 13,413 yr cal BP (11.320 ± 110 yr ^{14}C BP) (*Hippidion* sp.). These new dates indicate the age of death of at least three species of animals, corroborating previous dating of the site, and allow rejection of other dates (Steele and Politis 2009).These new dates also make it possible to consider different issues, such as the causes and reasons for the presence of different Pleistocene mammals at the site over a millennium.

As detailed above, the RS-I-50 and RS-Q-2 sites in southern Brazil are rejected as pre-Fishtail sites due to inconsistency of the data originally presented to justify them. The association of artifacts and *Glossotherium robust* in RS-I-50 was not demonstrated in an archaeological excavation, but derived from the stratigraphic profile in a natural exposure in a ravine. Several Brazilian authors have questioned the association of artifacts with extinct fauna at the RS-I-50 site. Dias (2004:258) argues that the remains of Pleistocene fauna (megafauna) are derived from fluvial entrainment, and the stone material is the product of "*flaking by natural processes*," so that the artifacts would actually be geofacts and not products of human activity. Something similar may have happened at the RS-Q-2 site (Paso de la Cruz 2) on the Quaraí river, where a date of 15,584 yr cal BP (12,690 ± 100 yr ^{14}C BP, SI-2351) would not be associated with cultural material or lithic artifacts (Dias, 2004, Dias and Jacobus, 2001; Milder 1995).

The site of La Moderna on the Pampa must also be discounted as a pre-Fishtail site. The date of 12,350 yr ^{14}C BP from La Moderna was rejected by the site investigator after other radiocarbon samples duplicated early Holocene ages around 7500 yr ^{14}C BP (Gutiérrez and Politis 1998), and the same results were obtained by dating organic matter from the oldest level and archaeological component of the site (Politis et al. 2003).The early component of La Moderna, then, is dated at 7500-7000 yr ^{14}C BP, or 8200 yr cal BP. In any case, the site is very important, as it allows us to observe the survival of Pleistocene fauna (glyptodonts) into the early Holocene ca. 7000 yr ^{14}C BP on the Pampa, data that are generally unknown to North American colleagues.

The Arroyo Vizcaíno site, due to the scarcity of robust and reliable data, must be rejected as a place where early humans coexisted with fauna of the Pleistocene. It may be considered an important paleontological site, but it has too many serious methodological, technical, and interpretive problems to be considered an early site with human occupation.

After evaluating the evidence, suspending judgment, and rejecting some other sites that present too serious problems to be accepted as valid for an occupancy exceeding 13,000 yr cal BP, there remain three human occupation sites of an age exceeding 14,000 yr cal BP for the Southern Cone. Monte Verde, Arroyo Seco 2, and Piedra Museo AEP-1 are accepted with an antiquity of over 14,000 yr cal BP. The following summarizes different aspects related to the flaked stone technology and subsistence economy of these three sites.

In relation to the stone material from Monte Verde II, Arroyo Seco 2, and Piedra Museo, no technological homogeneity can be recognized in the morphology of the artifacts. Except for several bifaces at Monte Verde II, the technology for these three sites apparently is relatively simple; no standardized morphology is recognized in the artifacts, which were made mostly of local rock of medium to good quality for flaking. The only diagnostic projectile points were recovered at Monte Verde II, points of a distinctive type that can serve to guide the search for similar artifacts, such as those recently discovered in Uruguay; and that may be examples of pre-Fishtail occupations in the Atlantic coast zone of South America.

There is a significant change in technology with the emergence of the Fishtail complex, when a new design of projectile point begins to expand across the Southern Cone by 13,000 yr ^{14}C BP. The evidence of high residential mobility during this period is associated with a larger number of sites, indicating an increase in population. These human groups shared such technological knowledge such as bifacial reduction and the same design in the style of the stem and blade of Fishtail points. Furthermore, people began to use colorful and better quality tool stone such as translucent agate, jasper, opal, chalcedony, chert, and silicified sandstone (see Figure 4), among others, with different strategies of lithic resource transfer from the sources to residential camps (Flegenheimer et al. 2003; Suárez 2011b). The Fishtail technology features high maintenance recycling, retooling, resharpening, and hafting (Politis 1991; Suárez 2003, 2011a). Three thinning sequences have been distinguished for the production of Fishtail points of the South Cone: full bifacial thinning of a large biface;

bifacial thinning retouch on flakes, and unifacial retouch of flakes used as performs (Suárez 2009).

As noted above, the emergence of the Fishtail complex in South America is simultaneous with the appearance of Clovis in North America by 13,100 yr cal BP (ca. 11,100 yr ^{14}C BP) indicate probably that the Clovis technological complex and the Fishtail complex had a common ancestor; and evolved in different ways in North and South America.

Monte Verde II, Arroyo Seco 2, and Piedra Museo share the use of Pleistocene fauna, although different species were identified in each site. In Monte Verde II, for example, were remains of mastodon and *Paleolama*; Arroyo Seco 2 yielded remains of giant sloth (*Megatherium* and *Glossotherium*) and horse (*Equus*); and at Piedra Museo were remains of horses and a small extinct guanaco species. With only three sites, it is difficult to try to draw any conclusions or generalizations about the economy of the groups occupying the Southern Cone ca. 14,500 yr cal BP. The excellent preservation of organic remains in the Monte Verde II site indicates the use of nine species of seaweeds (red, brown and green) as well as the edible or useful parts of some 60 species of land plants for both nutritional and medicinal purposes (Dillehay 1997; Dillehay et al. 2008). The use and consumption of different plant species could be supplemented with wildlife available, such as mastodon. Seaweeds must have been transported from the Pacific Ocean to the site. The other sites, Arroyo Seco 2 and Piedra Museo, did not have a good preservation of organic materials; only animal bones were preserved, indicating the species exploited during the exploration of the Southern Cone. The archaeological record indicates that remains of giant sloth and horse in Arroyo Seco 2 and Piedra Museo could be examples of hunting. Considering that at Monte Verde mastodons were hunted, ground sloths in Arroyo Seco 2, and horse in Piedra Museo, we can suggest that possibly at the end of the Pleistocene populations were not oriented towards specialization in hunting of a particular species of mammals. A generalized economy exploiting a broad spectrum of resources for the period between 12,000 to 7,000 yr ^{14}C BP has been proposed (Miotti and Salemme 1999; Gutierrez and Martinez 2008); however, for the period analyzed here the data are still fragmentary.

The population in the Southern Cone during the period between 15,000 and 14,000 yr cal BP must have been very small, with limited contact between groups and little social interaction. By 13.000 yr cal BP the Fishtail complex emerged, population density increased, and technological innovations (stemmed points) were introduced. This development is reflected in the increase in the number of sites for this period, with Fishtail points frequent in such different regions of South America as Ecuador, northern Peru, southern Brazil, northern and southern Chile, and the Pampa and Patagonia in Argentina and Uruguay. Technological innovation diversified in the pre-existing human populations in these regions; this development also coincides with improving climatic conditions.

In Uruguay and South of Brazil the post-Fishtail occupations are represented by two other different types of projectile points: Tigre and Pay Paso points (Figure 9 C and D).

Figure 9. Paleoamerican points from South Cone. A) Monte Verde II (ca. 14,600 yr cal BP), B) Fishtail point (ca. 13,000-12,300 yr cal BP), C) Tigre point (ca. 12,500-11,100 yr cal BP), D) Pay Paso point (ca. 11,000-9100 yr cal BP).

CONCLUSIONS

For more than fifteen years, it has been argued that the Fishtail complex was the earliest evidence of the first Americans in the Southern Cone. Now we know that the southern region of the continent was populated at least 2000 years before the emergence of the Fishtail complex in South America and the Clovis complex in North America. There are also post-Fishtail occupations to begin to recognize in Uruguay. Tigre and Pay Paso points are two examples of early cultures that were due adapt to environmental changes that occurred during the Pleistocene Holocene transition in the Southeast of South America.

There is conclusive evidence that the Southern Cone was populated by humans by 15,000 yr cal BP. The known main sites with evidence of pre-Fishtail occupations in southern South America are Monte Verde II (Chile) with a set of dates indicating 14,600 yr cal BP; Arroyo Seco 2 (Pampa Argentina), which has dates of ~ 14,000 yr cal BP; and Piedra Museo (Patagonia, Argentina) with a date of ~ 15,000 yr cal BP. Monte Verde is located 1200 km SW of Arroyo Seco 2 and 1000 km NW of Piedra Museo; Arroyo Seco 2 and Piedra Museo are located 1200 km away from each other.

The age of these three sites has been recognized and accepted by different researchers (e.g. Bonnichsen and Lepper 2005; Borrero 1999; Dillehay 2000; Gruhn 2005, Meltzer et al. 1997; Miotti 2006; Stanford et al., 2005; Steele and Politis 2009). The evidence is solid in archaeological, stratigraphic and chronological aspects. The chronology obtained for these three sites has a base of 24 radiocarbon dates, exceeding by at least a thousand years the age proposed for the beginning of Clovis (~ 13,000 yr cal BP). Fourteen dates from the three sites are more than 14,000 yr cal BP. The chronology of the three sites leads to dating them within one millennium, from 14,000 to 15,000 yr cal BP.

Current evidence, then, indicates that at least three small populations, with an unsophisticated flaked stone technology and a foraging economy were exploiting different ecosystems in the Southern Cone 15,000 years ago. The area related to the Pacific coastal plain, the open plains in Pampa, and the Patagonia steppe are highly productive regions, which should have attracted the people who explored the Southern Cone during the late Pleistocene.

Acknowledgments

I am sincerely grateful to Ruth Gruhn and Bruce Bradley for help in the translation from Spanish to English of the manuscript, and for comments and suggestions on this paper. Thanks also to the organizers of the Pre-Clovis Conference celebrated Washington D.C. for inviting me to present this paper, and to Bruce Bradley for his support over the years. Joaquin Mazarino edited the figures 3,6 and 9.

Reference cited

Adovasio, J.M.; Gunn, J.D.; Donahue, J. and R. Stuckenrath
1978 Meadowcroft Rockshelter, 1977: An Overview. *American Antiquity* 43 (4):632-651.

Ameghino, F.
1877 *Noticias sobre antigüedades indias en la Banda Oriental*. Imprenta de La Aspiración. Buenos Aires.

1880 *La antigüedad del Hombre en el Plata*. Editorial La Cultura Argentina. Buenos Aires.

Bird, J.
1969 A Comparison of South Chilean and Ecuadorian "Fishtail" Projectile Points. *The Kroeber Anthropological Society Papers* 40:52-71. Berkeley.

Bonnichsen, R.
1991 Clovis Origins. In *Clovis Origins and Adaptations,* edited by R. Bonnichsen and K.L. Turmire, pp. 309-329. Center for the Study of First Americans, Oregon State University, Corvallis.

Bonnichsen, R. and A.L. Schneider
1999 Breaking the Impasse on the Peopling of the Americas. In *Ice Age Peoples of North America,* edited by R. Bonnichsen and K. Turmire, pp: 497-519. Center for the Study of the First Americans. Oregon State University, Corvallis.

Bonnichsen, R. and Lepper, B.T.
2005 Changing Perceptions of Paleoamerican Prehistory. In *Paleoamerican Origins: Beyond Clovis*, edited by R. Bonnichsen, B.T. Lepper, D. Stanford, and M. Waters, pp: 9-19. Center for the Study of the First Americans. Texas A&M University Press, College Station.

Borrero, L.
1983 Distribuciones Discontinuas de Puntas de Proyectil en Sudamérica. Paper presented at the *11 International Congress of Anthropological and Ethnological Science Symposium "Early man in South America",* Vancouver, Canada.

1999 The Prehistoric Exploration and Colonization of Fuego-Patagonia. *Journal of World Prehistory* 13(3):321-355.

Bosch, A; Femenías, J. and A. Olivera
1980 [1974] Dispersión de las puntas de proyectil líticas pisciformes en el Uruguay. *III Congreso Nacional de Arqueología Uruguaya*. Centro Estudios Arqueológicos. Montevideo.

Bradley, B.; M. Collins, and A. Hemmings.
2010 *Clovis Technology*. International Monographs in Prehistory. Archaeological Series 17.Ann Arbor, Michigan.

Bryan, A.
1969 Early man in America and the Late Pleistocene Chronology of Western Canada and Alberta. *Current Anthropology* 10:339-365.

1973 Paleoenvironments and Cultural Diversity in the Late Pleistocene of South America. *Quaternary Research* 3: 237-256.

1979 The significance of the Taima-Taima site from the Perspective of America as a Whole. In *TaimaTaima: A Late Pleistocene Paleo-Indian Kill Site in Northernmost South America – Final Reports of 1976 Excavations,* edited by C. Ochsenius and R. Gruhn, pp. 111-119. South American Quaternary Documentation Program. Coro, Venezuela.

1986 Paleoamerican Prehistory as Seen from South America. In *New Evidence for the Pleistocene Peopling of the Americas,* edited by A.L. Bryan, pp: 1-14.Center for the Study of Early Man, University of Maine, Orono.

Bryan, A.L. and R. Gruhn
2003 Some Difficulties in Modeling the Original Peopling of the Americas. *Quaternary International* 109-110:175-179.

Cardich, A.; Cardich, L.A., and A. Hajduk
1973 Secuencia Arqueológica y Cronológica radiocarbónica de la Cueva 3 de Los Toldos (Santa Cruz, Argentina). *Separatas Relaciones* VIII, pp: 85-123. Sociedad Argentina de Antropología, Buenos Aires.

Dias, A.

2004 Diversificar para poblar: el contexto arqueológico brasileño en la transición Pleistoceno-Holoceno. *Complutum* 15:249-263.

Dias, A. and A. Jacobus

2001. The Antiquity of the Peopling of Southern Brazil. *Current Research in the Pleistocene* 18:17-19.

Dillehay, T.

1997 *Monte Verde. A Late Pleistocene Settlement in Chile*. Vol. 2. Smithsonian Institution Press, Washington D.C.

2000 *The Settlement of the Americas: A New Prehistory*. Basic Books. New York.

Dillehay, T. and Collins, M.B.

1991 Monte Verde, Chile: A comment on Lynch. *American Antiquity*, 56 (2):333-341.

Dillehay, T., and Pino, M.

1997 Site setting and stratigraphy. In Dillehay, T. (ed.), *Monte Verde. A Late Pleistocene Settlement in Chile, Vol. 2,* pp. 25-40. Smithsonian Institution Press, Washington, DC.

Dillehay, T.; Ramírez, C.; Pino, M.; Rossen, J. and D. Pino-Navarro

2008 Monte Verde: Seaweed, Food, Medicine and the Peopling of South America. *Science* 320:784-786.

Collins, M. B.

1997 The lithics from Monte Verde, a descriptive-morphological analysis. In T. Dillehay.(ed.), *Monte Verde. A Late Pleistocene Settlement in Chile, Vol. 2,* pp. 383-506.Smithsonian Institution Press, Washington, DC.

2002 *Clovis Blade Technology*. University of Texas Press.

2012 Preliminary Geographic Patterns in Older-than Clovis Assemblages of North America. Presentation 77th Annual Metting of SAA. Memphis, Tennessee.

Collins, M. and J.C. Lohse
2004 The Nature of Clovis Blades and Blade Cores. In *Entering America Northeast Asia and Beringia before the Last Maximum Glacial, Northeast Asia and Beringia before the Last Maximum Glacial*, edited by D.B. Madsen, pp. 159-183.University of Utah Press.

Corteletti, R.
2008 *Patrimonio Arqueológico de Caxias do Sul.* Editora Nova Prova. Porto Alegre.

Erlandson, J.
2002 Anatomically modern humans, maritime voyaging, and the Pleistocene colonization of the New World. In *The First Americans: the Pleistocene Colonization of the New World*, edited by N.G. Jablonski, pp. 59-42. Memoirs of the California Academy of Sciences, San Francisco.

Fariña, R,; Tambusso, P.S.; Varela, L.; Czerwonogora, A.; Di Giacomo, M.; Musso, M.; Bracco, R.; Gascue, A.
2014 Arroyo Vizcaíno, Uruguay: A fossil-rich 30-ka-old megafauna locality with cut-marked bones. Proceedings of The Royal Society B, 281:20132211.

Fiedel, S.
1999 Older than we thought: Implications of corrected dates for Paleoindians. *American Antiquity*, 64 (1): 95–116.

Figueira, J.H.
1892 *Los Primitivos Habitantes del Uruguay. Ensayo Paleoetnológico*. Imprenta Artística Dornaleche y Reyes. Montevideo.

Fladmark, K.R.
1979 Routes: Alternate Migration Corridors for Early Man in North America. *American Antiquity* 44 (1):55-69.

Flegenheimer, N.; Bayón, C.; Valente, M.; Baeza, J. and J. Femenías
2003 Long distance tool stone transport in the Argentine Pampas. *Quaternary International* Vol. 109-110:49-64.

Frison, G. C. and B.A. Bradley
1999 *The Fenn Cache: Clovis Weapons and Tools*. One Horse Land and Cattle Company, Albuquerque.

Goebel, T.; Waters, M.R. and D.H. O´Rourke
2008 The Late Pleistocene Dispersal of Moderns Humans in the Americas. *Science* 319:1497-1502.

Gruhn, R.
1991 Stratified Radiocarbon-dated Archaeological sites of Clovis Age and Older in Brazil. In *Clovis Origins and Adaptations,* R. Bonnichsen and K.L. Turmire, editors, pp. 283-287.Center for the Study of First Americans, Oregon State University, Corvallis.

1994 The Pacific Coast route of initial entry: an overview. In *Method and Theory for Investigating the Peopling of the Americas* edited by R. Bonnichsen and D. Gentry Steele, pp.249-256.Center for the Study of the First Americans, Oregon State University, Corvallis.

2004 Current Archaeological Evidence of Late-Pleistocene Settlement of South America. In *New Pespectives on the First Americans*, edited by B.T. Lepper and R. Bonnichsen, pp. 27-34.Center for the Study of the First Americans. Texas A&M University, College Station.

2005 The Ignored Continent: South America in Models of Earliest American Prehistory. In *Paleoamerican Origins: Beyond Clovis,* edited by R. Bonnichsen, B.T. Lepper, D. Stanford and M. Waters, pp. 199-208. Center for the Study of the First Americans. Texas A&M University, College Station.

2007 The Earliest Reported Archaeological Sites in South America. *Mammoth Trumpet* 23(1):14-18.

Gutíerrez, M. and G. Martínez

2008 Trends in the faunal human exploitation during the Late Pleistocene and Early Holocene in the Pampean region (Argentina). *Quaternary International* 191:53-68.

Haynes, C.V.
1973 The Calico Site: Artifacts or Geofacts? *Science* 181:305-310.

2005 Clovis, Pre-Clovis, Climate Change and Extinction. In *Paleoamerican Origins: Beyond Clovis,* edited by R. Bonnichsen, B.T. Lepper, D. Stanford and M. Waters, pp. 113-132. Center for the Study of the First Americans. Texas A&M University, College Station.

Haynes, G.
2002 *The Early Settlement of North America: The Clovis Era*. Cambrige University Press. Cambridge U.K.

Jackson, D.
2002 *Los instrumentos líticos de los primeros cazadores de Tierra del Fuego*. Colección Ensayos y Estudios. Ril Editores, Santiago.

Jackson, D. and A. Prieto
2005. Estrategias tecnológicas y conjunto lítico del contexto Paleoindio de Cueva de Lago Sofía 1, Última Esperanza, Magallanes. *Magallania* 33, 115–120.

Jennings, J.D.
1974 *Prehistory of North America*. Second edition. Mc.Graw-Hill Book Company.

Leipus, M.S. and M.C. Landini
2013 (in press). Materias primas y tecnología: un estudio comparativo del material lítico. In *Estado actual de las investigaciones en el sitio Arroyo Seco 2*, edited by G. Politis, M.A. Gutiérrez and C. Scabuzzo, pp: 149-187. Serie Monográfica INCUAPA 5. Olavarría.

Massone, M.
2002 El fuego de los cazadores Fell 1 a finales del Pleistoceno. *Anales Instituto Patagonia* 30:117-131.

2004 *Los Cazadores Después del Hielo*. Dirección de Bibliotecas, Archivos y Museos. Santiago.

Mayer-Oakes, W.J.

1986 El Inga: A Paleoindian Site in the Sierra of Northern Ecuador. *Transactions of the American Philosophical Society* Volume 76, Part 4. Philadelphia.

Meltzer, D.J.

1993 *Search for the First Americans*. Smithsonian Books. Washington, D.C.

2009 *First Peoples in a New World. Colonizing Ice Age America.* University of California Press. Berkeley and Los Angeles.

Meltzer, D.J.; Grayson, D.K.; Ardila, G.; Barker, A.; Dincauze, D.; Haynes, C. Mena, F.; Núñez, L. and D.J. Stanford

1997 On Pleistocene Antiquity of Monte Verde, Southern Chile. *American Antiquity* 62(4):659-663.

Meltzer, D.J.; Todd, L.C. y V.T. Holliday
2002 The Folsom (Paleoindian) Type Site: Past Investigations, Current Studies. *American Antiquity* 67(1):5-36.

Meneghin, U.

1977 Nuevas investigaciones en los yacimientos del "Cerro de los Burros". Edited by the author. Montevideo

2004 Urupez. Primer registro radiocarbónico (C-14) para un yacimiento con puntas líticas pisciformes del Uruguay. *Orígenes 2*, Fundación Arqueología Uruguaya, Montevideo.

2005 Yacimientos tempranos del Uruguay. *Orígenes 3*, Fundación Arqueología Uruguaya, Montevideo.

2006 Un nuevo registro radiocarbónico (c-14) en el Yacimiento Urupez II, Maldonado, Uruguay. *Orígenes 5*. Fundación Arqueología Uruguaya, Montevideo.

Milder, S.

1994 A fase Ibicuí: uma revisão arqueológica, cronológica e estratigráfica. *M.A. thesis.* Instituto de Filosofia e Ciências Humanas, PUCRS, Porto Alegre.

Miller, E.
1987 Pesquisas arqueológicas paleoindígenas no Brasil Ocidental. *EstudiosAtacameños* 8: 37–61. Chile.

Miotti, L.
1995 Piedra Museo Locality: a Special Place in the New World. *Current Research in the Pleistocene*, 12:34-40.

2006 La Fachada Atlántica como puerta de ingreso alternativa de la colonización de América del Sur durante la Transición Pleistoceno/Holoceno. In *II Simposio Internacional El Hombre Temprano en América*, editors J. C. Jiménez *et al.*, pp. 156-188. INAH. México D.F.

Miotti, L. and M. Salemme
1999 Biodiversity, taxonomic richness and specialists–generalists during Late Pleistocene/Early Holocene times in Pampa and Patagonia (Argentina, Southern South America). *Quaternary International* 53–54, 53–68.

Miotti, L.and M. Salemme
2005 Hunting and Butcering Events at the Pleistocene/Holocene Transition in Piedra Museo: An Example of Adaptation Strategies of the First Colonizers of Patagonia. In *Paleoamerican Origins: Beyond Clovis,* edited by R. Bonnichsen, B.T. Lepper, D. Stanford and M. Waters, pp. 209-218. Center for the Study of the First Americans. Texas A&M University, College Station.

Miotti, L.; Salemme, M. and J. Rabassa
2003 Radiocarbon Dating of Piedra Museo, Argentina. In*: Where the South Winds Blow, Ancient Evidence for Paleo South Americans,* edited by L. Miotti, M. Salemme, and N. Flegenheimer, pp. 99-104. Center for the Study of the First Americans, University of Texas A&M College Station.

Morrow, J. E. and T. A. Morrow

1999 Geographic variation in fluted projectile points: A hemispheric perspective. *American Antiquity* 64:215-231.

Nami, H.
2003 Experimentos para explorar la secuencia de reducción Fell de la Patagonia Austral. *Magallania* 31:107-138.

Paunero, R.
2003 The Cerro Tres Tetas Locality in the Central Plateau of Santa Cruz, Argentina. In *Where the South Winds Blow: Ancient Evidences From Paleo South Americans*. L. Miotti, M. Salemme and N. Flegenheimer editors, pp. 127-132. Center for the Study of the First American.Texas A & M University Press. College Station.

Politis, G.
1991 Fishtail Projectile Points in the Southern Cone of South America: An Overview. In *Clovis: Origins and Adaptations,* edited by R. Bonnichsen and K. Turnmire, pp. 287-301.Center for the Study of the First Americans. Oregon State University. Corvallis.

1999 La estructura del debate sobre el poblamiento de América. *Boletín de Arqueología* (2):25-51. Santa Fé de Bogotá.

2008 The Pampas and Campos of South America. *Handbook of South American Archaeology*, edited by H. Silverman and W. Isbelle, pp. 235-260.Springer, New York.

Politis, G. and M. Gutiérrez
1998 Gliptodontes y Cazadores-Recolectores de la Región Pampeana (Argentina). *Latin American Antiquity* 9 (2):111-134.

Politis, G.; Johnson, E.; Gutierrez, M. and W.T. Hartwell
2003 Survival of Pleistocene Fauna: New Radiocarbon Dates on Organic Sediments from la Moderna (Pampean Region, Argentina). In *Where the South Winds Blow: Ancient Evidences From Paleo South Americans*, edited by L. Miotti, M. Salemme and N. Flegenheimer pp. 45-50. Center for the Study of the First American and Texas A & M University Press. Corvallis.

Politis, G.; Messineo, P.G. and C.A. Kaufmann

2004 El Poblamiento temprano de las llanuras pampeanas de Argentina y Uruguay. *Complutum* 15:207-224.

Prieto, A.

1991 Cazadores tempranos y tardíos en Cueva de Lago Sofía 1. *Anales del Instituto de la Patagonia* 20, 75–99.

Stanford, D. and B. Bradley

2012 *Across Atlantic Ice. The Origin of America's Clovis Culture*. University of California Press.

Stanford, D.; Bonnichsen, R.; Meggers, B. and D. Steele

2005 Paleoamerican Origins: Models, Evidence, and Future Directions. In *Paleoamerican Origins: Beyond Clovis*, edited by R. Bonnichsen, B. Lepper, D. Stanford, and M. Waters, pp. 313-353. Center for the Study of the First Americans. Texas A&M University.

Steele, J. and G. Politis

2009 AMS ^{14}C dating of early human occupation of southern South America. *Journal of Archaeological Science* 36:419-429.

Suárez, R.

2000 Paleoindian Occupations in Uruguay. *Current Research in the Pleistocene* 17:78-80.

2001 Technomorphological Observations on Fishtail Projectile Points and Bifacial Artifacts from Northern Uruguay. *Current Research in the Pleistocene* 18:56-58.

2003 Paleoindian Components of Northern Uruguay: New data for Early Human Occupations of the Late Pleistocene and Early Holocene. In *Where the South Winds Blow: Ancient Evidences FromPaleo South Americans*. L Miotti, M. Salemme and N. Flegenheimer, editors, pp. 29-36. Center for the Study of the First Americans, Texas A & M University Press. College Station.

2006 Comments on South American Fisthail Points: Design, Reduction Sequences and Function. *Current Research in the Pleistocene* 23:69-72.

2009 Unifacial Fishtail Points and Considerations about the Archaeological Record of South Paleoamericans. *Current Research in the Pleistocene,* 26:12-15.

2011a *Arqueología durante la transición Pleistoceno Holoceno en Uruguay. Componentes Paleoindios, Organización de la Tecnología lítica y Movilidad de los Primeros Americanos.* British Archaeological Report, BAR International Series 2220. Archaeopress. Oxford, UK.

2011b Movilidad, acceso y uso de ágata traslucida por los cazadores-recolectores tempranos durante la transición Pleistoceno Holoceno en el Norte de Uruguay (ca.11,000-8500 AP. *Latin American Antiquity* 22(3):359-383.

Waters, M.R. and T.W. Stafford Jr.
2007 Redefining the Age of Clovis: Implications for the Peopling of the Americas. *Science* 315:1122-1126.

Plant Fiber Technologies and the Initial Colonization of the New World

J. M. Adovasio

Mercyhurst Archaeological Institute
Mercyhurst University
Erie, Pennsylvania 16546

Abstract

Ongoing research demonstrates that perishable industries—notably including the manufacture of textiles, basketry, cordage, netting, and sandals—were a well-established, integral component of the Upper Paleolithic technological milieu in many parts of the Old World. Moreover, extant data suggest that these technologies played a vital and, essentially unappreciated role in the ecological success of all late Pleistocene populations, notably including the first Americans. Late Pleistocene perishable assemblages from throughout this hemisphere are summarized including the most recent discoveries. Additionally, this paper explores the varied roles of early fiber technology in the New World and specifically examines the adaptive qualities, impact on social organization, and enhancements to food procurement strategies implicit in this critical series of interrelated industries. It is suggested that the manufacture of perishable plant fiber-derived artifacts was far more important in the successful colonization of this hemisphere than any of the more often recovered durable artifact classes, particularly stone.

Presented at the Conference "Pre-Clovis in the Americas"
Smithsonian Institution, Washington DC
9–10 November 2012

On many occasions over the past 44 years, I, alone (Adovasio 1970, 1974, 1977, 1997, 2011) or with colleagues, notably Olga Soffer (Adovasio, Hyland, And Soffer 1997, 2001, 2004, 2005; Adovasio, Hyland, Soffer, and Klima 1998a, 1998b, 2001; Adovasio and Illingworth 2004; Adovasio, Laub, Illingworth, McAndrews, and Hyland 2003; Adovasio, Soffer, Illingworth, and Hyland 2007), have stressed the critical roles of non-durable, so-called perishable technologies in the lifestyles and adaptive strategies of Late Pleistocene and Holocene populations around the world. As I have further observed, lamentably, even vexatiously, most of these "perishable pontifications" as David Madsen has called them, have fallen on deaf ears. Even the most positive reactions, at least on this side of the Atlantic, seldom exceed the often-heard "isn't that nice?" response. At the risk of over-preaching or, minimally, further boring the listener or reader, certain salient facts about perishable, especially plant-fiber based technologies, should be forcefully stressed or, at least, reiterated.

First and, again, as noted many times previously, throughout most of this as well as the last century, scholarly and popular interpretations of the technology, subsistence strategies, and the attendent social milieu of Paleoindians has stressed durable (i.e. stone) artifacts, the "central" role of big game hunting, and male-centered – if not male dominated – food procurement strategies. Until quite recently, this bias paralleled prevailing interpretations of Upper Paleolithic lifeways in the Old World which also focused on stones, large and mainly extinct mammals, and manly hunters to the virtual exclusion of non-durable technology, gathering or collecting, and communal hunting of small game. Indeed, as noted by Adovasio, Hyland, and Soffer (2004:157), "so ingrained is this emphasis on rocks, megafauna, and men in the reconstruction of Ice Age lifeways, especially in the northern hemisphere, that it is not inaccurate to state that this is still the prevailing image of the First Americans or first Nations and their Old World progenitors and counterparts."

Second, the tyranny of preservation and, until recently, the domination of Paleoindian and Upper Paleolithic studies by male scholars, have combined to render the persistence of the "man in furs sticking lethal stone-tipped spears into big animals (or anything else that moved!) mythology" perfectly understandable.

Indeed, paraphrasing Adovasio, Hyland, and Soffer (2004:157-158), given the ubiquity and, oftentimes, profusion of lithic artifacts in Late Pleistocene and Early Holocene archaeological sites, it is not surprising that a disproportionate amount of attention has been, and continues to be, devoted to their analysis and interpretations. This is particularly apparent when the distribution and frequency of stone tools and debitage are compared to the highly idiosyncratic occurrence and relative scarcity of non-durable artifacts. Unfortunately, despite the fact that this disparity is clearly the byproduct of differential preservation, it has helped perpetuate the idea that perishable artifacts, especially plant fiber derived ones, are somehow less important or significant than those produced in durable media. In short, their low recovery frequency is assumed to accurately reflect the incidence of their production and use in "life".

Third, in stark contrast to the aforementioned and still widely held view are these facts. In a still little appreciated quantification of excavated material from a dry cave with near perfect preservation in central Coahuila, Mexico, Taylor (1966:67, 73) noted that the *average* ratio of stone to wood to plant fiber artifacts per excavated square meter was 1:6:26. Put simply, perishable plant fiber artifacts were four times more common than artifacts fashioned of wood and twenty times more common than stone tools.

Rather than an anomaly, this seeming reversal of the relative importance of stone over non-durable items is paralleled in hundreds of dry caves, rockshelters, and other preservational contexts where conditions favor an accurate and representative characterization of the relative proportions of artifacts of all classes and compositional media. Moreover, as stressed almost ad nauseam previously (Adovasio, Soffer, Illingworth, and Hyland 2007; Adovasio 2011; Soffer and Adovasio 2004, 2007, 2010; Soffer, Adovasio, and Hyland 2000), Taylor's proportions apply not simply to archaic or formative groups, but to *all* time periods in *all* environmental situations (including the northern hemisphere!) and to *all* time periods including the ethnographic present.

The Antiquity of Perishable Fiber Artifacts in the Old World

It is now abundantly clear that, contrary to the received wisdom of three-quarters of a century ago (e.g., Childe 1936), the production of artifacts produced from plant fibers extends well back into the Pleistocene. Once defined as hallmarks of the Mesolithic or Neolithic in the Old World or the Archaic and Formative in the New World, the manufacture of cordage, cordage byproducts (like nets and snares), baskets, textiles, sandals, and their kin is now widely recognized as a Paleolithic innovation. The antiquity of plant-fiber-derived artifact production and use has been detailed in a long series of publications beginning in the mid-1990s through today. For those unfamiliar with this oftentimes arcane literature, a synopsis is provided below.

Continuing collaborative research with Czech scholars on Gravettian inventories from Dolni Vestonice I and II and Pavlov I in Moravia has documented the existence of highly diverse and sophisticated plant-fiber-based technologies that included the production of cordage, nets, basketry, and textiles (Adovasio, Hyland, Soffer, and Klima 2001; Adovasio, Soffer and Klima 1996; Adovasio et al. 1997, 1998a, 1998b, 1999, 2005; Soffer et al. 1998; Soffer and Adovasio 2004, 2007, 2010). The primary evidence for these industries derives from 36 negative impressions on small fragments of fired and unfired clay from Dolni Vestonice I, 1 from Dolni Vestonice II, and 41 from Pavlov I (Table 1). The inventory from these sites includes single-ply, multiple-ply, compound and braided cordage; knotted netting; plaited wicker-style basketry; and a surprisingly wide variety of non-heddle loom (i.e., hanging or ground frame) woven textiles. The textile assemblage includes simple and diagonal twined pieces as well as balanced plain weave. Some of the items exhibit structural decoration.

Because the Moravian fired and unfired clay assemblage is highly fragmentary and the impressions very diminutive, the configuration or dimensions "in life" of the fiber items represented cannot be specified with confidence. As has been argued elsewhere (Adovasio, Soffer, and Klima 1996; Adovasio et al. 1997, 1998a, 1998b; Soffer, Adovasio, and Hyland 2000), it is highly likely that the plaited basketry impressions represent containers or, less likely, mats. The relatively wide range of textile weaves and element gauges suggests that the textile impressions represent a variety of diverse forms. These may include mats, wall hangings, blankets, and flexible bags as well as a wide array of apparel forms such as shawls, shirts, skirts, and sashes. Additionally, the presence of sequentially spaced knots on some of the impressions directly suggests the production of netting either in the form of bags and/or hunting/fishing devices. Finally, the occurrence of seams conjoined by whipping stitches points to the sewing of fabric to produce more complex composite structures such as clothing and bags.

The variety of these Moravian inventories and the relative technical fineness of many of the products represented in the impressions clearly indicate, as has been stressed many times previously (e.g., Soffer, Adovasio, and Hyland 2000), that these are in no sense "primary essays in the craft." Rather, when coupled with the high degree of cordage ply, twined, or plain woven warp and weft processing, the collective evidence suggests considerable antecedent development both for these technologies, specifically, and for plant-fiber industries in general.

All of the Moravian materials summarized above derive from Upper Paleolithic sites assigned to the Eastern or Pavlovian variant of the Gravettian techno-complex and date between 29,000 and 24,000 14C yr BP. Significantly, and as hypothesized previously, it is now clear that the suite of interrelated plant-fiber-based crafts outlined above was not unique to some precocious Upper Paleolithic Moravians. Furthermore, although impressive in their diversity and sophistication, the Moravian materials are not the only such artifacts documented for the Paleolithic. Indeed, ample evidence exists for a widespread geographic distribution of this interrelated suite of industries.

As noted in Soffer, Adovasio, and Hyland (2000:514), more than a half century ago Cheynier (1967) published a textile impression from the Solutrean level at Badegoule in France. Recent reexamination of the collection confirmed his initial observation and indicated the presence of other plant-fiber-based impressions. Actual and somewhat younger cordage has been reported from the sites of Lascaux, France (LeRoi Gourhan and Allain 1979); Ohalo II, Israel (Nadel et al. 1994); Kosoutsy, Moldova (Adovasio et al. 1997, 1999); and Mezirich, Ukraine (Adovasio et al. 1997, 1999). Additionally, more-recent research demonstrates that plant-fiber artifact impressions are present at two sites in Russia, Kostenki I and Zaraisk. The Kostenki I assemblage includes multiple-ply cordage, perhaps employed to bind a bundle of reeds or stems, while the single specimen from Zaraisk documents a knotted net construction. Additionally, an impressed reindeer antler fragment from Gönnersdorf in Germany bears the impression of multiple-ply cords (Soffer, Adovasio, Illingworth, Amirkhanov, Praslov, and Street 2000). The Kostenki and Zaraisk specimens range in age from 21,000 14C yr BP to 19,000 14C yr BP, while the Gönnersdorf impression was recovered in a Magdalenian context dated to 15,000 14C yr BP.

Table 1. Representative Fiber Technology Identified on Negative Impressions on Fired and Unfired Clay Fragments from Upper Paleolithic Moravia

Fiber Technology Class and Type	Pavlov I	Dolni Vestonice I	Dolni Vestonice II
Cordage			
Single-Ply, Z-Spun	4	–	–
Single-Ply, S-Spun	8	6	–
Two-Ply, S-Spun, Z-Twist	2	2	–
Two-Ply, Z-Spun, S-Twist	4	3	–
Compound, Two-Ply, Z-spun, S-twist	1	–	–
Braided, Three-Strand	1	1	–
Untyped	–	1	–
Knotted Netting			
Weaver's Knotted	4	–	–
Twined Textiles			
Open, Simple, Z-twist Weft	2	–	–
Close, Simple, Z-twist Weft	2	1	–
Open, Simple, S-twist Weft	3	3	–
Close, Simple, S-twist Weft	–	1	–
Close, Diagonal, Z-twist Weft	1	–	–
Open, Diagonal, Z-twist Weft	4	–	–
Close, Diagonal, S-twist Weft	2	–	–
Open, Diagonal, S-twist Weft	3	1	–
Close, Simple, Z- and S-twist Weft	–	1	–
Open and Close, Simple, Z- and S-twist Weft	–	1	–
Close, Simple, Untyped-twist Weft	2	–	–
Open, Untyped, Z-twist Weft	2	–	–
Untyped, Simple, S-twist Weft	1	–	–
Untyped	7	1	1
Plaited Textiles			
Plain Weave (1/1 Balanced)	–	5	–
Plaited Basketry			
2/2 Twill	–	4	–
Untyped	–	6	–

Dramatic evidence for the deep antiquity of plant-fiber-based technologies was recently provided by the recovery of actual plant fibers and cordage from Dzudzuana Cave in Georgia (Kvavadze et al. 2009). This assemblage, which even includes apparent evidence of dyeing, is dated to ca. 30,000 14C yr BP (Kvavadze et al. 2009). By no later than 12,960 14C yr BP, two types of twining (Close Simple Twining, Z-Twist Weft; Open Simple[?] Twining, Z-Twist Weft) as well as cordage and/or cordage constructions have been identified in impressions on some of the world's oldest ceramics. These specifically originate from the very threshold of the Bering Platform as well as from the Amur River and Primorye regions of the Russian far east (Derevianko and Medvedev 1995; Hyland et al. 2002; Zhushchikhovskaya 1997).

In addition to the recovery of actual specimens of plant-fiber-based technologies, a variety of other lines of admittedly indirect evidence also attest to the deep antiquity and ubiquity of these interrelated industries. Soffer, Adovasio, and Hyland (2000, 2001) summarize the surprisingly extensive iconographic data, notably in the form of often incredibly detailed "dressed" Venus figurines from Gravettian contexts in Europe. Soffer (2004) and, more recently, Stone (2011), present formidable evidence for the presence of a variety of artifacts used in the processing and fashioning of plant fiber specimens, again in Upper Paleolithic European contexts.

The Antiquity of Perishable Fiber Artifacts in the New World

In both eastern and western North America, the oldest basketry, textile, or cordage materials are generally assignable to the mid-twelfth radiocarbon millennium B.P., though very few specimens have been recovered from well-dated contexts. As of this writing, the oldest bona fide basketry (or perishable fiber artifact) of any subclass or type from eastern North America derives from middle Stratum IIa at Meadowcroft Rockshelter in southwestern Pennsylvania (Andrews and Adovasio 1996; Stile 1982). The item is a wall fragment of a basket without selvage constituted of simple plaiting with single elements in a 1/1 interval. It is bracketed by radiocarbon dates of 12,800 ± 870 B.P. and 11,300 ± 700 B.P. and is associated with the Miller complex occupation at that site (Figure 1). According to Andrews and Adovasio (1996:39), a far older but more tentatively classified perishable from Meadowcroft Rockshelter derives from lowest Stratum IIa and is directly dated to 19,600 ± 2400 B.P. The specimen consists of a single element of intentionally cut birch-like (cf. *Betula* sp.) bark which is quite similar in overall morphology to the strips employed in all later Meadowcroft plaiting. If the specimen is a portion of a plaited basket and even if one sigma is subtracted from the date (i.e., 17,650 B.P.), it is at once the oldest basket fragment in North or South America. Basket fragment or not, it is the oldest anthropogenically modified perishable object in the hemisphere.

In the far west, recent radiometric research has revealed that plaiting recovered in the late 1960s from Spirit Cave, Nevada, is not ca. 1,500 or 2,000 years old as was originally reported (Wheeler and Wheeler 1969), but rather dates to 9415 ± 25 B.P. (Fowler et al. 1997). While not dating to Paleoindian times, plaited specimens of this age are unique in the Great Basin. Fowler and her colleagues (1997) report that plain or simple plaiting with paired two-ply, S-spun, Z-twist cordage wefts was used to construct burial shrouds for two interments. These shrouds, which consist of a head covering for one burial and a cremation bag for the other, are made of bulrush (*Scirpus acutus*) and most likely were produced with the aid of a three-bar upright frame. The warp edges of these plaited specimens are set with two rows of close plain (simple) twining, and one of the bags is decorated. Interestingly, all of the other textiles examined from this horizon are fully twined and represent three different forms of this production technique.

While plaiting is represented in very early contexts at Meadowcroft Rockshelter, Pennsylvania (Andrews and Adovasio 1996; Stile 1982) and somewhat later at Spirit Cave, Nevada (Fowler et al. 1997), most early North American basketry or textiles are twined. The ancient pedigree of twining in eastern North America is documented by a remarkable positive cast of a finely made twined bag fragment from the Hiscock Site in western New York (Adovasio et al. 2003). This unique specimen was produced with very fine gauge cordage warps and wefts via close diagonal twining with a Z-twist weft slant (Adovasio et al. 2003:274) (Figure 2). Recovered from the uppermost portion of the so-called Fibrous Gravely Clay Layer, the age of the specimen is somewhat ambiguous. With too little organic material to be directly dated even by accelerator mass spectrometry (AMS), the Hiscock twining, on stratigraphic grounds, is minimally 9,475 ± 95 to 9,205 ± 50 radiocarbon years old. An AMS date run on a twig of unidentifiable wood from the matrix of the Hiscock twining produced an uncorrected date of 10,180 ± 50 B.P. This determination is consistent with the late Pleistocene ascription of the deeper portions of this depositional unit, dates from which range in age from 11,450 ± 50 to 10,220 ± 120 B.P. Use of the Hiscock Site during this time range is attributed to itinerant fluted point makers of Clovis affinity. Whatever its exact age, the Hiscock twining is the oldest in eastern North America and if it is indeed of Clovis age, it is the oldest example of twining in this hemisphere. By ca. 9950–7950 B.P., perishable fiber artifacts are widely represented in eastern North America (Adovasio and Illingworth 2004:19–30; Andrews and Adovasio 1996:34–36) in the form of basketry, cordage, and sandals (Kuttruff et al. 1998:72–75).

Herein, mention should also be made of a possible square knot net impression on a partially fossilized extinct dugong (*Metaxytherium* sp.) bone from Turkey Point (8FR44) (Figure 3), a multicomponent open site or site complex on an active beach in the panhandle of Florida (Adovasio, Illingworth, and McKenzie 2012). While it is remotely possible that the very regular impression on this bone is *not* a net impression and, further, that the impression is more recent in age, it is also, at least, equally likely that the impression is penecontemporaneous with the bone and is of Late Pleistocene ascription. In any case, this is certainly not the only known case of the preservation of such a construction on odd or unusual media. Geographically furthest afield but obviously much closer in preservative media is the impressed antler fragment from a single component sealed cultural level at Gönnersdorf in Germany (Bosinski 1995). At this site of Magdalenian

Figure 1. Carbonized plaited basketry fragments with selvage from Meadowcroft Rockshelter.

age (ca. 12,800 cal B.P.), a segment of reindeer antler bears the impression of a cluster of 5 or 6 final S-twist cords (Soffer et al. 2000). Much closer to the study area, several specimens of bone from the Paleoindian site of Sloth Hole appears to contain negative cordage impressions which as of this writing have not been analyzed (Hemmings 2005).

West of the Mississippi River, the fifteen fiber artifacts recovered from Zones C1 and C2 at Pendejo Cave, New Mexico, were once thought to be among the oldest perishable artifacts ever recovered in the New World (Adovasio and Hyland 1993; Hyland 1997; Hyland and Adovasio 1995, 1998). Unfortunately, recent direct AMS assays suggest that most, if not all, of the Zones C1 and C2 perishable fiber artifacts are of intrusive Archaic origin (Hyland et al. 1998).

Stratum D1, Sand 1, at Danger Cave, Utah, has yielded the oldest definitive cordage and netting from the eastern reaches of the Great Basin (Jennings 1957:227–234). The small but informative collection includes single-ply, S twist cordage; a length of untwisted fiber; and more significantly, two segments of two-ply, Z spun, S twist cordage knotted together with a lark's head knot. Presumably, this specimen is the remnant of a section of knotted netting which, with the solitary and now highly unlikely exception of the possible Pendejo Cave netting, is the oldest such construction in North America. The Sand1, Level I netting fragment dates between $10,050 \pm 50$ B.P. (Beta 169848) and $10,310 \pm 40$ B.P. (Beta 168656 [Rhode et al. 2006]). While the specimen is not directly dated and there is some indication of disturbance, this net fragment is probably minimally 10,000 years old.

In the northern Great Basin, Cressman (1942) reports cordage of the two-ply, Z spun, S twist and single-ply, Z twist varieties from the bottom of Fort Rock Cave, Oregon. The cordage was apparently recovered with Fort Rock twined sandals and close simple twined basketry with Z twist wefts (Figure 4). Though the age of the basal deposits at Fort Rock Cave remain controversial, these perishable specimens are at least 11,000 years old (Andrews et al. 1986).

Recently, Tom Connolly, who unfortunately could not attend this symposium, provided me with descriptive data and illustrations of more early Northern Great Basin perishable material. In the recent re-excavations of Paisley Caves 1, 2, and 5, excavators have recovered a variety of knotted and unknotted, twisted and braided cordage and cordage constructions. Made of *Artemisia* sp., *Apocynum* sp., or untyped

Figure 2. Close-up of Close Diagonal Twining, Z-twist (paired) Wefts from Hicock, New York.

species, these directly dated items range in age from 10,030 ± 90 to 10,550 ± 40 B.P. (Connolly, personal communication, 2012).

Elsewhere in the northern Great Basin, specifically, and western North America, generally, the oldest basketry is invariably twined and includes open and close simple twined bags, mats, burden baskets, trays, and sandals of a variety of configurations. Though rarely directly dated, the age of these materials extends to at least 10,950 B.P. or slightly earlier (Andrews, Adovasio, and Carlisle 1986). Before turning to points south, mention should also be made of the remarkable set of micro-incised limestone pebbles and cobbles from the Gault Site, Texas. Recovered from Clovis contexts, these items appear to bear representations of plaited basketry, perhaps attesting to the importance of this perishable construction medium in the Clovis technology repertoire (Adovasio 2004) (Figure 5).

In South America, the production of textiles or basketry—again invariably twined—is evidenced in early tenth millennium B.P. contexts in the Peruvian highlands (Adovasio and Lynch 1973; Adovasio and Maslowski 1980), while the production of cordage and cordage byproducts is substantially more ancient. Presently, the oldest well-dated cordage in North or South America derives from Monte Verde in Chile

Figure 3. Dugong bone fragment from 8FR44 exhibitn an enigmatic, possible net, phenomena.

(Adovasio 1997). At that remarkable location, thirty-three individual specimens of cordage and at least eleven separate cordage impressions were identified in a context firmly dated between 13,565 ± 250 B.P. and 11,790 ± 200 B.P. The average age of the Monte Verde perishable materials is ca. 12,500–13,000 B.P. (Dillehay and Piño 1989:142), making this the longest and earliest perishable fiber assemblage (i.e., as opposed to individual specimens) from the entire New World. One structural type is apparently represented in the Monte Verde cordage assemblage, single-ply, S twist. However, regular indentations on several examples of this type strongly suggest that some of these specimens are plies of two-ply, S spun, Z twist cords which have become untwisted (Figure 6).

In upland South America, the oldest plant fiber-based artifacts derive from Guitarrero Cave in the north central highlands of Peru (Lynch 1980; 1999). Initially reported by Adovasio and Lynch (1973) then more fully described by Adovasio and Maslowski (1980), elements of this assemblage have very recently been directly redated (Jolie et al. 2011). A Two Ply, S-spun, Z-twist cord dates to 10,290 ± 45 B. P. while a square knotted leaf construction is only slightly younger at 10,230 ± 45. Significantly, in both the Monte Verde and Guitarerro cases, the recovered assemblages are presumed to be only a very small fraction of the non-durable repertoire of their respective populations.

Figure 4. Fort Rock Sandal type specimen from Fort Rock Cave, Oregon.

Figure 5. Incised stone specimen from Gault, Texas shown in: (a) standard photography and (b) PTM-enhanced.

Figure 6. Over-hand knotted fiber from Monte Verde, Chile.

Whatever the actual typological diversity of the Monte Verde and Guitarrero cordage and basketry industries, this brief overview indicates that the production of fiber perishables is well-documented in late Pleistocene contexts in both North and South America. As noted by Adovasio et al. (1996:533), perhaps significantly, twining is the earliest basketry or textile production technique known from virtually all of the areas enumerated above with the possible exception of eastern North America, where plaiting also exhibits a venerable antiquity. This seems to confirm, or at least strongly support, the hypothesis advanced some years ago (Adovasio 1970) that twining technology is at the heart of virtually all textile and basketry production, not simply as originally envisioned in North and South America, but apparently throughout the rest of the world.

The Implications of Perishable Fiber Artifacts

The relatively sophisticated character of the earliest New World perishable fiber artifacts, coupled with the demonstrable existence of plant-fiber-based technology in widely separated parts of the Old World by no later the 15,000 B.P., supports the hypothesis made long ago (Adovasio 1977:vii) that these items were part and parcel of the armamentarium of the first colonists to the New World. Indeed, given the dates on Dolní Vestonice I and II and Pavlov I (Adovasio et al. 1996, 2001, 2005; Soffer and Adovasio 2004; Soffer et al. 1998), even if the earliest percolations of humans across Berengia occurred more than 20,000 years ago, it is highly likely that perishable fiber technology was already an established and readily portable part of their techno-suite.

As with the Gravettian populations of Central Moravia, the presence of a plant-fiber-based technology—or, as Bettinger (personal communication, 1999) calls it, "soft technology"—has potentially profound implications for the elucidation of Paleoindian adaptations, behavior, and even social organization. First, as noted in the introduction, for many scholars, the mere existence of basketry, cordage, and related products in Upper Paleolithic or Paleoindian contexts is a revelation. This is particularly perplexing to perishable artifact analysts, especially given ample ethnographic evidence that perishable technologies form the bulk of hunter-gatherer material culture even in Arctic and sub-Arctic environments (e.g., Damas

1984; Helm 1981). Indeed, as alluded to above, archaeologists working with materials recovered from environmental contexts with ideal preservation confirm that the ubiquity of perishable artifacts observed in the ethnographic present also obtains in the past. For example, Croes (1997:531) reports that wet sites in the Pacific Northwest yield inventories where >95% of prehistoric material culture is made of wood and fiber, and Collins (1937) confirms the same for sites in Alaskan permafrost.

Second—perhaps more important than the production of baskets, textiles, cordage, and the like—are the producers of those items. Cross-cultural research shows that the production and use of textiles, basketry, and, of course, cordage, and cordage by-products (e.g., netting) is associated with both sexes (Murdock 1937; Murdock and Provost 1973). However, in pre-market societies these technologies, especially basketry manufacture and weaving, are usually associated with the labor of women (King 1991; Schneider and Weiner 1989). Given such patterns of production and use, the recovery of perishable fiber artifacts from Paleoindian (or Upper Paleolithic) sites attests directly to the presence of both men and women at such sites and reveals the heretofore virtually ignored labor of females.

Third, whichever sex made them, the manufacture of perishable fiber artifacts provides insights into previously unrecognized subsistence procurement choices, at least for the time period under consideration. For example, the identification of sheet bends or weaver's knots in the Gravettian impressions from Moravia and the even more dramatic recovery of very early nets (Adovasio et al. 2007; Frison et al. 1986) from the Americas suggests that net-hunting was, indeed, practiced in the late Pleistocene and on into the Holocene.

The possibility of net-hunting bears important implications for reconstructing subsistence behavior and attendant social organization. Hunting with nets is entirely different from the taking of individual animals or even the systematic predation of large aggregates of animals via lances, spears, spear throwers, and darts. Cross-cultural research indicates that net-hunting is a highly organized communal effort, which, because of the relative lack of individual expertise necessary for success as well as the minimal danger involved in such a non-confrontational harvesting technique, can and does utilize the labor of the entire co-residential social unit (Satterthwait 1986, 1987; Steward 1938; Turnbull 1965; Wilkie and Curran 1991). Net-hunting is also associated with mass harvests in short periods of time and, thus, with the production of a surplus (Satterthwait 1986, 1987). Although such surplus in some ethnographic cases is associated with participation in a market economy (e.g., the Ituri forest [Wilkie and Curran 1991]), in other cases (e.g., Aboriginal Australia [Satterthwait 1986, 1987] and New Guinea [Roscoe 1990, 1993]), it is associated with large gatherings, feasting, and ceremonialism. The implication of mass harvests have been discussed for the Gravettians of Central Europe (Soffer et al. 1998; Soffer and Adovasio 2004) but, as yet, have not been considered for Paleoindians in North or South America.

Fourth, whether it was used to generate surplus or to provisionally meet daily needs, we submit that net-hunting as an activity is less important than the physical appearance of nets, baskets, textiles, and other perishable artifacts as a technological class of objects. Although the manufacture of tools from non-durable media clearly extends well back into the Pleistocene, if not late Pliocene, and is associated with several grades of Homo, the systematic exploitation of plant fibers for diversified uses may well constitute a relatively unique event horizon as originally suggested by Barber (1991). Specifically, the development of plant-fiber-based industries may represent a technological signature of many late Pleistocene populations in much the same way as blade tool manufacture does, but with far more profound consequences. Moreover, the elaboration of such industries may also have facilitated adaptive success, both for populations who produced them in place as well as those on the move, including colonizing populations.

In this perspective, the successful colonization of this hemisphere probably owed more to broad spectrum hunting and collecting by all members of a given group employing non-durable technologies than it did to big game hunting by stone-wielding, big-game impaling, adult males. This is certainly the pattern suggested at Monte Verde (Dillehay 1989, 1997) and, even with their poorer preservation, at Meadowcroft (Adovasio et al. 1975, 1977a, 1977b, 1978, 1988; Carlisle and Adovasio 1982), Cactus Hill (McAvoy and McAvoy 1997), and Gault (Collins 2002); it is also reflected in the newly reported and quite early maritime-oriented sites on the Peruvian coast (Keefer et al. 1998; Sandweiss et al. 1998) and California coast (Connolly et al. 1995).

Fifth and finally, the documentation of perishable plant-fiber-based industries in an apparently broad spectrum economic milieu raises serious questions about the continuing utility of terms like Paleoindian and Archaic as valid archaeological constructs. If a series of interrelated industries like cordage, basketry, sandals, and textiles were being manufactured by a population or populations in the late Pleistocene when groups were previously posited to lack this technology and, further, if such groups were not exclusively or even predominantly big game hunters, what does the term Paleoindian mean? Similarly, if these same groups are, in fact, broad spectrum hunter-foragers, what does the term Archaic signify? We suggest, following Soffer et al. (1998), that these and all other similar labels, like Upper Paleolithic and Mesolithic, should be divorced from all pre-conceptions of prevailing technology and separated from posited food procurement patterns once and for all. In this reductionist perspective, if Paleoindian is retained as a culture stage or epoch diagnostic, it and the succeeding Archaic should be employed solely in chronological terms to designate populations living on either "side" of the Pleistocene/Holocene boundary, before the advent of food production. Whatever the utility, or lack thereof, of this suggestion, it is now abundantly clear that the origins of fiber artifact production, and cordage and basketry/textile manufacture, specifically, are millennia older than previously envisioned. Indeed, these crafts may have a potential antiquity that could extend to, and in a sense, help define the appearance of modern human behavior.

Acknowledgements

This paper was edited by David R. Pedler. Special thanks are also due to T. Connolly, who provided the data and images from the Paisley Caves.

References Cited

Adovasio, J. M.
- 1970 The Origin, Development and Distribution of Western Archaic Textiles and Basketry. *Tebiwa: Miscellaneous Papers of the Idaho State University Museum of Natural History* 13(2):1-40.
- 1977 *Basketry Technology: A Guide to Identification and Analysis*. Aldine Publishing, Chicago.
- 1997 Cordage and Cordage Impressions from Monte Verde. In *The Archaeological Context and Interpretation*, edited by T. D. Dillehay, pp. 221-228. Monte Verde: A Late Pleistocene Settlement in Chile, vol. 2. Smithsonian Institution Press, Washington, D.C.

Adovasio, J. M., R. L. Andrews, and J. S. Illingworth
- 2009 Netting, Net Hunting, and Human Adaptation in the Eastern Great Basin. In *Past, Present and Future Issues in Great Basin Archaeology: Papers in Honor of Don D. Fowler*, edited by B. Hockett, pp. 84-102. Cultural Resource Series No. 20, US Department of Interior, Bureau of Land Management, Elko.

Adovasio, J. M., A. T. Boldurian, and R. C. Carlisle
- 1988 Who Are Those Guys?: Some Biased Thoughts on the Peopling of the New World. In *Americans Before Columbus: Ice-Age Origins*, edited by R. C. Carlisle, pp. 45-61. Ethnology Monographs Number 12. Department of Anthropology, University of Pittsburgh, Pittsburgh.

Adovasio, J. M., J. D. Gunn, J. Donahue, and R. Stuckenrath
- 1977a Progress Report on Meadowcroft Rockshelter—A 16,000 Year Chronicle. In *Amerinds and their Paleoenvironments in Northeastern North America*, edited by W. S. Newman and B. Salwen, pp. 137–159. Annals of the New York Academy of Sciences 228.
- 1977b Meadowcroft Rockshelter: Retrospect 1976. *Pennsylvania Archaeologist* 47:2–3.
- 1978 Meadowcroft Rockshelter 1977: An Overview. *American Antiquity* 43:632–651.

Adovasio, J. M., and D. C. Hyland
- 1993 *Paleo-Indian Perishables from Pendejo Cave: A Brief Summary*. Paper presented at the 58th Annual Meeting of the Society for American Archaeology, April 14–18, St. Louis, Missouri.

Adovasio, J. M., D. C. Hyland, and O. Soffer
- 1997 Textiles and Cordage: A Preliminary Assessment. In *Pavlov I—Northwest: The Upper Paleolithic Burial and its Settlement Context*, edited by J. Svoboda, pp. 403–424. The Dolní Vestonice Studies, Vol. 4. Institute of Archaeology, Academy of Sciences of the Czech Republic, Brno, Czech Republic.
- 2001 Perishable Technology and Early Human Populations in the New World. In *On Being First: Cultural Innovation and Environmental Consequences of First Peopling*, edited by J. Gillespie, S. Tupakka, and C. De Mille, pp. 201–222. Chacmool, The Archaeological Association of the University of Calgary, Calgary.
- 2004 Perishable Fiber Artifacts and the First Americans: New Implications. In *New Perspectives on the First Americans*, edited by B. T. Lepper and R. Bonnichsen, pp. 157–164. Texas A&M University Press, College Station.
- 2005 Textiles and Cordage. In *Pavlov I Southeast: A Window Into the Gravettian Lifestyles*, pp. 432-444, edited by J. Svoboda. Academy of Science of the Czech Republic, Institute of Archaeology, Brno.

Adovasio, J. M., D. C. Hyland, O. Soffer, and B. Klíma
    1998a *Perishable Industries and the Colonization of the East European Plain*. Paper presented at the Fifteenth Annual Visiting Scholar Conference "Fleeting Identities: PerishableMaterial Culture in Archaeological Research," Carbondale, Illinois.
    1998b *Perishable Industries and the Colonization of the East European Plain*. Paper presented at the 14th International Congress of Anthropological and Ethnological Sciences, Williamsburg, Virginia.
    2001 Perishable Industries and the Colonization of the East European Plain. In *Fleeting Identities: Perishable Material Culture in Archaeological Research*, edited by P. B. Drooker, pp 285-313. Center for Archaeological Investigations Occasional Paper No. 28. Southern Illinois University, Carbondale.

Adovasio, J. M., and J. S. Illingworth
    2004 Prehistoric Perishable Fiber Technology in the Upper Ohio Valley. In *Perishable Material Culture in the Northeast*, edited by P. B. Drooker, pp. 19-30. The University of the State of New York, Albany.

Adovasio, J. M., J. S. Illingworth, and S. McKenzie
    2012 An Enigmatic Impression on a Dugong Bone. Paper presented at the 77th Annual Meeting of the Society for American Archaeology.

Adovasio, J. M., R. S. Laub, J. S. Illingworth, J. H. McAndrews, and D. C. Hyland
    2003 Perishable Technology from the Hiscock Site. In The Hiscock Site: Late Pleistocene and Holocene Paleoecology and Archaeology of Western New York. Proceedings of the Second Smith Symposium, held at the Buffalo Museum of Science, October 14–15, 2001, edited by R. S. Laub, pp. 272–280. Volume 37, Bulletin of the Buffalo Society of Natural Sciences. Buffalo, New York.

Adovasio, J. M., and T. F. Lynch
    1973 Preceramic Textiles from Guitarrero Cave, Peru. *American Antiquity* 38(1):84-90.

Adovasio, J. M., and R. F. Maslowski
    1980 Textiles and Cordage. In *Guitarrero Cave*, edited by T. F. Lynch, pp. 253–290. Academic Press, New York.

Adovasio, J. M., O. Soffer, D. C. Dirkmaat, C. A. Pedler, D. R. Pedler, D. Thomas, and M. R. Buyce
    1992 Flotation Samples from Mezhirich, Ukranian Republic: A Micro-View of Macro-Issues. Paper Presented at the 57th Annual Meeting of the Society for American Archaeology, Pittsburgh.

Adovasio, J. M., O. Soffer, and D. C. Hyland
    2005 Textiles and Cordage. In *Pavlov I Southeast: A Window into the Gravettian Lifesytles*, edited by J. Svobada, pp. 432-443. Academy of Sciences of the Czech Republic, Institute of Archaeology at Brno.

Adovasio, J. M., O. Soffer, D. C. Hyland, B. Klíma, and J. Svoboda
    1999 Textil, Kosíkárství A Síte V Mladém Paleolitu Moravy (Textiles, Basketry, and Nets inUpper Paleolithic Moravia). *Archeologicke Rozhledy* LI:58–94.

Adovasio, J. M., O. Soffer, J. S. Illingworth, and D. C. Hyland
    2007 Perishable Fiber Artifacts and Paleoindians: New Implications. Paper Presented a the Workshop "Weaving Together Two Ways of Knowing: Archaeological Organic Artifact Analysis and Indigenous Textile Arts". Metepenagiag Mi'kmaq Nation, Canada.

Adovasio, J. M., O. Soffer, and B. Klíma
    1995 *Paleolithic Fiber Technology: Data from Pavlov I, Czech Republic, ca. 27,000 YR B.P.* Paper presented at the 60th Annual Meeting of the Society for American Archaeology, Minneapolis.
    1996 Upper Paleolithic Fibre Technology: Interlaced Woven Finds From Pavlov I, Czech Republic, c. 26,000 Years Ago. *Antiquity* 70(269):526–534.

Andrews, R. L., and J. M. Adovasio
    1996 The Origins of Fiber Perishables Production East of the Rockies. In *A Most Indispensable Art: Native Fiber Industries from Eastern North America*, edited by J. B. Peterson, pp. 30–49. Knoxville: The University of Tennessee Press.

Andrews, R. L., J. M. Adovasio, and R. C. Carlisle
    1986 *Perishable Industries from Dirty Shame Rockshelter, Malheur County, Oregon*. Issued jointly as Ethnology Monographs No. 9 and University of Oregon Anthropological Papers No. 34. Department of Anthropology, University of Pittsburgh, Pittsburgh, and Department of Anthropology, University of Oregon, Eugene.

Barber, E. J. W.
    1991 *Prehistoric Textiles: The Development of Cloth in the Neolithic and Bronze Ages, with Special Reference to the Agean*. Princeton University Press, Princeton.

Bosinski, G.
    1995 The Earliest Occupation of Europe: Western Central Europe. In *The Paleolithic and Mesolithic of the Rhineland. Quaternary Field Trips in Central Euopre 15(2):967-971.*

Carlisle, R. C., and J. M. Adovasio (Editors)
　1982　*Meadowcroft: Collected Papers on the Archaeology of Meadowcroft Rockshelter and the Cross Creek Drainage*. Department of Anthropology, University of Pittsburgh, Pittsburgh.

Cheynier, A.
　1967　*Comment Vivait l'Home des Cavernes*. Robert Arnoux, Paris.

Collins, H. B., Jr.
　1937 *Archaeology of St. Lawrence Island, Alaska*. Smithsonian Miscellaneous Collections, Vol. 96. Smithsonian Institution Press, Washington, D.C.

Collins, M. B.
　2002 The Gault Site, Texas, and Clovis Research. *Athena Review* 3(2).

Connolly, T., J. M. Erlandson, and S. E. Norris
　1995　Early Holocene Basketry from Daisy Cave, San Miguel Island, California. *American Antiquity* 60:309-318.

Cressman, L. D.
　1942　*Archaeological Researches in the Northern Great Basin*. Carnegie Institution of Washington Publications No. 538. Washington, D.C.

Childe, V. G.
　1936　*Man Makes Himself*. Oxford University Press.

Croes, D. R.
　1997 The North-Central Cultural Dichotomy on the Northwest Coast of North America: Its Evolution as Suggested by Wet-Site Basketry and Wooden Fish-Hooks. *Antiquity* 71:594–615.

Damas, D. (Editor)
　1984 *Arctic*. Handbook of North America Indians, vol. 5. W. C. Sturtevant, general editor. Smithsonian Institution Press, Washington, D.C.

Derevianko, A. P., and V. E. Medvedev
　1995　The Amur River Basin as One of the Earliest Centers of Ceramics in the Far East. In *Toadzia Kekutonodoku No Kigaen (The Origin of Ceramics in Eastern Asia)*, edited by T. Fukushi, pp. 11–26. University Press, Sendai.

Dillehay, T. D. (Editor)
　1989　*Paleoenvironment and Site Context*. Monte Verde: A Late Pleistocene Settlement in Chile, vol. 1. Smithsonian Institution Press, Washington, D.C.
　1997　*The Archaeological Context and Interpretation*. Monte Verde: A Late Pleistocene Settlement in Chile, vol. 2. Smithsonian Institution Press, Washington, D.C.

Dillehay, T. D., and M. Piño
　1989　Stratigraphy and Chronology. In *Paleoenvironment and Site Context*, edited by T. D. Dillehay, pp. 133-145. Monte Verde: A Late Pleistocene Settlement in Chile, vol. 1. Smithsonian Institution Press, Washington, D.C.

Fowler, C. S., A. J. Dansie, and E. M. Hattori
　1997　*Plaited Matting from Spirit Cave, Nevada: Technical Implications*. Paper presented at the 62nd Annual Meeting of the Society for American Archaeology, Nashville.

Frison, G. C., R. L. Andrews, J. M. Adovasio, R. C. Carlisle, and R. Edgar
　1986　A Late Paleoindian Animal Trapping Net from Northern Wyoming. *American Antiquity* 51(2):352-361.

Glory, A.
　1959　Débris de corde paléolithique à la Grotte de Lascaux [Remains of a Paleolithic cord from the Cave of Lascaux]. *Mémoires de la Société Préhistorique Française* 5:135–169.

Helm, J. (Editor)
　1981 *Subarctic*. Handbook of North America Indians. vol. 6. W. C. Sturtevant, general editor. Smithsonian Institution Press, Washington, D.C..

Hemmings, C. A.
　2005　An Update of Recent Work at Sloth Hole (8JE121), Jefferson County, Florida. *Current Research in the Pleistocene* 22:47–49.

Hyland, D. C., and J. M. Adovasio
　1995　*Perishable Industries from Pendejo Cave, New Mexico*. Paper presented at the 60th Annual Meeting of the Society for American Archaeology, Minneapolis.

Hyland, D. C., and J. M. Adovasio, with J. S. Illingworth
　1998　The Perishable Artifacts. In *Excavations at Pendejo Cave*, edited by J. Libby and R. S. MacNeish, chapter 18. University of New Mexico Press, Albuquerque. in press.

Hyland, D. C., J. M. Adovasio, and R. E. Taylor
    1998   Corn Cucurbits, Coiling, and Colonization: An Absolute Chronology for the Appearance of Mesoamerican Domesticates and Perishables in the Jornada Basin, New Mexico. Presented at the 63rd Annual Meeting of the Society for American Archaeology, Seattle.

Hyland, D. C., I. S. Zhushchikhovskaya, V. E. Medvedev, A. P. Derevianko, A. V. Tabarev
    2002   Pleistocene Textiles in the Russian Far East: Impressions From Some of the World's Oldest Pottery. *Anthropologie* XL(1):1–10.

Jennings, J. D.
    1957   *Danger Cave*. Anthropological Papers No. 27. Department of Anthropology, University of Utah, Salt Lake City.

Jolie, E. A., T. F. Lynch, P. R. Geib, and J. M. Adovasio
    2011   Cordage, Textiles, and the Late Pleistocene Peopling of the Andes. *Current Anthropology* 52(2):285-297.

Kvavadze, E., O. Bar-Yosef, A. Belfer-Cohen, E. Boaretto, N. Jakell, Z. Matskevich, and T. Meshveliani
    2009   30,000-Year-Old Wild Flax Fibers. *Science* 325(5946):1359.

Keefer, D. K., S. D. Defrance, M. E. Moseley, J. B. Richardson III, D. R. Satterlee, and A. Day-Lewis
    1998   Early Maritime Economy and El Niño Events at Quebrada Tacahuay, Peru. *Science* 281(5384):1833–1835.

King, M. E.
    1991   The Perishable Preserved: Ancient Textiles from the Old World with Comparisons from the New World. *The Review of Archaeology* 13(1):2–11.

Kuttruff, J. T., S. G. Dehart, and M. J. O'Brien
    1998   7500 Years of Prehistoric Footwear from Arnold Research Cave, Missouri. *Science* 281(5373):72–75.

Leroi-Gourhan, A. and J. Allain
    1979   *Lascaux Inconnu*. Paris: CNRS.

Lynch, T. F.
    1980   *Guitarrero Cave: Early Man in the Andes*. Academic Press, New York
    1999. The earliest South American lifeways. In *The Cambridge history of native peoples of the Americas*, vol. 3. F. Salomon and S. B. Schwartz, eds. Pp. 188–263. Cambridge University Press, Cambridge.

McAvoy, J. M., and L. D. McAvoy
    1997   *Archaeological Investigations of Site 44SX202, Cactus Hill, Sussex County Virginia*. Research Report Series No. 8. Commonwealth of Virginia Department of Historic Resources, Richmond.

Murdock, G. P.
    1937Comparative Data on the Division of Labor by Sex. *Social Forces* 15:551–553.

Murdock, G. P., and C. Provost
    1973   Factors in the Division of Labor by Sex: A Cross-Cultural Analysis. *Ethnology* 12:203–226.

Nadel, D., A. Danin, E. Werkerc, T. Schick, M. E. Kislev, and K. Stewart
    1994   19,000-Year-Old Twisted Fibers from Ohalo II. *Current Anthropology* 35:451–457.

Rhode, D., T. Goebel, K. E. Graf, B. S. Hockett, K. T. Jones, D. B. Madsen, C. G. Ovlatt, D. N. Schmitt
    2006   Latest Pleistocene – Early Holocene Human Occupation and Paleoenvironmental Change in the Bonneville Basin, Utah – Nevada. *Geological Society of America, Field Guide 6*.

Roscoe, P. B.
    1990   The Bow and Spreadnet: Ecological Origins of Hunting Technology. *American Anthropologist* 92:691–701.
    1993   The Net and the Bow in the Ituri. *American Anthropologist* 95:153–155.

Sandweiss, D. H., H. Mcinnis, R. L. Burger, A. Cano, B. Ojeda, R. Paredes, M. C. Sandweiss, and M. D. Glascock
    1998   Quebrada Jaguay: Early South American Maritime Adaptations. *Science* 281(5384):1830-1832.

Satterthwait, L.
    1986   Aboriginal Australian Net Hunting. *Mankind* 16:31–48.
    1987   Socioeconomic Implications of Australian Aboriginal Net Hunting. *Man* (N.S.) 22:613–636.

Schneider, J. and A. Weiner
    1989Introduction. In *Cloth and Human Experience*, edited by A. B. Weiner and J. Schneider, pp. 1–15. Smithsonian Institution Press, Washington, D.C.

Soffer, O.
    2004   Recovering Perishable Technologies through Use Wear on Tools: Preliminary Evidence for Upper Paleolithic Weaving and Net Making. *Current Anthropology* 45(3):407–413.

Soffer, O., and J. M. Adovasio
    2004   Textiles and Upper Paleolithic Lives. A Focus on the Perishable and Invisible. In *The Gravettian Along the Danube*, edited by J. Svoboda and L. Sedláková, pp. 270–282. Archeologický ústav AVR, Brno.

    2007   Роль собирательства И технологий обработки растительного сырья в верхнем палеолите. In *Проблемы археологии каменного века (к юбилею М.Д. Гвоздовер)*, pp. 62-79, edited by Н.Б. Леонова and Е. В. Леонова. Авторы статй, Москва.

    2010   Not By Stone Alone: Perishable Technologies and Upper Paleolithic Lifeways. In *Proceedings of the Gintsi Site Conference*, edited by I. Gavrylenko (in press).

Soffer, O., J. M. Adovasio, and D. C. Hyland
    2000   The "Venus" Figurines: Textiles, Basketry, Gender, and Status in the Upper Paleolithic. *Current Anthropology* 41(4): 517–537.

Soffer, O., J. M. Adovasio, D. C. Hyland, and B. Klíma
    1998   *Perishable Industries from Dolní Vestonice: New Insights into the Origin of the Gravettian*. Paper presented at the 63rd Annual Meeting of the Society forAmerican Archaeology, Seattle, Washington, 25–29 March 1998.

Soffer, O., J. M. Adovasio, D. C. Hyland, B. Klíma, and J. Svoboda
    1998   Textiles and Basketry in the Paleolithic: What Then is the Neolithic? In *Historical-Cultural Contacts Between Aborigines of the Pacific Coast of North Western America and North Eastern Asia*, edited by A. R. Artemiev, pp. 311–321. Russian Academy of Sciences, Far Eastern Branch, Institute of History, Archaeology and Ethnography of the Peoples of the Far East, Vladivostok.

Soffer, O., J. M. Adovasio, J. S. Illingworth, H. A. Amirkhanov, N. D. Praslov, and M. Street
    2000   Palaeolithic Perishables Made Permanent. *Antiquity* 74:812–821.

Steward, J. H.
    1938   *Basin-Plateau Aboriginal Sociopolitical Groups*. Bureau of American Ethnology Bulletin 120.

Stile, T. E.
    1982   Perishable Artifacts from Meadowcroft Rockshelter, Washington County, Southwestern Pennsylvania. In *Meadowcroft: Collected Papers on the Archaeology of Meadowcroft Rockshelter and the Cross Creek Drainage*, edited by R. C. Carlisle and J. M. Adovasio, pp. 130–141. Prepared for the Symposium "The Meadocroft Rockshelter Rolling Thunder Review: Last Act." 47th Annual Meeting of the Society for American Archaeology, Minneapolis.

Stone, E. A.
    2011   Through the Eye of the Needle: Investigations of Ethnographic, Experimental, and Archaeological Bone Tool Use Wear from Perishable Technologies. Unpublished Ph.D. dissertation, Department of Anthropology, University of New Mexico, Albuquerque.

Taylor, W. W.
    1966   Archaic Cultures Adjacent to the Northeastern Frontiers of Mesoamerica. In *Handbook of Middle American Indians* 4, pp. 59-94. University of Texas Press, Austin.

Turnbull, C. M.
    1965   *Wayward Servants*. Natural History Press, Garden City, New York.

Wheeler, S.N., and G.N. Wheeler
    1969   Cave Burials Near Fallon, Churchill County, Nevada. *Nevada State Museum Anthropological Papers* 14:70-78.

Wilkie, D. C., and B. Curran
    1991   Why Do Mbuti Hunters Use Nets? Ungulate Hunting Effiiency of Archers and Net-Hunters in the Ituri Rain Forest. *American Anthropologist* 93:680–689.

Zhushchikhovskaya, I. S.
    1997   Current Data on Late-Pleistocene/Early-Holocene Ceramics in the Russian Far East. *Current Research in the Pleistocene* 14:89–91.

ORIGIN AND ANTIQUITY OF A WESTERN NORTH AMERICAN STEMMED POINT TRADITION: A PRE-CLOVIS PERSPECTIVE

David G. Rice[1]

My interest in the First Americans grew out of early training and work experience from my mentors Douglas Osborne, Alan Bryan, B. Robert Butler, Roy Carlson and Charles Borden. In college I was a student of Alex Krieger, and conducted independent fieldwork for Richard Daugherty, Earl Swanson, and Claude Warren. And in graduate school I worked under the guidance and professional experience of Richard Daugherty and Roald Fryxell to define the Windust Phase in lower Snake River Region Prehistory (Rice 1972). Early colleagues were Charles M. Nelson and Robson Bonnichsen. These comments stem from the mix of ideas, thoughts and perspectives that I derived from these colleagues and mentors, along with some of my own reflections. In addition, I have had rare opportunities for discussions with other influential thinkers involved in the sharp clash of differing opinions, theories, definitions and paradigms during the late twentieth century for what constitutes evidence. These individuals include Luis Aveleyra Arroyo de Anda, Jose Luis Lorenzo, Harold Malde, Howard Powers, Gordon Willey, Scotty MacNeish, Karl Butzer, Tom Dillehay, and Robson Bonnichsen. Some of these ideas and perspectives point to a new paradigm that requires a broader understanding of the near and far pre-Clovis cultural possibilities, and their environmental contexts in New World Prehistory. This focus is prerequisite to sorting out revisions to the scientific belief system we collectively use to discover the truths and the realities that guide our research. My intent is to stress the need for open-mindedness and imagination to achieve a new vision of New World prehistory, to identify some recent documentation, and to identify a few directions that need to be intensified to pursue that vision.

For many years Richard Daugherty's (1962) Intermontane Western Tradition, based on the Lind

---

[1] Tkwinat Twati Anthropological Services, 1300 Miller Dr., Kelso, WA. Email: windust1@yahoo.com

Coulee Site, stood alone as an anomalous context of stemmed lanceolate points associated with *Bison antiquus* in eastern Washington State. Dated at 8,700 radiocarbon years B.P., its true significance was not appreciated for nearly half a century. Now, over the past 40 years, an accumulation of early archaeological evidence for a stemmed point tradition from diverse settings in the American West has gradually made investigators aware that these finds may not follow Clovis, as previously assumed, but may parallel and predate Clovis, as a separate co-tradition (Bryan 1980; Hanes 1988; Willig and Aikens 1988; Braje et. al. 2011; Davis, Willis, and Mcfarlan 2012: Jenkins et. al. 2012). Initially, much of the evidence was dismissed because it was anomalous with the prevailing view that Clovis represented the earliest New World people. Now new possibilities for New World origins seem more feasible and realistic, than previous putative observations. The mechanics for realization of possible colonization along the coastal shelf of North America makes the concept feasible in light of modeling of sea level changes during the Pleistocene (Fladmark 1978), better understanding of local and regional details of eustacy and isostacy, improvements in geochronology and dating techniques, enabling details of the timing of such changes; and LIDAR, satellite telemetry and GIS and to reveal its form, location, and areal extent. These developments in the advancement of science and scientific method now allow the interdisciplinary exploration of what seemed impossible just a few years ago (Braje et. al. 2011; Davis 2011). New possibilities are emerging in our awareness that now seem almost *self-evident* they are so obvious! But this is only so once you are able to connect the relationships that create a new awareness in the eye of the beholder. Albert Einstein once said, "Logic will get you from A to B. Imagination will take you everywhere." To unravel the story of the First Americans takes both.

New obsidian hydration studies, use of thermoluminescence for dating, and more refined radiocarbon dating from an increased number of sites from California (Budinger 2004:21; Parker 2012), Idaho (Gruhn 1961; Davis 2011), Oregon (Jenkins et. al. 2012), Texas (Waters et. al. 2011), and Washington (Waters et.al. 2011) present unequivocal evidence for pre-Clovis existence. Not all of these finds have been recent. Some evidence has been present over the past half century of archaeological discovery and investigation, but lacked a "context" to give it standing and validity (Krieger 1964:31). For example, George Carter's hearths exposed along southern California sea cliffs (Krieger1964:46), or Phil Orr's discovery of hearths with extinct faunal remains off the California coast on the Channel Islands were examined and accepted by Krieger (1964:44), but discredited as evidence by others in his time. The southern California coast presented a different kind of investigation problem because the high relief of the sea cliffs does not produce large horizontal erosion exposures deeply buried sediment horizons that represent the late Pleistocene-Holocene transition, allowing them to keep their secrets. New discoveries in the Channel Islands, and new methods, along with much sustained and intensive fieldwork (Braje et. al. 2011), however, have established a surprising and unexpected maritime context strengthening Gruhn's (1994) arguments for coastal migrations for the earliest Americans. This context reflects a localized diversified subsistence adaptation emphasizing sea birds, mammals, and fish. Charcoal and bird bones from San Miguel Island have been radiocarbon dated to between 11,800 and 11,500 years ago (Braje et. al. 2011). Of particular importance here is that these dated faunal remains are directly associated with two styles of stemmed lithic projectile points and chipped stone crescents. These finely made artifacts are illustrated in Smith (2011:28-31) and appear to represent a new aspect of the Western Stemmed Point Tradition. Often, determined efforts to discover the earliest material have been thwarted because researchers could not conceive where to find it. Prevailing attitudes discouraged investigation of locations and environments that has produced new finds as old, and older than Clovis (Waters and Stafford 2007). Also, new evidence includes not only formed artifacts, but associated material, such as human coprolites (Jenkins et. al. 2012). Advances in molecular DNA research, electron microscopes, XRF analysis, small sample biological and chemical analysis, and AMS radiocarbon dating refinements now enable much of new detailed micro-analytical information in the most recent studies, and they provide a definitive level of precision not previously obtainable in Krieger's time.

A random short list of recently dated Pre-Clovis finds from the Intermontane West includes:

Budinger, Jr. (2004): Rock Wren Locality, Calico Hills Site, California
- Chalcedony, Biface, dated by sediment thermoluminescence to 14,400+/-2,200 years B.P.

Davis (2011): Cooper's Ferry, Idaho
- Mammal bone associated with Western Stemmed Points, radiocarbon dated at 14,062+/-315 Cal years B.P.
- Charcoal associated with Western Stemmed Points, radiocarbon dated at 13,131+/-131 Cal B.P.

Jenkins et al. (2012): Paisley Caves, Oregon
- Human coprolite associated below cave deposits containing Western Stemmed Points, AMS radiocarbon dated at 14,725+/-302 Cal B.P.
- Western Stemmed Points in cave deposits, AMS radiocarbon dated at 12,959+/-30 Cal B.P.

Parker (2012): Clear Lake Basin, California
- Obsidian, Flat-base Widestem with single hydration rim of 12.2 microns, indicating an age of 14,200 years B.P.
- Obsidian, reworked Mendocino Side-notched with multiple hydration bands, the oldest of which is 12.2 microns, indicating an age of 14,200 years B.P.

Waters et.al (2011): Manis Mastodon Site, Washington
- Formed mastodon bone shaft imbedded in mastodon rib, AMS radiocarbon dated at 13,860 Cal years B.P.

Waters et. al. (2011): Buttermilk Creek, Texas
- An assemblage of 15,528 artifacts that define the Buttermilk Creek Complex, which stratigraphicly underlies a Clovis assemblage, and dates between 13,200 and 15,500 years ago.

As additional 14,000 year dates mount, predating Clovis finds by hundreds to a thousand years, a new paradigm needs to be built to consider prospects for new locations that should be investigated, and to identify available new methods necessary to investigate them (Bonnichsen and Steele 1994). Since potential currently submerged coastal shelf sites face the challenge of high cost to develop innovative methods to investigate them, other more accessible settings need to be reexamined. The interactive boundary along the southern extent of the Laurentide and Cordilleran ice sheets from Montana to Washington during the late Pleistocene presented unusual, now almost unimagined habitats. These existed along this stressful fluctuating boundary during the Younger Dryas and may have been more productive to prehistoric hunter-gatherers than what has so far been considered. The rugged mountain boarder along the Cascade and Sierra Nevada and California North Coast ranges to the west appear to have been another attraction to prehistoric hunters in settings like active, deglaciating high mountain cirque lakes, kame terraces, precipitous mountain valleys, and lake basins, such as at Clear Lake in the north coast ranges of California (DeGeorgey 2004), or at Beech Creek in the southern Washington Cascade Mountains (Mack, Chatters, and Prentiss 2010). Investigations in Idaho and Washington along similar locations have unexpectedly revealed points of the Western Stemmed Point Tradition (Davis 2011). In places, they may turn out to be as old or older than Clovis in the West. These seemingly marginal areas appear to contain better preserved evidence than much of the Great Basin heartland and suggest new settings to investigate. In the late Pleistocene-Holocene transition period, even the Washington and Oregon coasts west of these ranges are beginning to show a

Great Basin flavor in new unpublished findings.

In my current view, diverse stemmed points developed within separate regions along the coast, and around the pluvial lakes that existed in the basin-plateau, and in the surrounding mountains, connecting valleys, and riverine alluvial terraces. Many distinct stylistic variations evidently emerged within a common Western Stemmed Point Tradition, accounting for separate scattered geographic clusters of point types named Lind Coulee, Cougar Mountain, Haskett, Windust, Silver Lake, Lake Mojave, and others (Jenkins et. al. 2004). In 1972 my belief was that the inspiration for this broader tradition came from the south, and was much older than we now understand. Subsequent observations in the 1980s sustained that view (Bryan 1980; Hanes 1988; Willig and Aikens 1988). It was also my initial belief that San Dieguito may represent an aspect of this older source, based on Campbell et.al. (1937), Rogers (1939), and Warren (1967). Ultimately, there may be some connection to substantially older archaeological finds in the now unsorted Calico Hills Lake Mohave-Lake Manix region (Budinger 2004; Hardaker 2009) that is presently discredited by many American prehistorians. Such possible associations are hard to imagine because the hydrology, vegetation, and bio-populations of mammals, birds, and fish that comprised that broad landscape is now almost totally gone, but for localities like Tule Springs, Nevada.

Lack of preserved artifacts alone, however, may be insufficient grounds to rule out this possibility, if other contextual environmental and geological information suggest otherwise (Krieger 1964:31). That is really the problem for much of Great Basin-California desert archaeology – lack of preservation of the context. When I examined the work of Giambastiani and Bullard (2009) at China Lake, CA, I was struck by the fact that the only recoverable artifacts from the Pleistocene-Holocene lake terraces were stone. In my mind this presented enormous difficulty in imagining all the possible elements of that past environmental context that could not be considered or properly imagined for accurate archaeological interpretation.

A recent extra-terrestrial comet strike in Russia, impacting hundreds of people and causing extensive property damage, raises *catastrophism* as a new consideration in the search for the earliest Americans. Apart from earthquakes, Pleistocene era floods, and volcanic eruptions that have produced catastrophic events that are customary and easily imaginable, the recent, more heavenly cosmic events have become relevant, and really are reminders of unexpected situations out of the past. Firestone et. al. (2007) suggested that a meteor strike about 12,900 years ago might account for the sudden end of the Clovis Fluted Point Tradition in North America. Toner (2010) discusses the controversy over this possibility. If continued research supports this hypothesis, many of our current thoughts regarding, late Pleistocene extinction, gaps in cultural sequences, human adaptations, processes of culture change, and the effects of climate change, may need to be revised. The possible environmental devastation of a meteor strike is random and unexpected, so not usually considered in New World human prehistory. The Tunguska incident that occurred in Siberia in 1908 may have been the breakup of a comet. It has taken 100 years to understand, and represents a rare historical example of the widespread environmental devastation that occurred there. That incident helps to imagine what might have happened in the glaciated Canadian shield landscape 12,900 years ago. The possibility is bigger than detonation nanodiamonds! In the more distant past, Phillips et. al. (1991) report on the origin and revised age of Meteor Crater in northern Arizona. Using thermoluminescence and independent analysis of cosmogenic radio-nuclides his research team has determined the age for the Meteor Crater at 49,000+/-3,000 years B.P. and 49.7+/-0.85 ka, respectively. Meteor Crater is the largest terrestrial crater (1.2 km in diameter and 170 m deep) whose impact origin is proven by meteorite fragments. If researchers are to consider the interpretations of the earliest cultural antiquity given in Burdinger (2004) and Hardaker (2009) at the Calico Hills, then Meteor Crater might become a relevant context problem in looking for possible preserved evidence for humans during Pleistocene times earlier than 50,000 years ago in North America. If hominins did reach North America before 50,000 years ago, then they may not have survived the explosions from the impact of Meteor Crater. The actual devastation of this single event, and its areal extent, are presently unimaginable. There is now a new dimension to the meaning of catastrophism in North American prehistory.

If the Calico Hills region of 15,000 to 30,000 years ago did have any cultural counterparts, they too, were likely to the south, in Mexico. I suspect that during the late Pleistocene a natural system of luxuriant cienegas extended from southern California and Nevada into Mexico, especially including the major valleys of Mexico, Puebla, Tehuacan, and Oaxaca. The early evidence in Mexico for these relationships has been long obscured there by thousands of years of horticulture and agriculture. There, too, factors of longer day-length, greater moisture and humidity than is evident today, and more intensive solar radiation due to the southerly latitude, would have expanded foraging subsistence opportunity and hunting productivity for humans in the tropical uplands of the Central Plateau of Mexico. It was a long marshy pluvial phase in the late Pleistocene there. These environmental conditions presented a much more productive big game landscape for hunter-gatherers than we can clearly see today. These conditions may have supported a still earlier phase of regional human existence, with environmental cross-ties extending ultimately into South America and back, as some fossil biota suggest (Aveleyra 1964).

The idea of much earlier occupations during the Pleistocene in Mexico is not new and was recognized since the 1960s at sites in the Valsequillo reservoir area like Tequixquiac (Aveleyra 1964; Krieger 1964; Armenta 1978). The Santa Isabel Iztapan mammoth kills in the Valley of Mexico were found with projectile points described as similar to Angostura-Agate Basin and Scottsbluff at the time of the reported find in 1954. Aveleyra (1964: 402), however, mentions that these artifact finds may be related to two wood dates of >16,000 radiocarbon years B.P., and a peat date of 11,003+/- 500 radiocarbon years ago [This would be 12,785+/-625 Cal B.P.] from the Armenta Horizon of the Becerra Formation, which is the characteristic upper Pleistocene phase in the Valley of Mexico. This association is not firm, but might illustrate an example of how the *possible* age of the mammoths might have been prejudged based on existing scientific beliefs and erroneous artifact cultural assignments. Forty years ago when I was a graduate student, I did not know what to make out of these odd stemmed point associations with the Mexican mammoths. Accepted thought was that mammoths were associated with the Fluted Point Tradition. Now, I would say that the stemmed points found with the Mexican mammoths could well be part of an earlier Western Stemmed Point Tradition. By extension, if the actual age of the Santa Isabel Iztapan mammoths is at least 12,785+/-625 Cal B.P., then the associated points would be older than the point styles they were ascribed to from the northern American plains. What once was perceived as the earliest Paleoindian stemmed points in America, the Cody Complex may actually be a more localized and specialized development out of an older, broader, more diverse and generalized stemmed point tradition.

Many stemmed points may have been found in various parts of Mexico, but not necessarily viewed in Mexico as being associated with the late Pleistocene-Holocene transition, or necessarily southern in origin. Current and future work in Mexico will be able to judge. Researchers will, however, have to look in areas of older alluvial sediment accumulation in terraces, lacustrine basins, bogs, caves and rockshelters in locations that were in tropical uplands of that time period, such as the Mesa Central of Mexico (West 1964: 46). Much archaeological work has been done there, but it has focused on agricultural cultures whose more recent pursuits have erased and obscured much of the evidence. Hydrology is another important limiting factor in Mexico since that is where much of native North American agriculture developed. Although important for agriculture, the importance of water at the time of the Pleistocene-Holocene transition lies in the lacustrine sediments that may have preserved early period evidence along the higher terraces or strand lines. Perhaps the most promising drainage to investigate would follow the Rio Lerma from Lake Chapala eastward towards the large interior lake basins in the Central Plateau of Mexico. Many of these are mapped and the hydrology generally described in Tamayo and West (1964:108-113). The uplands surrounding the large lake basins at the upper limits of agriculture, and following high terraces from receding late Pleistocene lake levels should expose areas of sediments that would have potential for preserved botanical and faunal remains, and artifacts, to answer some of these questions.

In conclusion, new scientific methods and conceptual paradigms to identify now unknown locations to search, will refocus our thinking to the reality of multiple Pre-Clovis archaeological traditions in New World prehistory. Part of the challenge is to be open to situations of apparent *lack of physical evidence*, and to search, instead, for more subtle circumstantial evidence based on broader situational awareness. This means gathering new information about now unknown environmental contexts, before making judgments about their possible potential as subsistence and habitat settings, or about their content or age. This calls for a longer fuse for contemplation in hypothesis development. As research advances it is not unreasonable to consider that potential human activities in the New World could well be 20,000 and 30,000 years or more. In the 1960s Krieger often spoke of potential for more than 30,000 years of New World prehistory. Now, the evidence in South America suggests that possibility at Monte Verde in Chile (Dillehay and Manosa 2004). The lesson of Monte Verde in Chile is that some archaeological sites there are predominately represented by vegetal fabric, bone and wooden artifacts and features, and that lithic artifacts seem not to be as common. It is possible that lithic technology there was not as advanced as the developed cultural dexterity with reed, wood, and fiber source materials. Physical preservation of early cultures there potentially might be a larger problem, except at wet sites, and recognition of such artifact types as well. Krieger (1964) recognized this problem very early, in his proposed Pre-Projectile Point developmental stage. He remarked about the technical difficulty in *thinning* and *flattening* stone into useful tools. Both Krieger (1964) and Bryan (1978: 305-327) argue for use of bone, antler by early Pleistocene hunters, and state that they could have used alternative materials to make needed tools more easily than using stone tipped thrusting javelins. A North American example would be use of a mastodon bone shaft found imbedded in the Manis mastodon rib in the Pacific Northwest of North America 14,000 years ago (Waters et. al.2011). Bryan (1978) gave momentum to Krieger's (1964) views through a compilation of contributed papers revisiting many sites he discussed, by reviewing new evidence, and by expanding the evidence in South America at sites like Taima-Taima and Pedra Furada.

The problem with building a new paradigm with no physical evidence is that there will be no fore-knowledge of the early subsistence base, site and feature types, or technological traditions for tool manufacture and use. For site context, preservation will be the key problem. For artifacts, it will be pattern recognition. It is possible, however, to hypothesize what some of the future evidence might be. To admit its possibility. For instance, there is the existing body of geological information, scientifically derived and analytically studied, recovered from Hueyatlaco at Valsequillo reservoir in Mexico that is not presently accepted because scientific findings exceed present views regarding human antiquity in the New World (Armenta 1978; Hardaker 2007). Collaborative work is needed now to reevaluate these and other old findings for their scientific merit in light of more recent discoveries and new analytical methods. These finds, representing years of intensive research by many multidisciplinary scholars should not be dismissed out of hand by personal belief or inflexible professional positions. As a memorial tribute to Alfred Kroeber, Krieger (1961) spoke about the responsibilities of the critical mind, and the importance of being open to new possibilities and different interpretations. This open quality of mind is needed to resolve some present professional conflicts, and may help to facilitate an enlarged scope for research that will set us back on track in the search for the First Americans.

REFERENCES CITED

Armenta, Juan (1978) Vestigios de Labor Humana en Huesos de Animales Extintos de Valsequillo, Puebla. Publications of the Editorial Council of the State of Puebla.

Aveleyra, Luis Arroyo de Anda (1964) The Primitive Hunters, pp.. In *Handbook of Middle American Indians 1, Natural Environment and Early Cultures*:384-412, Robert Wauchope (general editor). University of Texas Press: Austin.

Bonnichsen, Robson and D. Gentry Steele, Editors. (1994) Method and Theory for Investigating the Peopling of the Americas. Center for the Study of the First Americans, Oregon State University.

Braje, Todd et. al. (2011) Archaeology and Historical Ecology of California Sea Mammals, pp. 273-296. In *Resiliance and Reorganization*, Todd Braje, (ed.). University of California: Berkeley.

Bryan, Alan L., (ed.) (1978) Early Man in America from a Circum-Pacific perspective. *Occasional Papers No. 1 of the Department of Anthropology, University of Alberta* (Edmonton).

Bryan, Alan L. (1980) The Stemmed Point Tradition: an early technological tradition in western North America. In Anthropological Papers in Memory of Earl H.Swanson, Jr., L.B.Harten, C.N.Warren, and D.R. Tuohy, (eds.), pp. 77-107. *Special Publication of the Idaho Museum of Natural History.* Pocatello.

Budinger, Fred E., Jr. (2004) Middle and Late Pleistocene Archaeology of the Manix Basin, San Bernardino County, California. In *New Perspectives on the First Americans*, B.T. Lepper and R. Bonnichsen, (eds.), pp.13-25. Center for the Study of the First Americans, Texas A&M University.

Campbell, E.W.C. et. al. (1937) The Archeology of Pleistocene Lake Mohave: A Symposium. *SW Museum Papers* 11.

Daugherty, Richard D. (1962) The Intermontane Western Tradition. *American Antiquity* 28:144-150.

Davis, L.G., S. C. Willis, S. J. Macfarlan. (2011) Lithic technology, cultural transmission, and the nature of the Far Western Paleoarchaic-Paleoindian Co-tradition. In *Meetings at the Margins: Prehistoric Cultural Interactions in the Intermountain West,* D.Rhode, (ed.), pp. 47-64. University of Utah Press.

Davis, Loren G. (2011) Return to the Cooper's Ferry site: studying cultural chronology, geoecology, and foragers in context. *Idaho Archaeologist* 34(1):1-5.

Davis, Loren G. (2011) The Paleocoastal concept reconsidered. In *Trekking the Shore: Changing Coastlines and the Antiquity of Coastal Settlement,* N. Bicho, J. Haws and L.G. Davis, (eds.), pp. 3-26. Springer Publishing Company, New York.

DeGeorgey, Alex (2004) A Single Component Paleo-Indian Site in the North Coast Ranges, California. *Proceedings of the Society for California Archaeology* 17: 35-39.

Dillehay, Tom D. & Cecilia Manosa (2004) *Monte Verde: Un Asentumiento del Pleistoceno en el Sur de Chile.* Santiago: LOM Ediciones.

Edgar, Blake (2007) Going Coastal: The Channel Islands may hold the key to New World prehistory. *Archaeology* 60 (3): 55-62.

Firestone, R.B. et. al. (2007) Evidence for an extraterrestrial impact event 12,900 years ago that led to megafaunal extinctions and the onset of Younger Dryas cooling. *Proceedings of the National Academy of Science* 104: 16016-16021.

Fladmark, K. R. (1978) The feasibility of the Northwest Coast as a migration route for early man. In Early Man in America from a Circum-Pacific perspective, A.L. Bryan, (ed.), pp. 119-128. *Occasional Papers No. 1 of the Department of Anthropology, University of Alberta* (Edmonton).

Giambastiani, Mark and T.F. Bullard (2009) Terminal Pleistocene—Early Holocene Occupations on the Eastern Shoreline of China Lake, California. *Pacific Coast Archaeological Society Quarterly* 43 (1&2): 50-70.

Gonzalez, Silvia and David Huddart (2008) The Late Pleistocene Human Occupation of Mexico. *Publico da Fundaco Museu do Homem Americano (FUMDHAM), Mentos* VII: 237-259.

Gruhn, Ruth (1961) The Archaeology of Wilson Butte Cave, south-central Idaho. *Occasional Papers of the Idaho State College Museum* 6.

Gruhn, Ruth (1994) The Pacific Coast route of initial entry: An overview. In *Method and Theory for Investigating the Peopling of the Americas,* Robson Bonnichsen and D.Gentry Steele, (eds.), pp. 249-256. Center for the Study of the First Americans, Oregon State University.

Hanes, Richard C. (1988) Lithic Assemblages of Dirty Shame Rockshelter: Changing Traditions in the Northern Intermontane. *University of Oregon Anthropological Papers* 60.

Hardaker, Christopher (2007) *The First Americans: The Suppressed Story of the People Who Discovered the New World.* New Page Books: Franklin Lakes, N.J.

Hardaker, Christopher (2009) Calico Redux: Artifacts or Geofacts? *Society for California Archaeology Proceedings* 22: 1-18.

Jenkins, Dennis et. al. (2004) Early and Middle Holocene Archaeology in the Northern Great Basin: Dynamic Natural and Cultural Ecologies. *University of Oregon Anthropological Papers* 62: 1-20.

Jenkins, Dennis L. et.al. (2012) Clovis Age Western Stemmed Projectile Points and Human Coprolites at the Paisley Caves. *Science* 337:223-228.

Krieger, Alex D. (1961) On Being Critical. *Kroeber Anthropological Society Papers* 25: 19-23.

Krieger, Alex D. (1964) Early Man in the New World, pp. 23-81. In *Prehistoric Man in the New World*, Jesse Jennings and Edward Norbeck, Editors. University of Chicago Press.

Mack, C.A., J.C. Chatters, A.M. Prentiss (2010) Archaeological Data Recovery at the Beech Creek Site (45LE415), Gifford Pinchot National Forest, Washington. U.S. Forest Service, Heritage Program, Trout Lake, WA.

Parker, John W., web-master. (2012) Lake County Archaeology: 20,000 Years of Cultural Change in the Clear Lake Basin. Parker & Associates Archaeological Research (www.wolfcreekarcheology.com). Lucerne, CA.

Phillips, F.M. et. al. (1991) Age and Geomorphic History of Meteor Crater, Arizona, from Cosmogenic CL-36 and C-14 in Rock Varnish. *Geochimica et Cosmochimica Acta* 55: 2695-2698.

Rice, David G. (1972) The Windust Phase in lower Snake River region prehistory. *Washington State University, Laboratory of Anthropology, Reports of Investigations* 50.

Rogers, Malcom J. (1939) Early Lithic Industries of the Lower Basin of the Colorado River and Adjacent Desert Areas. *San Diego Museum Papers* 3.

Smith, Julian (2011) First American Seafarers. *American Archaeology* 15(2): 26-31.

Tamayo, J. L. and R.C.West (1964) The Hydrography of Middle America. In *Handbook of Middle American Indians 1, Natural Environment and Early Cultures*: 85-121, Robert Wauchope, (general ed.) University of Texas Press: Austin.

Toner, Mike (2010) The Clovis Comet Controversy. A*merican Archaeology* 14(3):12-18.

Warren, Claude N. (1967) The San Dieguito complex: a review and hypothesis. *American Antiquity* 32(2): 168-185.

Warren, Claude N. (1980) Pinto points and problems in Mojave Desert archaeology. In Anthropological Papers in Memory of Earl H.Swanson, Jr., L.B.Harten, C.N.Warren, and D.R. Tuohy, (eds.), pp. 67-76. *Special Publication of the Idaho Museum of Natural History.* Pocatello.

Waters, M. R. et. al. (2011) Pre-Clovis mastodon hunting 13,800 years ago at the Manis site, Washington. *Science* 334: 351-353.

Waters, M. R. et. al. (2011) The Buttermilk Creek Complex and the Origins of Clovis at the Debra L. Friedkin Site, Texas. *Science* 331: 1599-1603.

Waters, M. R. and Stafford, T. W. (2007) Redefining the age of Clovis: Implications for the peopling of the Americas, *Science* 315: 1122-1126.

West, Robert C. (1964) Surface Configuration and Associated Geology of Middle America. In *Handbook of Middle American Indians 1, Natural Environment and Early Cultures*: 33-83, Robert Wauchope, (general ed.). University of Texas Press: Austin.

Willig, Judith and C. Mel Aikens (1988) The Clovis-Archaic Interface in Far Western North America. In *Early Human Occupation in Far Western Northern America: The Clovis-Archaic Interface,* J. A. Willig, C. M. Aikens, and J. L. Fagan, (eds). *Nevada State Museum Anthropological Papers* 21:1-40.

PALEOENVIRONMENTS AND PALEOINDIANS IN EASTERN SOUTH AMERICA

Astolfo Gomes de Mello Araujo[1]

In the last decades, the number of paleoenvironmental studies in Eastern South America, in what today is Brazilian territory, has grown considerably. The same can be said about archaeological data, whose exponential growth is due to management activities of both academic and cultural resources. This made a real integration between archaeological and paleoenvironmental data possible, allowing the understanding of regional occupation and abandonment scenarios, as well the construction of models of site preservation that take into account the response of geomorphic agents. In this chapter I will initially present the state of knowledge about paleonvironmental data for Eastern South America, showing its points of accordance and discrepancies. Based on this scenario, I will explore the contribution of paleoenvironmental data for the understanding of the Late Pleistocene and Early Holocene archaeological record of eastern South America, be it in terms of human occupation (expansion routes) or in terms of preservation and /or destruction of this same archaeological record.

PALEOENVIRONMENTS IN EASTERN SOUTH AMERICA

Many publications on Eastern South America paleoenvironments have been published in the last five years, but the most important development is perhaps the growing corpus of data regarding speleothem oxygen isotopic studies. There was also a general increase in the number of articles and dissertations whose main paleoenvironmental proxies are palinology, soil carbon isotopes, and speleothems, together with studies using sedimentary, geochemical, and soil phytolith data.

The characteristics of the different proxies in terms of resolution, sensitivity to climatic factors and the associated dating methods are sometimes responsible for incongruities that impart some dubiousness in the paleoenvironmental inferences. In such cases, archaeology

---

[1] Museum of Archaeology and Ethnology, University of São Paulo (MAE/USP). E-mail: astwolfo@usp.br

can help in the establishment of the more likely scenarios (Araujo et al.2005; Araujo et al. 2006).

Proxies, Convergences and Divergences.

When one tries to make sense of the paleoenvironmental scenarios depicted during the following sessions, the lack of correspondence that some of them present is sometimes striking. Such discrepancies can happen both inside the same proxy, or between different proxies, but are not uncommon in science, especially in climate-related sciences. An example of disagreement between different proxies happens in NE Brazil, where some authors see a dry event between 15,100 and 13,200, when speleothems stopped growing (Cruz et al. 2009a), while other authors see the opposite trend, of an increase in moisture based on pollen data (Ledru et al 2006). In SE Brazil the same happens, when the LGM is viewed as cold and wet based on speleothems (Cruz et al. 2006), but cold and dry when based on pollen (Behling 2002; Ledru et al. 2009). However, disagreement can occur inside the same proxy, and we can mention, for instance, the contention between palynologists about the meaning of poaceae (grasses) in the palynological record. Poaceae can spread both in dry conditions (forest regression, open landscapes, endurance to water stress) and in wet conditions (aquatic grasses, bamboos inside forested environments, or near lake shores – Racza 2009). Changes in the concentration of CO_2 in the atmosphere can also change the poaceae signal (Jolly and Haxeltine 1997).

Sensitivity of different proxies in relation to climatic factors and the dating methods involved are other possible sources of disagreement. In the case of speleothems, the analysis is done on stable isotopes, whose concentrations are extremely variable and very sensitive to environmental factors. Their growth can be traced to annual layers, and calcite can be dated by U-series, which allows precision on a decadal order (Cruz et al. 2009b). This means that speleothems can potentially be dated in time slices that are relevant in a human time-frame.

In the case of pollen, the accuracy of the environmental reconstructions is related not only to the analyst and his methods, but to the degree of sensitivity that different plants have in relation to climatic changes. Also, pollen samples are collected at intervals between 1 and 5 cm, which can represent centuries, depending on the rate of sedimentation. In addition, since lake sediments are dated by radiocarbon, it is not possible to reach a high degree of chronological resolution. Also, pollen is often collected in wet, marshy environments, and therefore the floral signature of a savannah, for instance, can be blurred by the presence of gallery forest (Berrio et al. 2000).

The analysis of carbon isotopes in soils, incorporated by means of plant decay, is another important climatic proxy, much used in Brazil. Plant metabolism can be divided into C3 or C4 paths, C3 being more common in trees, and C4 more common in grasses (Pessenda et al. 2001). Hence, a strong C3 signal in soil organic matter would mean a former forested environment, whereas a C4 signal would mean a grassy, open environment. This approach has some problems. First, it is not possible to know which plants (taxa) are responsible for the isotopic signal, although this shortcoming can be circumvented by coupled soil phytolith analysis. Second, there is some evidence that CO_2 atmospheric concentrations can favor C4 plants over C3 ones, which means that an increase in grasses could be driven by low CO_2 concentrations, and not necessarily by dry climates. Third, radiocarbon ages obtained in soils are always more problematic due to biological action (in the case of charcoal) and organic matter turnover (mixture of old and new organic matter). Of course, this problem can be minimized by OSL dating. It is worth noting also that even the same dating method can give different results. Radiocarbon ages obtained in charcoal fragments are different from those obtained in bulk organic matter, be it in soil or in peat. In the first case, what is being dated is a single individual, a tree or a branch. In the second case, what is being dated is a pool of organic matter coming from several different individuals that happened to be deposited in the same place. This means that bulk organic matter is always a mixture of carbon from different ages. For instance, Pessenda et al. (2001) compared radiocarbon ages from soil profiles at eight localities in Brazil, and showed that ages obtained from total soil organic matter (SOM) were significantly younger than the humin fraction (considered to be the oldest organic component of the soil) or than ages obtained from charcoal. Differences in ages between total SOM and humin obtained in the Brazilian samples ranged between 13 % and 209 %. In general, charcoal ages were similar and/or older than the humin ages. This means that a radiocarbon age obtained at a lake sediment can be considerably older than it appears.

By pointing to these facts, we are not saying that one method is "better" than another, but that paleoenvironmental evidence, as any other scientific data, has to be approached with caution.

The Last Glacial Maximum (23,000 to 19,000 cal yrs BP)

The main climatic events that are of interest for our discussion occurred since the global decrease in Earth's temperature called the Last Glacial Maximum (LGM). During this period, between 23,000 and 19,000 cal BP[2] (Mix et al. 2001), Earth's mean temperature was 5°C lower than today, CO_2 concentrations half those of today (Mayle et al. 2009) and the imprisonment of water in glaciers and icesheets was responsible for the removal of 3% of the ocean's volume (Wright 2009), causing a 120 m drop in the sea level around 18,000 cal yrs BP (Clark and Mix 2002). The factors that led to the LGM are still under debate, being probably a confluence of Earth's orbital parameters and complex climatic feedback mechanisms.

The LGM has been deeply studied since the 1970s, starting with the CLIMAP Project (Climate Long-range Investigation Mapping and Prediction – CLIMAP 1981), and countless papers on the theme have been published since then. The more consensual aspects point to a cooler and, in most parts, drier climate in South America. However, several issues still need to be clarified, among them the role that the LGM conditions played in the tropics. The traditional view of a "cold-dry" LGM in tropical settings, based on computational simulations, is changing rapidly as more data is amassed, but the scenario seems not only more complex than previously thought, but also less consensual.

The Late Glacial (19,000 to 11,500 cal yrs BP)

The period between the LGM and the beginning of the Holocene is well known in the Northern Hemisphere for short-term, drastic climate changes, such as the cold spell called the Heinrich Event 1 (H1), around 16,500 cal BP, that was followed by a major increase in temperature between 14,500 cal BP and 13,000 cal BP, called Bolling-Allerod interstadial,

---

[2] I will use the notation "cal BP" for calibrated radiocarbon ages, and only "BP" for radiocarbon ages too old to be calibrated, or other methods that do not need calibration (e.g. OSL, TL, U-Th).

and then going back again to a short and extremely cold period between 12,600 cal BP and 11,800 cal BP, known as the Younger Dryas.

In the Southern Hemisphere, however, the signals left by these extreme climatic changes are somewhat dubious. Perhaps only the Younger Dryas can be recognized, especially in marginal environments, such as the Andean Region (Baker et al. 2001) or Patagonia (Hadjas et al. 2003). However, Ledru et al. (2002) and Jacob et al. (2007) suggest that the Younger Dryas can also be recognized in Caçó Lake, Northern Brazil. New data available from speleothem oxygen isotopes suggest that climatic effects related to the H1 event and the Younger Dryas can also be observed in NE and SE Brazil (Cruz Jr. et al. 2005, 2006, 2009a,b).

The Holocene (11,500 cal yrs BP to present)

The Holocene marks the end of the globally cold and dryer conditions that reigned since 130,000 years ago when the last glacial cycle was established. Oxygen isotope data from Arctic and Antarctic ice cores show clearly that the Holocene is a period of extreme mild climatic conditions when compared to the Pleistocene (Watanabe et al. 2003; North Greenland Ice Core Project Members 2004). Burroughs (2005) even dubbed the Holocene "the end of the reign of chaos", with good reason. Temperatures rose about 20°C since the LGM, and climate fluctuations were much less severe. In fact, there is strong evidence that large-scale agriculture would not be possible at all in full and late-glacial conditions (Richerson et al. 2001). In the mid-Holocene, between 8300 and 5000 BP, there was a period of global increase in temperature called "hypsithermal", or "climatic optimum" (Fairbridge 2009). This latter term is erroneous, since the climate did not become "optimal" globally, i.e., in some places the increase in temperature was accompanied by a decrease in precipitation, leading to the mid-Holocene dry climatic events recognized in many portions of the world.

Once agriculture spreads and population booms (Diamond 1987), the effects of minor climatic fluctuations, even in Holocenic overall mild conditions, can be extremely deleterious (eg. deMenocal 2001). Hence, the Holocene epoch, its climatic fluctuations, and their impact on humans is a topic of major importance.

Holocene climatic conditions for Eastern South America were discussed at length elsewhere (Araujo et al. 2005, 2006), and here I will provide a panorama of the main issues.

Summing Up Eastern South American Paleoclimates

For the sake of our discussion, it is not only important to address the paleoenvironments in different time frames, but also to pay attention to signals of climatic disruptions, or rapid climate changes ("RCCs", following Mayewski et al. 2004) along the records. Several instances of disagreement between authors and proxies can be solved if we take into account that different proxies respond differently, that local conditions can influence local deposition modes and local vegetation, and also that some sites are very distant from their neighbors. Figure 1 shows the locations where paleoenvironmental data was gathered. However, if we want to have an overall picture of the paleoenvironments of Eastern South America since the LGM, a possible one would be:

1) Amazon Basin: this huge area, bigger than Europe, is subsumed under a single label more through our lack of knowledge rather than due to any unity, be it geomorphologic or biotic. In any case, there is consensus about the drop in temperature during the LGM, some consensus about a decrease in precipitation, and no consensus as to the degree of forest fragmentation. Figure 2 shows a comparison between different sites and authors for the Amazon basin (data based on Absy et al. 1991; Behling 2001; Carneiro Filho et al. 2002; Colinvaux et al. 1996; Colinvaux et al. 2000; Freitas et al. 2001; Haffer & Prance 2001; Hooghiemstra & van der Hammen 1998; Jacob et al. 2004; Latrubesse & Kalicki 2002; Sarges et al. 2009; Siffedine et al. 2003), where one can perceive that a hiatus in sedimentation in Carajás, a time of eolian dune reactivation at Rio Negro, and low lake levels at Six Lakes converge in some way. The same can be said for the Late Glacial, and it is important to note that Carajás is 1500 km from Rio Negro and 2000 km from Six Lakes. For more recent times, another emerging pattern is the existence of a mid-Holocene drought (Bush et al. 2004; Bush et al. 2007).

2) Northeastern Brazil: Auler and Smart (2001) point to the importance of regional differences in climate that can be superimposed over general atmospheric circulation patterns. Nimer (1989) observes that today the rain patterns in NE Brazil

are extremely conditioned by orography. The distribution of the dry season is very variable, with mountains being responsible for the abbreviation of the dry season, and plains responsible for its extension. Even so, there is a relatively good agreement between different proxies (Figure 3; based on Auler et al. 2006; Cruz et al. 2005; Cruz et al. 2006; Czaplewski & Cartelle 1998; Tsoar et al. 2009;), signaling a wet mid-Holocene in contrast with increasing aridity in the late Holocene. The LGM and Late Glacial records are more variable, but it is expected that the Icatu and Bahia Central records, very inland, would be different from Caçó, near the coast and inside the fluctuation zone of the ITCZ (Inter-Tropical Convergence Zone), a low-pressure belt characterized by abundant precipitation and turbulence.

3) Central Brazil: For central Brazil some authors believe, based on pollen records, that a very dry Last Glacial Maximum and Late Glacial was followed by an increase in moisture beginning ca. 7000 cal BP, leading to modern conditions (Barberi *et al.* 2000; Salgado-Laboriau *et al.* 1997). In the same region, however, other researchers found evidence of a somewhat different scenario, with a drier and cooler LGM followed by a cool and moist Late Glacial between ca.12,000 and 9500 cal BP, followed again by an arid period between ca. 9500 and 4500 cal BP, and a trend towards modern conditions since then (Behling 2002; Ledru 1993; Ledru et al. 1996). Possible reasons for these differences were discussed elsewhere (Araujo et al. 2005), but Figure 4 shows that there are some points in common (based also on Ferraz-Vicentini & Salgado-Laboriau 1996; Ledru et al. 1998; Parizzi et al. 1998), such as the climatic instability that can be perceived in the mid-Holocene. Some authors tend to differentiate this Central Brazil into two climatic zones: the "core" savanna area would show the trend of increasing moisture during the Holocene, whereas the "peripheric" area, towards the east and the south, would show more oscillations in moisture, probably due to incursions of Antartic polar fronts (Barberi 2001:145) and periods of aridity in the Late Holocene due to increasing ENSO activity or to a northward shift of the ITCZ (De Oliveira *et al.* 1999:335).

4) Southeastern Brazil: most paleoenvironmental data for SE Brazil comes from the Serra do Mar mountain range, and we simply do not know how useful these data are from an archaeological standpoint, since the mountain heights were probably drier and colder than the valley bottoms. The few studies in the lowlands (Turcq et al. 1997; Behling et al. 1998) suggest a major period of instability, marked by strong fluvial erosion (Figure 5; data based also on Behling et al. 2007; Behling & Lichte 1997; Behling & Safford 2010; Coelho Netto 1999; Coelho Netto & Fernandes 1990; Cruz et al. 2009a; Cruz et al. 2009b; Ledru et al. 2005; Modenesi-Gautieri 2000; Moura & Mello 1991; Pessenda et al. 2009; Saia et al. 2008; Wang et al. 2006; Wang et al. 2007). Stevaux (2000) found evidence of a dry LGM in the Paraná River, further inland. As already mentioned, the main contention nowadays is over the precipitation at the LGM (Cruz et al. 2006; Ledru et al. 2009).

5) Southern Brazil: The southern states have the best paleoenvironmental coverage, and the scenario seems to be well understood (Figure 6; data based on Behling 1997, 1998, 2007; Behling et al. 2001; Behling & Negrelle 2001; Behling et al. 2004; Behling et al. 2005). It is somewhat consensual that the LGM was dry and cold, with a dominance of *campos* or grassy savannah, subject to a gradual change towards warmer and moister climates throughout the Late Glacial and Holocene (Behling 2002).

The Impact of Climate Change in the Landscape: A Tropical Regard

Tropical and sub-tropical areas are subject to geomorphic processes that are intimately linked to heavy rainfalls, high temperature and strong biological activity. These processes are responsible for the formation of deep soils and thick weathering mantles that were never subject to scouring by glacial processes. The understanding and recording of tropical geomorphic processes is a relatively new addition to the general field of geomorphology, since most theories and studies were related to temperate or desert settings (Thomas 1994). Important contributions to the formation of tropical landscapes in a evolutionary framework came from the work of Henri Erhart (1956), who proposed the "biorhesistatic theory", in which a given landscape with stable vegetal cover (or in "biostasy") will be subject mainly to chemical leaching, releasing soluble minerals and retaining insoluble ones. When there is a disruption in climate, however, the biological equilibrium is lost ("rhesistasy"), vegetation retracts and the clastic and clayey insoluble component that was stored under it is released, causing valley infilling and deposition of several meters of sediment in short intervals. This idea was applied by Ab´Sáber (1969) and Bigarella (Bigarella et al. 1965) in Brazilian colluvial deposits, whose rudaceous pavements or "stone lines" were considered testimonies of dry climates, or correlative deposits, since they could be theoretically linked to specific environmental conditions. Another important contribution for tropical geomorphology came from Bigarella and Mousinho (Bigarella et al. 1965) who defined the concept of *rampa complexes*, or anphiteatre-like hollows filled by both colluvial and alluvial material, showing epidodes of aggradation and degradation that were probably linked to climatic factors. The definition of *rampa* complexes allowed a more dynamic and spatially heterogeneous view of tropical soil processes, since it presupposes that erosion and sedimentation occur simultaneously over different sectors of the slope, in different directions and at different rates, converging to the longitudinal axis, or hollow (Moura and Silva 2003).

A parallel development of Erhart´s biorhesistatic theory was proposed by Knox (1972), who devised a model of biogeomorphic response to abrupt climate changes. In spite of being originally applied to eastern United States, the model was used by Roberts and Barker (1993) in tropical Africa, and by Thomas (2008; Thomas and Thorp 1995) in interpreting the signal of climatic changes in several tropical settings. The model predicts that an increase in precipitation causes an increase in relative vegetation cover, and therefore a decrease in the potential for hillslope erosion, and vice-versa (Figure 7). However, the transitions from humid to dry and from dry to humid are asymetrical in relation to a fourth variable, namely the relative geomorphic work, or sediment yield. In normal conditions, the sediment yield in vegetated areas is lower than in semi-arid areas, due to the lack of soil protection against torrential rains in the latter. The transition from a dry to humid period will produce a peak in sediment yield, because the soil is unprotected, and rainfall would be high. This situation will last until the vegetation adjusts to the new conditions. The opposite situation, from humid to dry, tends to produce a depression in the sediment yield, since the soil is covered, and the precipitation is low. Again, this situation will last until the new vegetation (or lack thereof) prevails, and the normally high sediment yield of semi-arid settings is established.

To better grasp the actual implications of these models, we can provide some empirical evidence. In the humid highlands of Southeastern Brazil, in a broad stripe parallel to the coast that incorporates the Serra do Mar mountain range and adjacent areas in the Doce and Paraíba do Sul river basins, several studies coupling pedology and geomorphology were undertaken by Meis, Moura and colleagues (Meis & Moura 1984; Moura & Mello 1991) using the concept of allostratigraphy, that involves stratigraphic interpretation, correlation, and mapping which uses discontinuities and surfaces to subdivide the sedimentary section. These discontinuities and surfaces are assumed to have time-stratigraphic significance (NACSN 1983), being related to episodes of environmental instability (Moura 2003). In the Bananal region, between São Paulo and Rio de Janeiro, Moura and Mello (1991) recognized nine alloformations placed on the slopes and valley bottoms. The obtained chronology brackets the deposits in a late Pleistocene – Holocene time slice. The basal alloformation, Santa Vitoria, is a reddish colluvium formed directly over the basal rock, probably of late Pleistocene age. Above it, Rio do Bananal alloformation is a yellowish colluvial deposit, more than 8 m thick, ill-sorted and with unstructured clastic material, suggesting climatic instability, with episodes of strong slope erosion filling the valley bottoms. Its upper portion exhibits an A horizon suggesting a stable surface at 11,300 cal BP. Above Rio do Bananal, the authors defined another seven alloformations. Of interest here is alloformation Manso, a sedimentary sequence encompassing interfinged fluvial and slope deposits subdivided in three facies, probably associated with another major period of instability, when slope erosion reached the rock basement in some places, causing an extreme valley infill and drowning drainages that can be seen today. These deposits can reach thicknesses of 10 m or more. Figure 8 (modified from Moura et al. 1991) exemplifies the complex spatial relationship between the allostratigraphic subsurface deposits and present relief, the different paleotopographies and, more importantly, the difficulty in reaching older surfaces. In fact, later work in the area (Coelho Netto 1999) showed that Pleistocene sediments were almost totally eroded. They appear in isolated spots on some slope terraces, and never in the valley bottoms, washed away by a regional episode of erosion-deposition that filled the landscape between 11,000 and 9,600 cal BP.

Another example of climatic disruption was described by Modenesi-Gauttieri (2000) on the top of the Serra da Mantiqueira range, where at least three episodes of colluvial accretion occurred since the LGM, ages bracketed between 22,000 - 17,300, 17,300 - 11,500, and 11,500 - 8,300 cal BP.

HUMANS ENTER THE STAGE

After our discussion about paleoenvironments and their implications on the landscape, it is time to put humans in this scenario. However, an initial question has to be posed: since when? To ask such a question is somewhat rhetorical because we will probably never have a correct answer, but we can start by at least dismissing some incorrect ones.

The detection of sites that are older than the Clovis horizon as defined in North America, or sites that are slightly newer than Clovis but totally unrelated from a cultural point of view, suggest that the "Clovis First" model is inadequate to explain the South American empirical evidence. What remains to be clarified is the timing and routes of expansion of the first human inhabitants of the continent, a theme that has been addressed by a great number of researchers, and will not be discussed here (e.g. Anderson & Gillam 2000; Dillehay 2000; Dixon 2001; Fladmark 1979). My contribution starts with a somewhat misty scenario, grounded in admittedly scant evidence, but that at the same time cannot be entirely dismissed: the possibility of a pre-LGM human arrival in the Americas. Besides the Pedra Furada evidence (Parenti 2001), there are other sites that should be considered, such as Sitio do Meio, with ages older than 24,200 cal BP (Felice 2002; Santos et al. 2003), Santa Elina, with ages up to 23,000 BP (Vilhena-Vialou 2003, 2011), and Monte Verde, with an upper occupation of 14,000 BP and a possible lower occupation of 33,000 BP (Dillehay & Collins 1988), not to mention a site in Colombia where obsidian and chert flakes were found underneath Pleistocene volcanic ash, together with wood dated at 19,500 cal BP (Cooke 1998). But why do we have to take this scant evidence into consideration? The simple answer is that because it explains better what happened later. Otherwise, it is very difficult to explain that by 12,000 BP South America was suddenly and completely filled by totally different cultural groups, with distinct lithic technologies (Araujo & Pugliese 2009; Bate 1990; Dillehay 2000), showing adaptations to the most extreme environments, from Amazonian rainforest (Roosevelt et al. 1996) to the Patagonian steppes (Borrero et al.

1998), all this a mere 1,250 calendar years after the earliest Clovis sites, or 800 calendar years after the latest Clovis sites (Waters & Stafford 2007). So, as an exercise, let us consider for a while that this occupation happened in pre-LGM times, allowing sufficient room for the expansion and cultural differentiation of human groups until 12,000 BP. Continuing our exercise, let us make a cursive examination of the radiocarbon age distribution for South American sites older than 10,000 14C BP, or 11,500 cal BP. Figure 9a shows a plot of 202 radiocarbon ages obtained at 59 archaeological sites in seven countries (Argentina, Brazil, Chile, Colombia, Ecuador, Peru, and Venezuela). Even if we dismiss one or other site, the pattern suggests a continuum of ages that goes up to 14,700 14C BP, or 18,000 cal BP, extending back to the end of the LGM. When we reach the LGM, coincidence or not, the ages become scattered. Is this because the ages are wrong, the sites are fake, or the archaeological signal is weak?

If we close the focus in Eastern South America, our sample is comprised of 81 radiocarbon ages coming from 37 sites in Eastern South America (shown in Figure 10). Figure 9b shows a plot of the calibrated ages, where it is equally possible to perceive the "break" around the LGM. Even more intriguing is the pattern we can observe when the calibrated ages are lumped into 1000 calendar years intervals (Figure 9c). Instead of a progressive increase in the number of ages, as we would suppose if only the vagaries of sampling and the decay of charcoal were in action, it is possible to perceive a peak of ages around 17,000 BP. Finally, in Figure 9d the calibrated ages were lumped in three equal-time, 6000 years slices: Late Glacial (12 to 18 ka; n = 34); LGM (18 to 23 ka; n = 4); Pre-LGM (24 to 30 ka; n = 6). Again, the results suggest a decrease in ages during the LGM. In this case, a question remains: what happened between these early sites and the "explosion" of archaeological sites all over South America by 12,000 BP? The first and easy answer is to say that "what happened" was the LGM. But we can go further and think about (at least) three possibilities: one is related to the number of objects left by these populations, assuming that they were present in the landscape during the whole period; the second is related to the impact that extreme climatic events such the LGM could have on the structure of small hunter-gatherer bands, causing local extinctions, lowered reproduction rates as a response to a much lower environmental carrying capacity, or population displacement towards coastal areas, that happen to be submerged by now; and the third is related the role of natural processes over the archaeological signal.

In terms of the *archaeological signal*, it is almost a truism to say that the detection of physical (archaeological) evidence of any given human behavior is always delayed in relation to the actual onset of such behavior, by a simple matter of probability. The probability of finding a site is totally correlated with the number of sites, and the same can be said for individual artifacts. Hence, the archaeological signal, here defined as the quantity of material evidence of human behavior accumulated in a given segment of the landscape, is a function of net artifact discard. If we assume that in pre-capitalist societies artifact discard is correlated to population size, low population means low artifact production and discard, which means a low archaeological signal. However, it is difficult to separate low archaeological signal from actual absence of humans, and this can be the case if small hunter-gatherer bands become too separated, or population numbers decreased in response to rapid climate changes. Given the already mentioned cultural variability that emerged in South America around 12,000 cal BP, the most parcimonious scenario would not involve the total extinction of the first settlers, but the survival of the early human groups during the LGM and Late Glacial, perhaps with a low archaeological signal. However, if this is the case, why does the archaeological signal not increase exponentially with time, instead of presenting an "LGM depression"? A possible explanation has to do with the impact that geomorphic processes can impart in the archaeological record.

The geomorphic models presented in the preceding session enable us to better explore the role of abrupt climatic changes in the archaeological record, summed up in Figure 11. If we have a cold and humid LGM, for instance (box 1), and climate changes towards drier and warmer conditions, there will be a floristic change without a significant change in the relative vegetation cover. Hence, the rates of erosion and sedimentation (sediment yield) will be low (box 2), which increases the chances of preservation of any pre-LGM archaeological site. If the same system undergoes a dry period during the mid-Holocene, it will probably suffer a major disruption (box 3), potentially causing the erosion and reworking of any site older than 5000 BP. If the mid-Holocene is moist, on the other hand, the system will move towards a maintenance of the low sediment yield (box 6). In this case, preservation of sites is optimal. On the other hand, if we have a cold and dry LGM (box 4), if climate changes rapidly towards humid and warmer, there will be a lag until the vegetation colonizes the environment, the soil will be unprotected, a high erosion and deposition rate will occur (box 5), and this will decrease in the probability of pre-LGM site preservation. Again, during the mid-Holocene, this system can undergo either dry or moist conditions. Of course, this same reasoning can be applied to any abrupt climate change along the period of interest, or the last 23,000 years. If we go back to the environmental data presented earlier, we can recognize several periods where proxies suggest strong and rapid climate oscillations in several parts of Eastern South America, leading to "rhesistatic" conditions; for instance, to cite a few: the transition from a fully developed rainforest in pre-LGM times towards a major forest reduction between 23,000 and 12,000 cal BP, interrupted by a peak of moisture around 17,000 cal BP at Colonia (Ledru et al. 2009); the sedimentation hiatuses that are apparent in several LGM records across Central and SE Brazil (Ledru et al. 1998); the widespread slope erosion and valley infill events present at the Tamanduá river between 20,000 and 12,100 cal BP (Turcq et al. 1997), and at Bananal between 11,000 and 9,600 cal BP (Coelho Netto 1999), both affecting the LGM deposits; the three episodes of colluvial accretion detected by Modenesi-Gauttieri (2000) between 22,000 - 17,300, 17,300 - 11,500, and 11,500 - 8,300 cal BP; a episode of strong slope denudation detected by Melo et al. (2003) at Ponta Grossa, Southern Brazil, at 19,000 cal BP; the strong climatic oscillations detected between 17,300 and 16,400 BP on speleothems at Rio Grande do Norte (Cruz et al. 2009a); the sand layer inside the organic clay at Caçó Lake between ca. 18,600 cal BP and 16,300 cal BP (Ledru et al. 2002), and so on. When we enter the Holocene, these rapid climate changes (RCC) do not stop. Mayewski et al. (2004), based on more than 50 records spread worldwide, recognized at least six RCC events placed at ca. 9000 to 8000 cal BP, 6000 to 5000 cal BP, 4200 to 3800 cal BP, 3500 to 2500 cal BP, 1200 to 1000 cal BP, and since 600 cal BP. The signature and strength of each event is different around the globe, but they can impart considerable changes in the landscape. At least in some portions of Brazil, the mid-Holocene dryness (Araujo et al. 2005; Behling 2002; Bush et al. 2007; Ledru et al. 1996; Ledru et al. 2009) or climatic instability (Meggers 2007; Racza 2009) is becoming more evident.

The Impact of Climate Change on Humans: A Central Brazilian Example

After exploring the consequences of climate changes in the archaeological record, let us move our attention to the consequences that climate imparted on humans. The subject is vast, and was treated by several authors in different parts of the world, with different degrees of detail, sometimes carrying an aura of determinism and heated debate (Diamond 2005; McAnany & Yoffee 2010). Nevertheless, my personal position is that climate should always be taken into consideration when we start thinking about anything archaeological, not because humans are driven by climate in a deterministic way, but simply because climate-based explanations and their subsequent expectations are a good starting point. And, of course, if humans are not deterministically driven by climate, the artifacts they leave on the ground are very much so, as we have just seen.

Five years ago, we noted that there was a depression in the archaeological signal for the Lagoa Santa region during the mid-Holocene (Araujo et al. 2005, 2006). Radiocarbon ages obtained by our project[3] and by other authors showed a remarkable behavior when plotted in 500 year intervals: they showed a frequency peak between 10,300 and 7,000 cal BP, and another peak between 1960 and 900 cal BP, without any ages in between, which would constitute a major gap, of about 5,000 years, in human occupation. We then expanded our dataset and perceived that the same pattern was present in several other states, encompassing an estimated area of 920,000 km² in Central Brazil. On the other hand, the northeastern, southeastern and southern portions of the country did not show the same pattern. This lead us to seek paleoenvironmental literature and, not surprisingly, there was good evidence of dry periods in Central Brazil in several records, which led us to propose that climatic deterioration (dryness) drove the Lagoa Santa Paleoindians away from the area. After this, we started to look for Paleoindian sites in open-air settings, to be sure that our depleted mid-Holocene archaeological signal was not an artifact of rockshelter excavations. We also sent many more samples from the rockshelters themselves, and started a collaborative work with palynologists in order to refine the Lagoa Santa paleoenvironmental data.

---

[3] Headed by Walter A. Neves, from University of São Paulo, and funded by FAPESP, grants no. 99/00670-7 and 04/01321-6.

Starting with the lithic open-air sites, we found at least three of them, named Sumidouro, Coqueirinho, and Lund, and the evidence gathered suggests that their ages cluster into two intervals: 12,300 to 9,300 cal BP for Sumidouro and Coqueirinho, and 2,200 cal BP for Lund. For Sumidouro we also obtained OSL ages for the Paleoindian levels between 12,500 and 9900 BP (Araujo and Feathers 2008). Hence, still no evidence of a mid-Holocene occupation in open-air settings. This view changed when we sent more samples from the three rockshelters we excavated, named Lapa das Boleiras (32 radiocarbon ages), Lapa do Santo (63 radiocarbon ages), and Lapa Grande de Taquaraçu (13 radiocarbon ages). We perceived that there was indeed a very rapid reoccupation in the mid-Holocene, bracketed between 5,300 and 4,000 cal BP, or during approximately 1300 calendar years. Thus, instead of a large interval of 5,000 years without any archaeological signal, we have two hiatuses, one between 7,000 and 5,300 cal BP, and the other between 4,000 and 2,000 cal BP. Judging from the similarity in terms of lithics, the same human population came back at 5,300 cal BP. In fact, if it were not for the radiocarbon ages, we would never say that there was a hiatus of 1,700 years in the stratigraphy.

Finally, pollen data gathered in two lakes, Lagoa Olhos d'Água and Lagoa dos Mares by De Oliveira and Racza (Racza 2009) did not show any clear signal of dryness, and the lakes seemed to be perennial, without hiatuses in sedimentation. After a cool and humid LGM, vegetation was gradually substituted by tropical taxa, which suggests an increase in temperature since the beginning of the Holocene. However, the authors detected a strong oscillation of several taxa, suggesting an instable climate during the mid-Holocene, with strong rainfall periods followed by long dry periods in an erratic manner. Hence, the area abandonment could be triggered not by a pervasive dryness, but by a very inconstant and, therefore, unreliable environment. It is worth noting that Lagoa Santa Paleoindians relied strongly on vegetal foodstuff, as suggested by the high incidence of dental caries, related to carbohydrate consumption (Neves & Cornero 1997), and also by the botanical remains found at the rockshelters (Nakamura et al. 2010). Hence, the landscape floral composition would be as important as the game. I believe that the brief (in geological terms) human reoccupation of 1300 years is linked to a short period of climate stability. It is possible that these periods of stability and instability, that are obviously perceived by humans, do not always leave a clear signal in the lake sediments. Soils and geomorphic features can help in illuminating this issue.

It is important to note that archaeological gaps are present not only at Lagoa Santa but in other portions of Brazil, such as Amazonia (Neves 2007), as well in several parts of South America, such as the Pampas (Araujo et al. 2006), NW Argentina (Neme & Gil 2009) and Chile (Nuñez et al. 2001). These gaps are related to a *low archaeological signal*. It does not mean that people did not foray in those areas: it means that they placed the focus of their occupation elsewhere. Another important consequence of the Lagoa Santa data is that it showed the abandonment of an area that is far from being a marginal environment, such as deserts or high altitude settings.

Be that as it may, Lagoa Santa provides a very good example of change in the focal occupation of Paleoindians, who moved elsewhere, while retaining the region inside their territory. The clear continuity in the lithic industry after 1,700 years of virtual absence is a good clue, and suggests that the Lagoa Santa Paleoindians retained their cultural characteristics well beyond the early Holocene.

Where do We Go from Here? Towards Geo-Informed Approaches

To speak about the importance of what we call nowadays "geoarchaeology" is perhaps unnecessary. In fact, geoarchaeology is so ingrained in archaeology that it should be regarded not as a specialty, but simply as "well done archaeology", both from the theoretical and methodological points of view (Araujo 1999). In order to approach a given site, a given region or problem, we ought to have a geoarchaeological theoretical framework operating in our minds (Butzer 1980), because no matter what we do, from Paleoindian to Historical, we have to face that most contemporary, relevant questions can only be answered by means of materials, techniques and machines that are outside "business as usual".

One interesting aspect about Paleoindian archaeology is that it has to rely even more on geoarchaeology, not only because the sites generally have only stones, but also because they have been part of the lithosphere for a long time. The necessity of using a geoarchaeological approach in dealing with Paleoindian sites was acknowledged since the 1930's (Holliday 2009). However, it is even more pressing that we use a geoarchaeological-driven rationale for survey, or what I will call a *geo-informed approach*. Very few Paleoindian sites in the Americas were found by means of a geo-informed approach, or *survey procedures explicitly based on geomorphic / pedological / geological approaches in order to find Paleoindian sites* (but see Stafford 1995 and Mandel 2008).

In the preceding sections I have tried to show that it is necessary to have some clues about what kind of paleoenvironment reigned in a given area, and also which geomorphic processes should be taken into consideration, both in the past and today. I also tried to show some aspects of the relationship among paleoenvironments, humans and the materials they left on the landscape, but we should try to go beyond the recognition of shortcomings and necessities, and adopt a more proactive stance.

One of the major problems we face in Paleoindian archaeology is to solve the question of the late Pleistocene sites. They are around, and even when dismissed by a group of scholars, they are still around. The gathering of scholars in order to verify the validity of a given site does not work as it used to in the XIX century (Borrero 1995). A good example is Pedra Furada, dismissed in the mid-1990s (Meltzer et al. 1994), but fully published only later by Parenti (2001), and now undergoing support by French reasearchers (e.g. Fogaça and Boeda 2006). Another, perhaps better, example is Meadowcroft Rockshelter, in the Eastern United States. As the excavation coordinator stated, "there is no prospect for an end" to the controversy (Adovasio 1999). We can sit and wait until the day these sites will be "solved" (which will never happen) or, alternatively, we can develop a geo-informed approach and go after the places and methods suited to our needs. Our needs are presently related to increase the archaeological signal and, therefore, the sample size of possible Pleistocene sites. Only a *pattern* of sites can solve the problem, inside a "siteless survey" (Dunnell & Dancey 1983) framework. Most Paleoindian sites in the Americas were found by chance, either because they were being eroded and contained easily recognizable artifacts, or because they were underneath some younger deposit, and somebody dug a bit deeper.

Perhaps it is unnecessary to say that there is no recipe, and that each team has to find the best ways to seek Paleoindian sites.

Final Remarks

My objective was to show some of the implications that paleoenvironments could have imparted on humans and in the materials they left. As archaeologists, we can only hope to address the former issue by means of addressing the later.

Many questions regarding the peopling of the Americas are within this inquiry. There is no prospect for an end, as Adovasio (1999) said, unless we approach the "problem" of the old sites in South America from a regional perspective, less centered on sites and diagnostic artifacts, and more centered (or non-centered) on lithic scatters over the landscape. Given our present state of knowledge, nobody is really wrong about the timing of the peopling of the Americas, and conversely nobody is really right. As scientists, we should try to make our data collection an enterprise that is greater than our prejudices.

REFERENCES

Ab´Sáber, A.N. 1969. Um conceito de geomorfologia a serviço das pesquisas do Quaternário. Geomorfologia 18, University of São Paulo.

Absy, M.L.; Cleef, A.; Fournier, M. Martin, L.; Servant, M.; Siffedine, A.; Ferreira da Silva, M.; Soubies, F.; Suguio, K., Turcq, B.; van der Hammen, T. 1991. Mise en évidence de quatre phases d'ouverture de la forêt dense dans le sud-est de l'Amazonie au cours des 60,000 dernières années. Première comparaison avec d'autres régions tropicales. Comptes Rendus d'Academie des Sciences, Paris, Serie II, 312: 673-678.

Adovasio. J. 1999. No vestige of a beginning nor prospect for an end: Two decades of debate on Meadowcroft Rockshelter. In Bonnichsen, R.; Turmire, K.L. (eds) Ice Age Peoples of North America: Environments, Origins, and Adaptations of the First Americans. Oregon State University Press, Corvallis, pp. 416–31.

Anderson, D.G.; Gillam, C.J. 2000. Paleoindian colonization of the Americas: Implications from an examination of physiography, demography, and artifact distribution. American Antiquity 65:43-66.

Araujo, A.G.M. 1999. As geociências e suas implicações em teoria e métodos arqueológicos. Revista do Museu de Arqueologia e Etnologia, Anais da I Reunião Internacional de Teoria Arqueológica na America do Sul. Suplemento 3: 35-45.

Araujo, A.G.M.; Feathers, J.K. 2008. First notice of open-air Paleoamerican sites at Lagoa Santa: Some geomorphological and paleoenvironmental aspects, and implications for future research. Current Research in the Pleistocene 25: 27-29.

Araujo. A.G.M.; Neves, W.A.; Piló, L.B.; Atui, J.P. 2005. Holocene dryness and human occupation in Brazil during the "Archaic Gap". Quaternary Research, 64: 298-307.

Araujo. A.G.M.; Neves, W.A.; Piló, L.B.; Atui, J.P. 2006. Human occupation and paleoenvironments in South America: Expanding the notion of an 'Archaic Gap'. Revista do Museu de Arqueologia e Etnologia, 15/16: 3-35.

Araujo, A.G.M.; Pugliese, F. 2009. The use of non-flint raw materials by Paleoindians in Eastern South America: A Brazilian perspective. In: F. Sternke, L. Eigeland and L-J. Costa (eds) Non-Flint Raw Material Use in Prehistory: Old prejudices and New Directions. BAR Series 1939, Oxbow, Oxford, pp 169-175.

Auler, A.S.; Piló, L.B.; Smart, P.L.; Wang, X.; Hoffmann, D.; Richards, D.A.; Edwards, R.L.; Neves, W.A.; Cheng, H. 2006. U-series dating and taphonomy of Quaternary vertebrates from Brazilian Caves. Palaeogeography, Palaeoclimatology, Palaeoecology 240: 508–522.

Auler, A.S.; Smart, P.L. 2001. Late Quaternary paleoclimate in semiarid Northeastern Brazil from U-Series dating of travertine and water-table speleothems. Quaternary Research 55: 159–167.

Baker, P.A.; Seltzer, G.O.; Fritz, S.C.; Dunbar, R.B.; Grove, M.J.; Tapia, P.M.; Cross, S.L.; Rowe, H.D.; Broda, J.P. 2001. The history of South American tropical precipitation for the past 25,000 years. Science 291: 640-643.

Barberi, M. 2001. Mudanças Paleoambientais na Região dos Cerrados do Planalto Central Durante o Quaternário Tardio: O Estudo da Lagoa Bonita, DF. Ph.D. Dissertation, Universidade de São Paulo.

Barberi, M.; Salgado-Labouriau, M.L.; Suguio, K. 2000. Paleovegetation and paleoclimate of "Vereda de Águas Emendadas", central Brazil. Journal of South American Earth Sciences 13: 241-254.

Bate, L.F. 1990. Culturas y modos de vida de los cazadores recolectores en el poblamiento de America del Sur. Revista de Arqueología Americana 2:89-153.

Behling, H. 1997. Late Quaternary vegetation, climate and fire history of the *Araucaria* forest and campos region from Serra Campos Gerais, Paraná State (South Brazil). Review of Palaeobotany and Palynology 97: 109-121.

Behling, H. 1998. Late Quaternary vegetational and climatic changes in Brazil. Review of Palaeobotany and Palynology 99: 143-156.

Behling, H. 2001. Late Quaternary environmental changes in the Lagoa da Curuçá region (eastern Amazonia, Brazil) and evidence of Podocarpus in the Amazon lowland. Vegetation History and Archaeobotany 10: 175-183.

Behling, H. 2002. South and southeast Brazilian grasslands during Late Quaternary times: a synthesis. Paleogeography, Palaeoclimatology, Palaeoecology 177: 19-27.

Behling, H. 2007. Late Quaternary vegetation, fire and climate dynamics of Serra do Araçatuba in the Atlantic coastal mountains of Paraná State, southern Brazil. Vegetation History and Archaeobotany 16:77-85.

Behling, H.; Bauermann, S.; Neves, P.C.P. 2001. Holocene environmetal changes in the São Francisco de Paula region, southern Brazil. Journal of South American Earth Sciences 14: 631-639.

Behling, H. Dupont, L.; Safford, H.D.; Wefer, G. 2007. Late Quaternary vegetation and climate dynamics in the Serra da Bocaina, southeastern Brazil. Quaternary International 161: 22–31.

Behling, H.; Lichte, M. 1997. Evidence of dry and cold climatic conditions at glacial times in tropical Southeastern Brazil. Quaternary Research 48: 348-358.

Behling, H.; Lichte, M.; Miklós, A. 1998. Evidence of a forest free landscape under dry and cold climatic conditions during the last glacial maximum in the Botucatu region (São Paulo State), Southeastern Brazil. In: J. Rabassa and M. Salemme (eds.) Quaternary of South America and Antarctic Peninsula. A.A. Balkema, Rotterdam, pgs. 99-110.

Behling, H.; Negrelle, R.B. 2001. Tropical rain forest and climate dynamics of the Atlantic Lowland, Southern Brazil, during the Late Quaternary. Quaternary Research 56: 383-389.

Behling, H.; Pillar, V.; Orlóci, L.; Bauermann, S. 2004. Late Quaternary Araucaria forest, grassland (Campos), fire and climate dynamics, studied by high-resolution pollen, charcoal and multivariate analysis of the Cambará do Sul core in southern Brazil. Palaeogeography, Palaeoclimatology, Palaeoecology 203:277-297.

Behling, H.; Pillar, V.; Bauermann, S. 2005. Late Quaternary grassland (Campos), gallery forest, fire and climate dynamics, studied by pollen, charcoal and multivariate analysis of the São Francisco de Assis core in western Rio Grande do Sul (southern Brazil). Review of Palaeobotany and Palynology 133: 235-248.

Behling, H.; Safford, H.D. 2010. Late-glacial and Holocene vegetation, climate and fire dynamics in the Serra dos Órgãos, Rio de Janeiro State, southeastern Brazil. Global Change Biology 16: 1661–1671.

Berrio, J.C.; Hooghiemstra, H.; Behling, H.; Borg, K. 2000. Late Holocene history of Savanna gallery forest from Carimagua area, Colômbia. Review of Palaeobotany and Palynology 111: 295-308.

Bigarella, J.J.; Mousinho, M.R.; Silva, J.X. 1965. Pediplanos, pedimentos e seus depósitos correlativos no Brasil. Boletim Paranaense de Geografia 16/17: 117-151.

Borrero, L.A. 1995. Human and natural agency: some comments on Pedra Furada (Brazil). Antiquity 69: 602.

Borrero, L.A.; Zárate, M.; Miotti, L., Massone, M. 1998. The Pleistocene-Holocene transition and human occupations in the southern cone of South America. Quaternary International 49/50:191-199.

Burroughs, W.J. 2005. Climate Change in Prehistory – The End of the Age of Chaos. Cambridge University Press, Cambridge, 356pp.

Bush, M.B.; De Oliveira, P.E.; Colinvaux, P.A.; Miller, M.C.; Moreno, J.E. 2004. Amazonian paleoecological histories: one hill, three watersheds. Palaeogeography, Palaeoclimatology, Palaeoecology 214: 359–393.

Bush, M.B.; Silman, R.; Toledo, M.; Listopad, C.; Gosling, W.; Williams, C.; De Oliveira, P.; Krisel, C. 2007. Holocene fire and occupation in Amazonia: records from two lake districts. Phylosophical Transctions of the Royal Society B 362: 209-218.

Butzer, K.W. 1980. Context in archaeology: an alternative perspective. Journal of Field Archaeology 7: 417-422.

Carneiro Filho, A.; Schwartz, D.; Tatumi, S.; Rosique, T. 2002. Amazonian paleodunes provide evidence for drier climate phases during the Late Pleistocene-Holocene. Quaternary Research, 58:205-209.

Clark, P. U.; Mix, A. C. 2002. Ice Sheets and sea level of the Last Glacial Maximum. Quaternary Science Reviews 21:1-7.

CLIMAP Project Members. 1981. Seasonal Reconstructions of the Earth's Surface at the Last Glacial Maximum. Boulder, CO: Geological Society of America, Map and chart Series, 36, 36pp.

Coelho Netto, A.L. 1999. Catastrophic landscape evolution in a humid region (SE Brazil): Inheritances from tectonic, climatic and land use induced changes. Annals of the Fourth Conference on Geomorphology, Italy. Supplementi di Geografia Fisica e Dinamica Quaternaria III: 21-48.

Coelho Netto, A.L.; Fernandes, N.F. 1990. Hillslope erosion, sedimentation, and relief inversion in SE Brazil: Bananal, SP. Research Needs and Applications to Reduce Erosion and Sedimentation in Tropical Steeplands (Proceedings of the Fiji Symposium, June 1990: IAHS-AISH Publ. No.192: 174-182.

Colinvaux, P.A.; De Oliveira, P.E.; Moreno, J.E.; Bush, M.B. 1996. A long pollen record from lowland Amazonia: Forest and cooling in glacial times. Science, 274: 85-88.

Colinvaux, P.A.; De Oliveira, P.E.; Bush, M.B. 2000. Amazonian and neotropical plant communities on glacial time-scales: The failure of the aridity and refuge hypotheses. Quaternary Science Reviews 19: 141-169.

Cooke, R. 1998. Human settlement of Central America and northernmost South America (14,000 – 8000 BP). Quaternary International 49/50: 177-190.

Cruz Jr, F.W.; Burns, S.J.; Karmann, I.; Sharp, W.D.; Vuille, M.; Cardoso, A.O.; Ferrari, J.A.; Dias, P.L.; Viana Jr, O. 2005. Insolation-driven changes in atmospheric circulation over the past 116,000 years in subtropical Brazil. Nature 434: 63-66.

Cruz Jr, F.W.; Burns, S.J.; Karmann, I.; Sharp, W.D.; Vuille, M. 2006. Reconstruction of regional atmospheric circulation features during the late Pleistocene in subtropical Brazil from oxygen isotope composition of speleothems. Earth and Planetary Science Letters 248: 495-507.

Cruz Jr, F.W.; Vuille, M.; Burns, S.J.; Wang, X.; Cheng, H.; Werner, M.; Edwards. L.R.; Karmann, I.; Auler, A.S.; Nguyen, H. 2009a. Orbitally driven east–west antiphasing of South American precipitation. Nature Geoscience 2: 210-214.

Cruz Jr, F.W.; Wang, X.; Auler, A.S.; Vuille, M.; Burns, S.J.; Edwards. L.R.; Karmann, I.; Cheng, H. 2009b. Orbital and millennial-scale precipitation changes in Brazil from speleothem records. In: Vimeux, F.; Sylvestre, F.; Khodry, M. (eds.) Past Climate Variability in South America and Surrounding Regions. Springer: 29-60.

Czaplewski, N.J.; Cartelle, C. 1998. Pleistocene Bats from Cave Deposits in Bahia, Brazil. Journal of Mammalogy 79: 784-803.

De Oliveira, P.E.; Barreto, A.M.; Sugio, K. 1999. Late Pleistocene/Holocene climatic and vegetational history of the Brazilian caatinga: the fossil dunes of the middle São Francisco River. Palaeogeography, Palaeoclimatology, Palaeoecology 152: 319-337.

deMenocal, P.B. 2001. Cultural responses to climate change during the Late Holocene. Science 292: 667-673.

Diamond, J. 1987. The worst mistake in the history of the human race. Discover Magazine, May: 64-66.

Diamond, J. 2005. Collapse. How Societies Choose to Fail or Succeed. Viking Press: New York.

Dillehay, T. 2000. The Settlement of the Americas – A New Prehistory. Basic Books, New York.

Dillehay, T.; Collins, M.B. 1988. Early cultural evidence from Monte Verde in Chile. Nature 332: 150-152.

Dixon, E.J. 2001. Human colonization of the Americas: timing, technology and process. Quaternary Science Reviews 20: 277-299.

Dunnell, R.C.; Dancey W.S. 1983. The siteless survey: a regional scale data collection strategy. Advances in Archaeological Method and Theory 6: 267-287.ethod and Theory 6:267-287.

Erhart, H. 1956. La genèse des sols en tant que phénomène géologique. Esquisse d'une théorie géologique et géochimique. Biostasie et rhéxistasie. Masson, Paris.

Fairbridge, R.W. 2009. Hypsithermal. In: Gornitz, V. (ed) Encyclopedia of Paleoclimatology and Ancient Environments. Springer: Dordretch pp. 451-453.

Ferraz-Vicentini, K.R.; Salgado-Labouriau, M.L. 1996. Palynological analysis of a palm swamp in Central Brazil. *Journal of South American Earth Sciences, 9*: 207-219.

Fladmark, K.R. 1979. Routes: alternate migration corridors for Early Man in North America. American Antiquity 44: 55-69.

Felice, G.D. 2002. A controvérsia sobre o sítio arqueológico Toca do Boqueirão da Pedra Furada. Revista Fundhamentos 2: 144-178.

Fogaça, E.; Boeda, E. 2006. A antropologia das técnicas e o povoamento da América do Sul pré-histórica. Habitus 4: 673-684.

Freitas, H.A.; Pessenda, L.C.; Aravena, R.; Gouveia, S.E.; Ribeiro, A.S.; Boulet, R. 2001. Late Quaternary vegetation dynamics in the Southern Amazon basin inferred from carbon isotopes in soil organic matter. *Quaternary Research* 55:39-46.

Hadjas, I.; Bonani, G.; Moreno, P.; Ariztegui, G. 2003. Precise radiocarbon dating of Late-Glacial cooling in mid-latitude South America. Quaternary Research 59: 70-78.

Haffer, J.; Prance, G.T. 2001. Climate forcing of evolution in Amazonia during the Cenozoic: on the refuge theory of biotic differentiation. Amazoniana 16:579–607.

Holliday, V.T. 2009. Geoarchaeology and the search for the first Americans. Catena 78: 310–322.

Hooghiemstra, H.; van der Hammen, T. 1998. Neogene and Quaternary development of the neotropical rain forest: the forest refugia hypothesis, and a literature overview. Earth-Science Reviews 44: 147–183.

Jacob, J.; Disnar, J-R.; Boussafir, M.; Sifeddine, A.; Turcq, B.; Albuquerque, A.S. 2004. Major environmental changes recorded by lacustrine sedimentary organic matter since the last glacial maximum near the equator (Lagoa do Caçó, NE Brazil). Palaeogeography, Palaeoclimatology, Palaeoecology 205: 183– 197.

Jacob, J.; Huang, Y.; Disnar, J-R.; Sifeddine, A.;Boussafir, M.; Albuquerque, A.S.; Turcq, B. 2007. Paleohydrological changes during the last deglaciation in Northern Brazil. Quaternary Science Reviews 26: 1004–1015.

Jolly, D.; Haxeltine, A. 1997. Effect of low glacial atmospheric CO2 on tropical African montane vegetation. Science 276: 786-788.

Knox, J.C. 1972. Valley alluviation in southwestern Wisconsin. Annals of the Association of American Geographers 62: 401-410.

Latrubesse, E.M.; Kalicki, T. 2002. Late Quaternary paleohydrological changes in the upper Purus basin, southwestern Amazonia, Brazil. Zeitschrift fur Geomorphologie 129: 41-59.

Ledru, M.P. 1993. Late Quaternary environmental and climatic changes in Central Brazil. Quaternary Research 39: 90-98.

Ledru, M.P.; Bertaux, J.; Siffedine, A.; Suguio, K. 1998. Absence of Last Glacial Maximum records in lowland tropical forests. Quaternary Research 49: 233-237.

Ledru, M.P.; Braga, P.S.; Soubies, F.; Fournier, M.; Martin, L.; Suguio, K.; Turcq, B. 1996. The last 50,000 years in the Neotropics (Southern Brazil): Evolution of vegetation and climate. Palaeogeography, Palaeoclimatology, Palaeoecology 123: 239-257.

Ledru, M.P.; Mourguiart, P.; Ceccantini, G.; Turcq, B.; Siffedine, A. 2002. Tropical climates in the game of two hemispheres revealed by abrupt climatic change. Geology 30: 275-278.

Ledru, M.P.; Mourguiart, P.; Riccomini, C. 2009. Related changes in biodiversity, insolation and climate in the Atlantic rainforest since the last interglacial. Palaeogeography, Palaeoclimatology, Palaeoecology 271: 140-152.

Ledru, M.P.; Rousseau, D.D.; Cruz Jr., F.W.; Riccomini, C.; Karmann, I.; Martin, L. 2005. Paleoclimate changes during the last 100,000 yr from a record in the Brazilian Atlantic rainforest region and interhemispheric comparison. Quaternary Research 64: 444-450.

Mandel, R. D. 2008. Buried paleoindian-age landscapes in stream valleys of the central plains, USA. Geomorphology 101: 342–361.

Mayewski, P.A.; Rohling, E.; Stager, J.C.; Karlén, W.; Maasch, K.; Meeker, L.; Meyerson, E.; Gasse, F.; van Kreveld, S.; Holmgren, K.; Lee-Thorp, J.; Rosqvist, G.; Rack, F.; Staubwasser, M.; Schneider, R.; Steig, E. 2004. Holocene climate variability. Quaternary Research 62: 243–255.

Mayle, F.E.; Burn. M.J.; Power, M.; Urrego, D. 2009. Vegetation and fire at the Last Glacial Maximum in tropical South America. In: Vimeux, F.; Sylvestre, F.; Khodry, M. (eds.) Past Climate Variability in South America and Surrounding Regions. Springer: 89-112.

McAnany, P.A.; Yoffee, N. (eds.) 2010. Questioning Collapse. Human Resilience, Ecological Vunerability, and the Aftermath of Empire. Cambridge University Press: Cambridge.

Meggers, B.J. 2007. Mid-Holocene climate and cultural dynamics in Brazil and the Guianas. In: Anderson, D.; Maasch, K.; Sandweiss, D. (eds). Climate Change and Cultural Dynamics: A Global Perspective on Mid-Holocene Transitions. Elsevier: 117-155.

Meis, M.R.; Moura, J.R. 1984. Upper Quaternary sedimentation and hillslope evolution; southeastern Brazilian Plateau. American Journal of Science 284: 241-254.

Melo, M.S.; Medeiros, C.; Giannini, P.C.; Garcia, M.J.; Pessenda, L.C. 2003. Sedimentação quaternária no espaço urbano de Ponta Grossa, PR. Geociências UNESP 22: 33-42.

Meltzer, D.J.; Adovasio, J.M.; Dillehay, T. 1994. On a Pleistocene human occupation at Pedra Furada, Brazil. Antiquity 68: 695-714.

Mix, A. C.; Bard, E.; Schneider, R. 2001. Environmental processes of the ice age: land, oceans, glaciers (EPILOG). Quaternary Science Reviews 20: 627-657.

Modenesi-Gauttieri, M.C. 2000. Hillslope deposits and the Quaternary evolution of the *altos campos* - Serra da Mantiqueira, from Campos do Jordão to the Itatiaia massif. Revista Brasileira de Geociências 30: 508-514.

Moura, J.S. 2003. Geomorfologia do Quaternário. In: Guerra, A.J.T.; Cunha, S. B. (org.) Geomorfologia: uma atualização de bases e conceitos. 3rd.ed. Bertrand Brasil, Rio de Janeiro, pp. 335-364.

Moura, J.S.; Mello, C.L. 1991. Classificação aloestratigráfica do Quaternário superior na região de Bananal (SP/RJ). Revista Brasileira de Geociências 21:236-254.

Moura, J.S.; Peixoto, M.O.; Silva, T.M. 1991. Geometria do relevo e estratigrafia do Quaternário como base à tipologia de cabeceiras de drenagem em anfiteatro – Médio vale do Rio Paraíba do Sul. Revista Brasileira de Geociências 21: 255-265.

Moura, J.S.; Silva, T.M. 2003. Complexo de rampas de colúvio. In: S.B. Cunha and A.T. Guerra (eds.) Geomorfologia do Brasil. Bertrand, Rio de Janeiro, pp. 143-180.

NACSN - North American Commission on Stratigraphic Nomenclature. 1983. North American stratigraphic code, American Association of Petroleum Geologists, Bulletin 67: 841-875.

Nakamura, C.; Melo Jr, J.C.; Ceccantini, G.T. 2010. Macro-restos vegetais: uma abordagem paleoetnobotânica e paleoambiental. In: Araujo, A.G.M.; Neves, W.A. (eds.) Lapa das Boleiras: Um Sítio Paleoíndio do Carste de Lagoa Santa, MG, Brasil. Annablume / FAPESP, São Paulo: 159-187.

Neme, G.; Gil, A. 2009. Human Occupation and Increasing Mid-Holocene Aridity. Current Anthropology 50: 149-163.

Neves, E.G. 2007. El Formativo que nunca terminó: la larga história de la estabilidad en las ocupaciones humanas de la Amazonía Central. Boletín de Arqueología PUCP 11: 117-142.

Neves, W.A.; Cornero, S. 1997. What did South American paleoindians eat? Current Research in the Pleistocene 14:93-96.

Nimer, E. 1989. Climatologia do Brasil. 2nd ed. Rio de Janeiro: IBGE.

North Greenland Ice Core Project Members. 2004. High-resolution record of Northern hemisphere climate extending into the last interglacial period. Nature 431: 147–151.

Nuñez, L.; Grosjean, M.; Cartajena, I. 2001. Human dimensions of Late Pleistocene/Holocene arid events in southern South America. In: Markgraf, V. (Ed.) Interhemispheric Climate Linkages. San Diego, Academic Press: 105-117.

Parenti, F. 2001. Le Gisement Quaternaire de Pedra Furada (Piauí, Brésil) - Stratigraphie, Chronologie, Évolution Culturelle. Éditions Recherche sur les Civilisations. Ministére des Affaires Étrangères, 325 pp.

Parizzi, M.G.; Salgado-Labouriau, M.L.; Kohler, H.C. 1998. Genesis and environmental history of Lagoa Santa, southeastern Brazil. The Holocene 8: 311–321.

Pessenda, L. C.; De Oliveira, P.E.; Mofatto, M.; Medeiros, V.B.; Garcia, R.F.; Aravena, R.; Bendassoli, J.A.; Leite, A.Z.; Saad, A.R.; Etchebehere, M.L. 2009. The evolution of a tropical rainforest/grassland mosaic in southeastern Brazil since 28,000 14C yr BP based on carbon isotopes and pollen records. Quaternary Research 71: 437–452.

Pessenda, L.C.; Gouveia, S.E.M.; Aravena, R. 2001. Radiocarbon dating of total soil organic matter and humin fraction and its comparison with C14 ages of fossil charcoal. Radiocarbon 43:595-601.

Racza, M.F. 2009. Mudanças paleoambientais quaternárias na região de Lagoa Santa, MG, Brasil: A palinologia como subsidio para o entendimento do padrão de ocupação humana. Unpublished M.A. Thesis, University of Guarulhos, São Paulo.

Richerson, P.J.; Boyd, R.; Bettinger, R.L. 2001. Was agriculture impossible during the Pleistocene but mandatory during the Holocene? A climate change hypothesis. American Antiquity 66: 387-411.

Roberts, N., Barker, P.,1993. Landscape stability and biogeomorphic response to past and future climate shifts in intertropical Africa. In: Thomas, D.S.G., Allison, R.J. (Eds.), Landscape Sensitivity. John Wiley, New York, pp. 65–82.

Roosevelt, A.C.;Costa, M.L.; Machado, C.L.; Michab, M.; Mercier, N.; Valladas, H.; Feathers, J.; Barnett, W.; Silveira, M.I.; Henderson, A.; Sliva, J.; Chernoff, B.; Reese, D.S.; Holman, J.A.; Toth, N.; Schick, K. 1996. Paleoindian cave dwellers in the Amazon: The peopling of the Americas. Science 272:373-384.

Salgado-Labouriau, M. L.; Casseti, V.; Ferraz-Vicentini, K.; Martin, L.; Soubies, F.; Suguio, K.; Turcq, B. 1997. Late Quaternary vegetational and climatic changes in cerrado and palm swamp from Central Brazil. Palaeogeography, Palaeoclimatology, Palaeoecology 128: 215–226.

Santos, G.M.; Bird, M.I.; Fifield, L.K.; Guidon, N.; Parenti, F. 2003. A revised chronology of the lowest occupation layer of Pedra Furada Rock Shelter, Piauí, Brazil: the Pleistocene peopling of the Americas. Quaternary Science Reviews 22: 2303-2310.

Saia, S.E.; Pessenda, L.C.; Gouveia, S.E.M.; Aravena, R.; Bendassoli, J.A. 2008. Last glacial maximum (LGM) vegetation changes in the Atlantic Forest, southeastern Brazil. Quaternary International 184: 195-201.

Sarges, R.R.; Nogueira, A.R.; Riccomini, C. 2009. Sedimentação coluvial pleistocênica na região de Presidente Figueiredo, nordeste do estado do Amazonas. Revista Brasileira de Geociências 39: 350-359.

Sifeddine, A., Albuquerque, A. S., Ledru, M.-P., Turcq, B., Knoppers, B., Martin, L., Mello, L. Z., Passenau, H., Dominguez, J. L., Cordeiro, R. C., Abrão, J. J., Bittencourt, A. P. 2003. A 21000 cal years paleoclimatic record from Caçó Lake, northern Brazil: vidence from sedimentary and pollen analysis. Palaeogeography, Palaeoclimatology, Palaeoecology 189: 25-34.

Stafford, C.R. 1995. Geoarchaeological perspectives on paleolandscapes and regional subsurface archaeology Journal of Archaeological Method and Theory 2:69-104.

Stevaux, J.C. 2000. Climatic events during the Late Pleistocene and Holocene in the Upper Parana River: correlation with NE Argentina and South-Central Brazil. Quaternary International 72: 73-85.

Thomas, M.F. 1994. Geomorphology in the Tropics - A Study of Weathering and Denudation in Low Latitudes. John Willey & Sons, New York.

Thomas, M.F. 2008. Understanding the impacts of Late Quaternary climate change in tropical and sub-tropical regions. Geomorphology 101: 146-158.

Thomas, M.F.; Thorp, M.B. 1995. Geomorphic response to rapid climatic and hydrologic change during the Late Pleistocene and Early Holocene in the humid and subhumid tropics. Quaternary Science Reviews 14:193–207.

Turcq, B.; Pressinoti, M.M.N.; Martin, L. 1997. Paleohydrology and paleoclimate of the past 33,000 years a the Tamanduá river, Central Brazil. Quaternary Research 47: 284-294.

Tsoar, H.; Levin, N.; Porat, N.; Maia, L.P.; Herrmann, H.J.; Tatumi, S.; Claudino-Sales, V. 2009. The effect of climate change on the mobility and stability of coastal sand dunes in Ceará State (NE Brazil). Quaternary Research 71: 217–226.

Vilhena-Vialou, A.V. 2003. Santa Elina Rockshelter, Brazil: Evidence of the coexistence of man and *Glossoterium*. In: Miotti, L., Salemme, M. and Flegenheimer, N. (eds.) Where the south winds blow. Ancient evidence of paleo South Americans. Center for the Study of the First Americans, Texas, A&M, pp. 21-28.

Vilhena-Vialou, A.V. 2011. Occupations humaines et faune éteinte du Pléistocène au centre de l'Alamerique du Sud: L'abri rupestre Santa Elina, Mato Grosso, Brésil. In: Vialou, D. (ed) Peuplements et Prehistoire en Amériques. Comité des Travaux Historiques et Scientifiques, Collection Documents Préhistoriques 28: 193-208.

Wang, X.; Auler, A.S.; Edwards, R.L.; Cheng, H.; Ito; Solheid, M. 2006. Interhemispheric anti-phasing of rainfall during the last glacial period. Quaternary Science Reviews 25: 3391–3403.

Wang, X.; Auler, A.S.; Edwards, R.L.; Cheng, H.; Ito, E.; Wang, Y.; Kong, X.; Solheid, M. 2007. Millennial-scale precipitation changes in southern Brazil over the past 90,000 years. Geophysical Research Letters 34: L23701.

Watanabe, O.; Jouzel, J.; Johnsen, S.; Parrenin, F.; Shoji, H.; Yoshida, N. 2003. Homogeneous climate variability across East Antarctica over the past three glacial cycles. Nature 422: 509–512.

Waters, M.R.; Stafford, T.W. 2007. Redefining the Age of Clovis: Implications for the peopling of the Americas. Science 315: 1122-1126.

Wright, J. D. 2009. Cenozoic climate change. In: Gornitz, V. (ed.). Encyclopedia of Paleoclimatology and Ancient Envrionments. Springer, pp. 148-155.

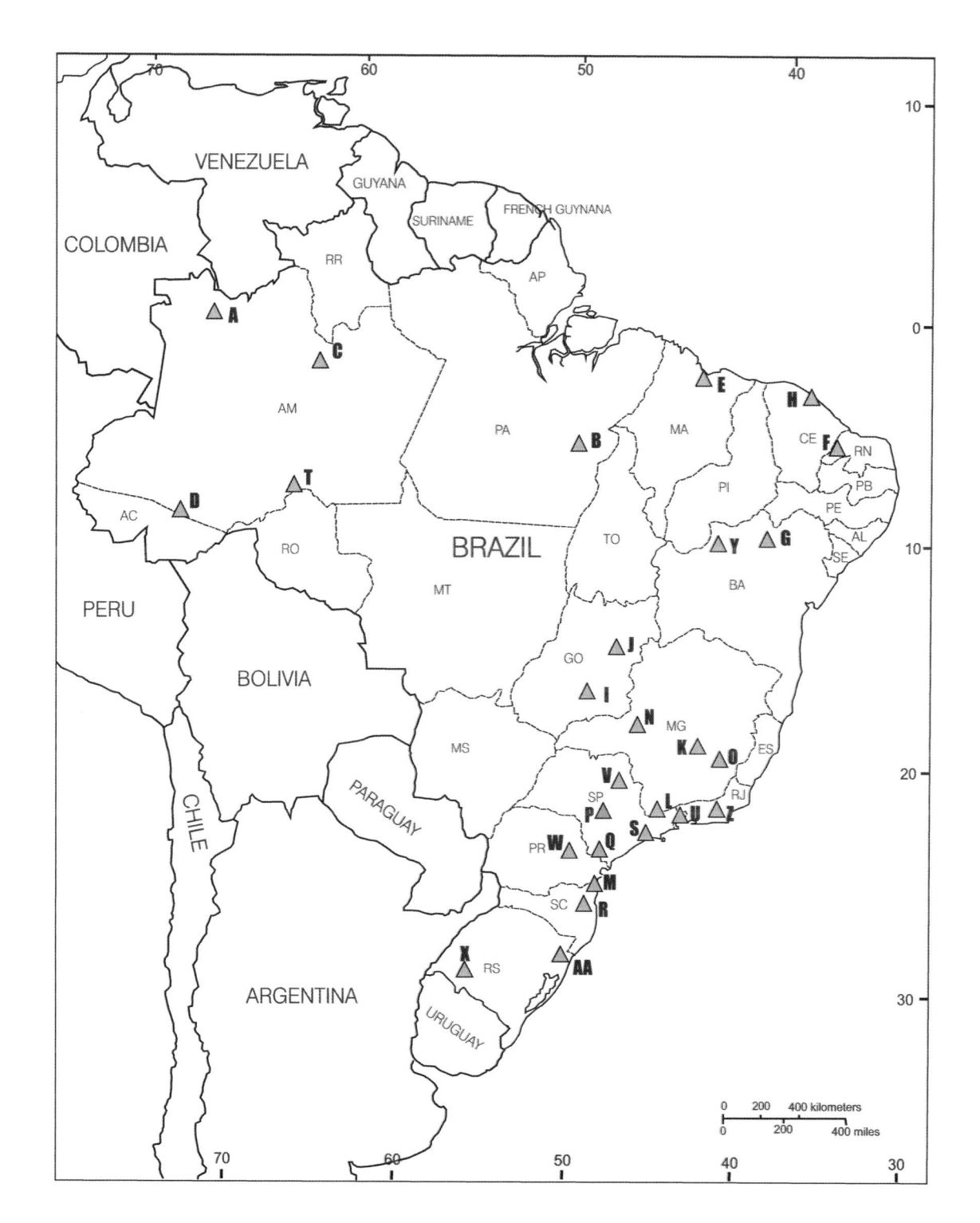

Figure 1. Paleoenvironmental studies cited in the text.

A= Six Lakes (Pata, Dragão, Verde); B= Carajás; C= Rio Negro Dunes; D= Purus River; E= Caçó Lake; F= Rio Grande do Norte; G= Toca Boa Vista; H= Ceará Dunes; I = Cromínia; J = Águas Emendadas and Lagoa Bonita; K=Lagoa Santa, Lagoa Olhos, Lagoa Mares; L= Morro Itapeva; M = Volta Velha and Serra Araçatuba; N= Salitre and Serra Negra; O= Catas Altas; P = Botucatu; Q= Santana Cave and Petar; R = Botuverá Cave; S= Curucutu and Colonia; T= Humaitá; U= Serra da Bocaina; V= Tamanduá River; W= Serra Campos Gerais; X= São Francisco de Assis; Y = Icatu Dunes; Z = Serra Orgãos; AA= São Francisco de Paula and Cambará do Sul.

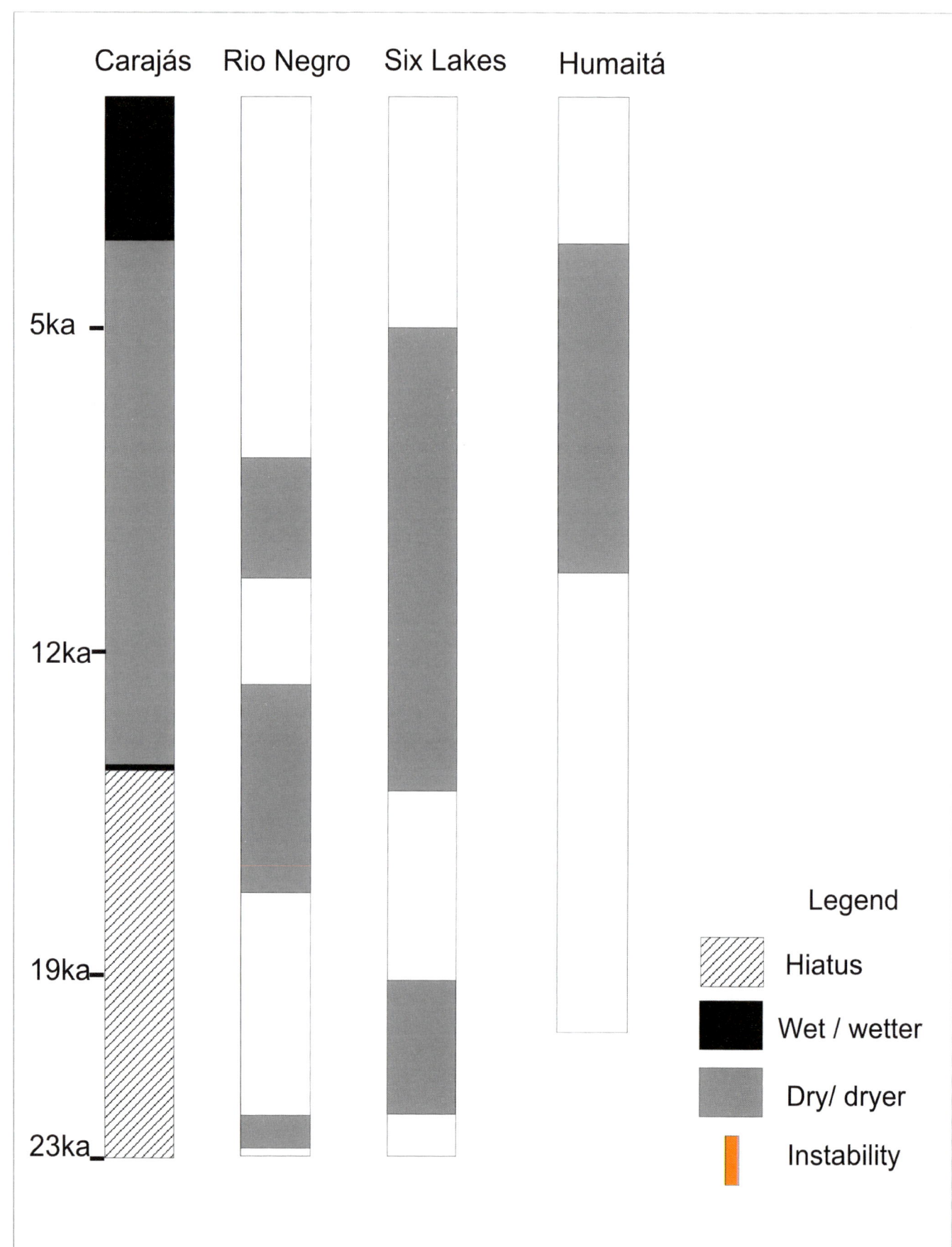

Figure 2: Scheme showing the main climatic shifts occurring in the Amazon Basin.

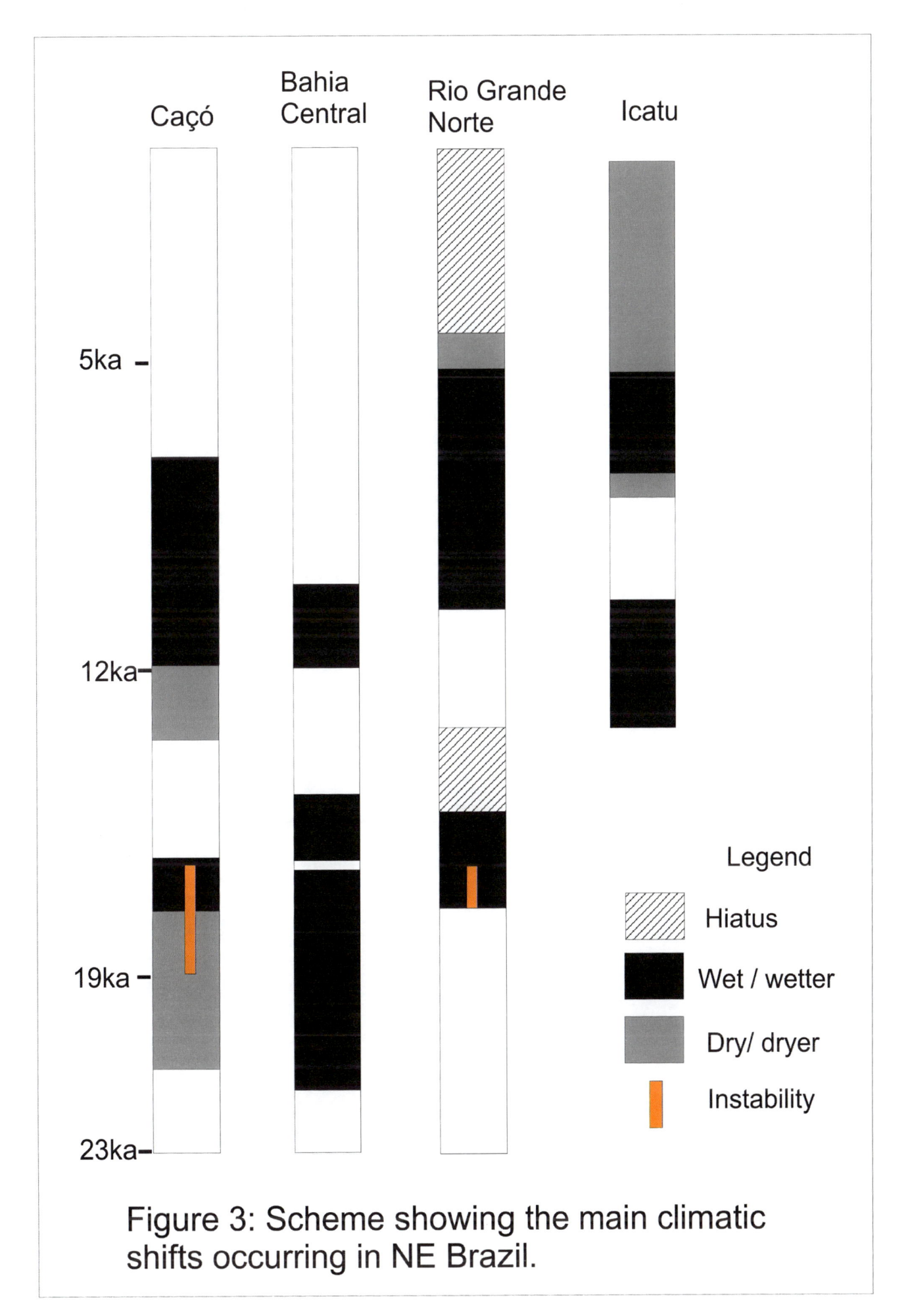

Figure 3: Scheme showing the main climatic shifts occurring in NE Brazil.

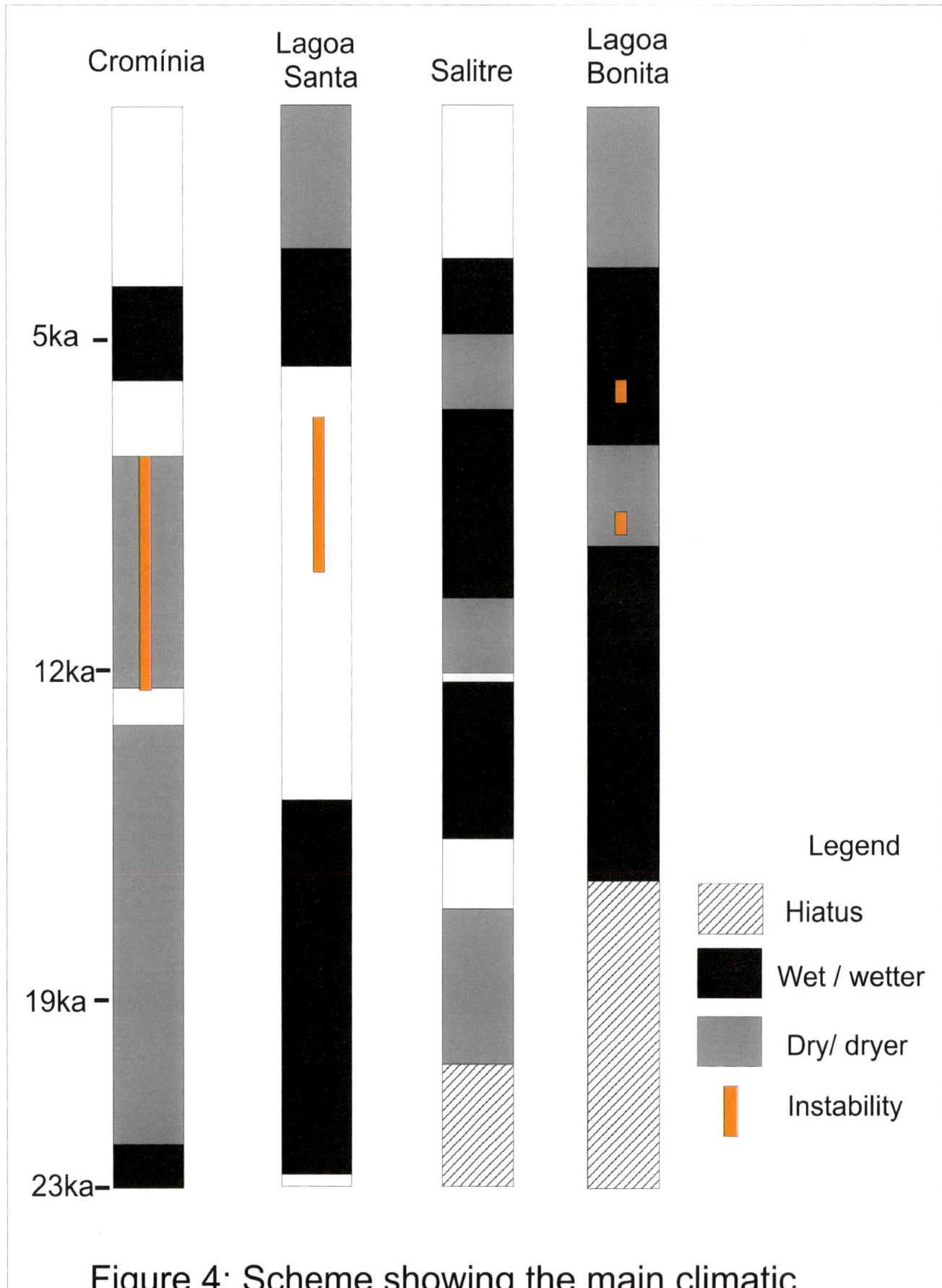

Figure 4: Scheme showing the main climatic shifts occurring in Central Brazil.

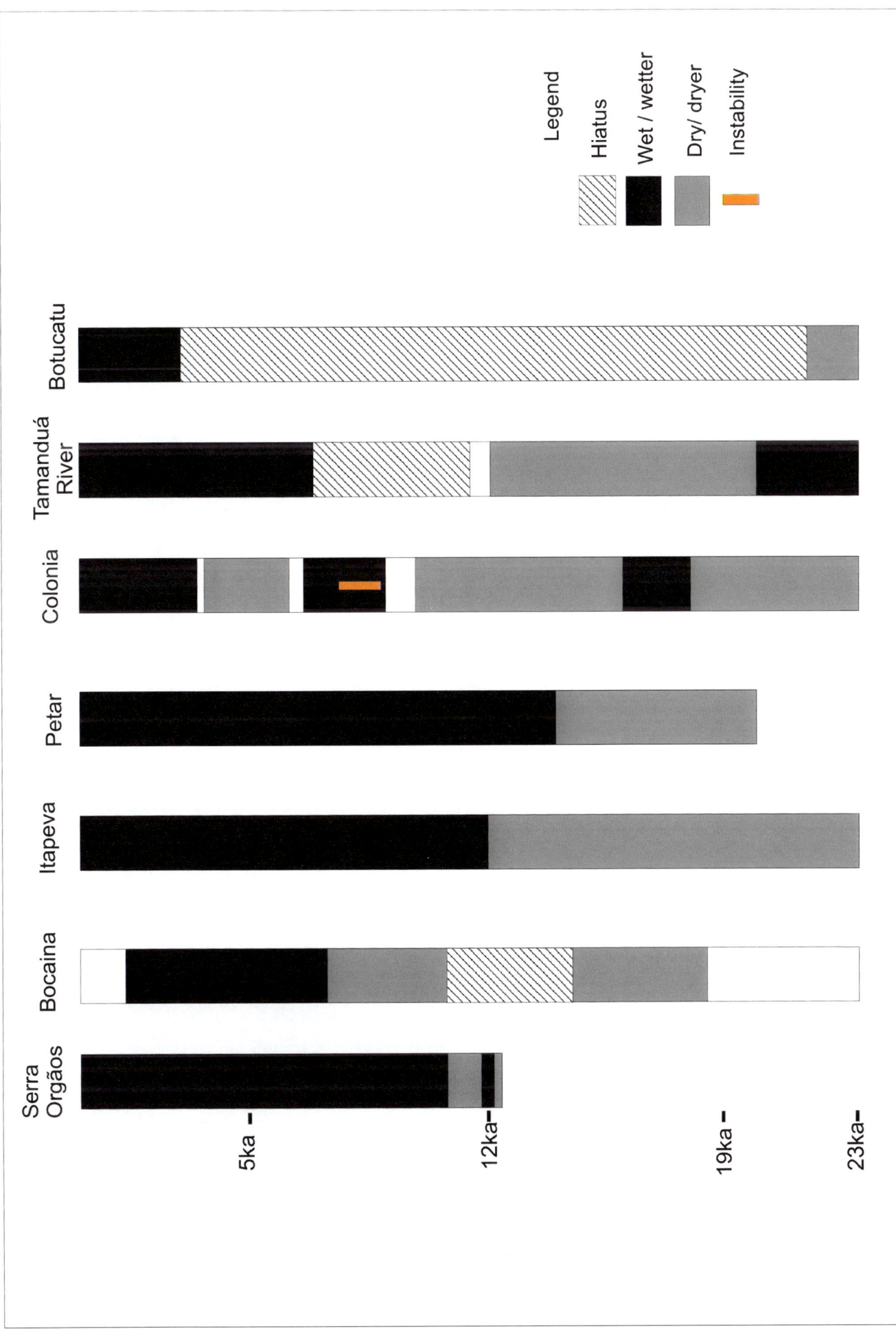

Figure 5: Scheme showing the main climatic shifts occurring in Southeastern Brazil.

Figure 6: Scheme showing the main climatic shifts occurring in Southern Brazil.

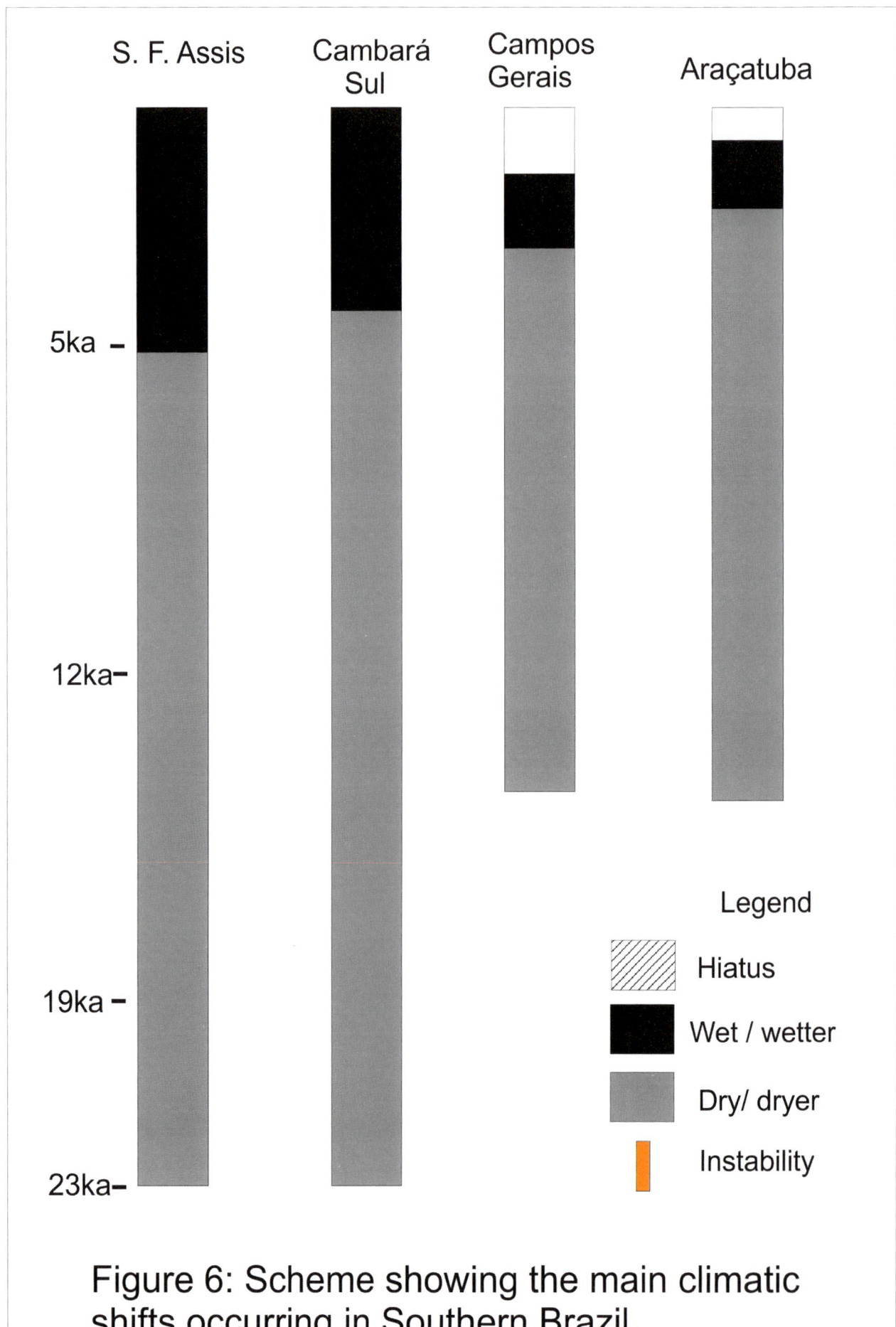

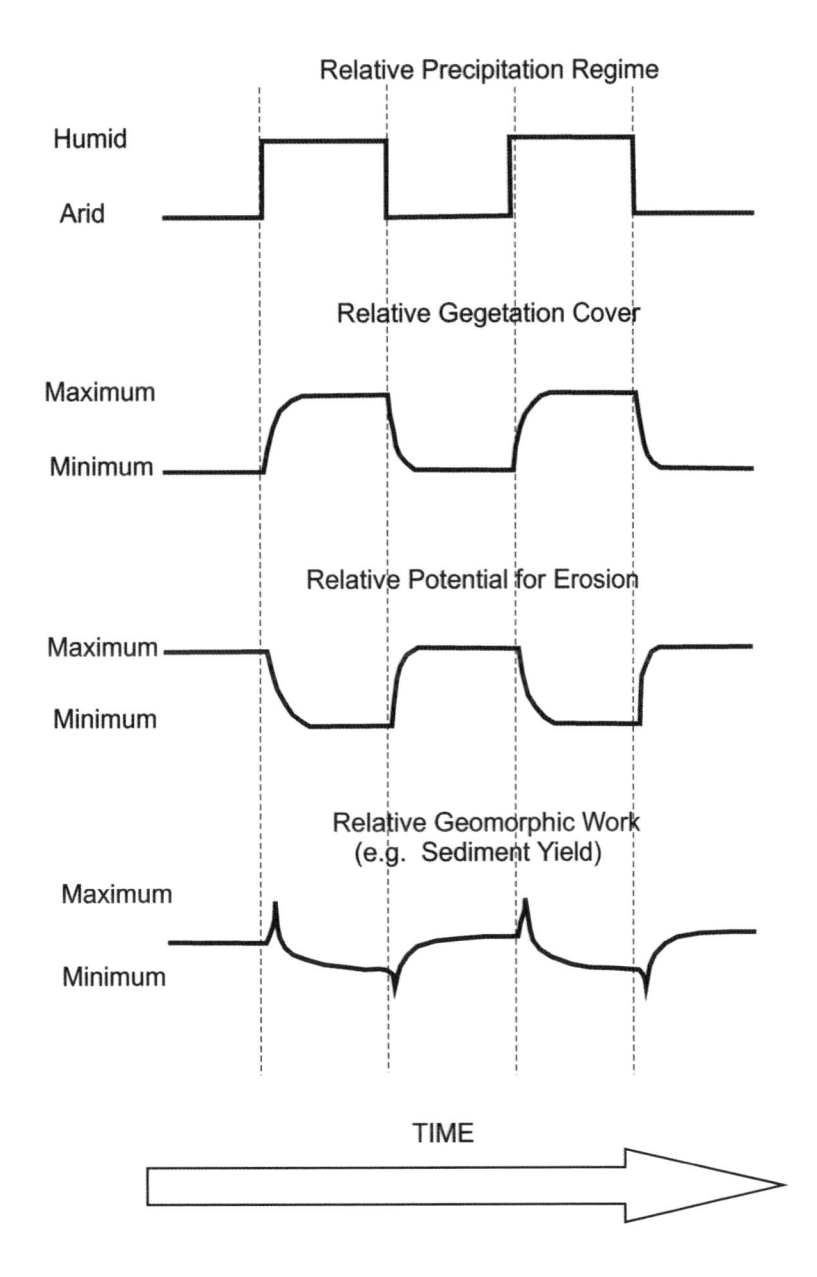

Figure 7: Model of geomorphic response to climate changes modified from Knox (1972).

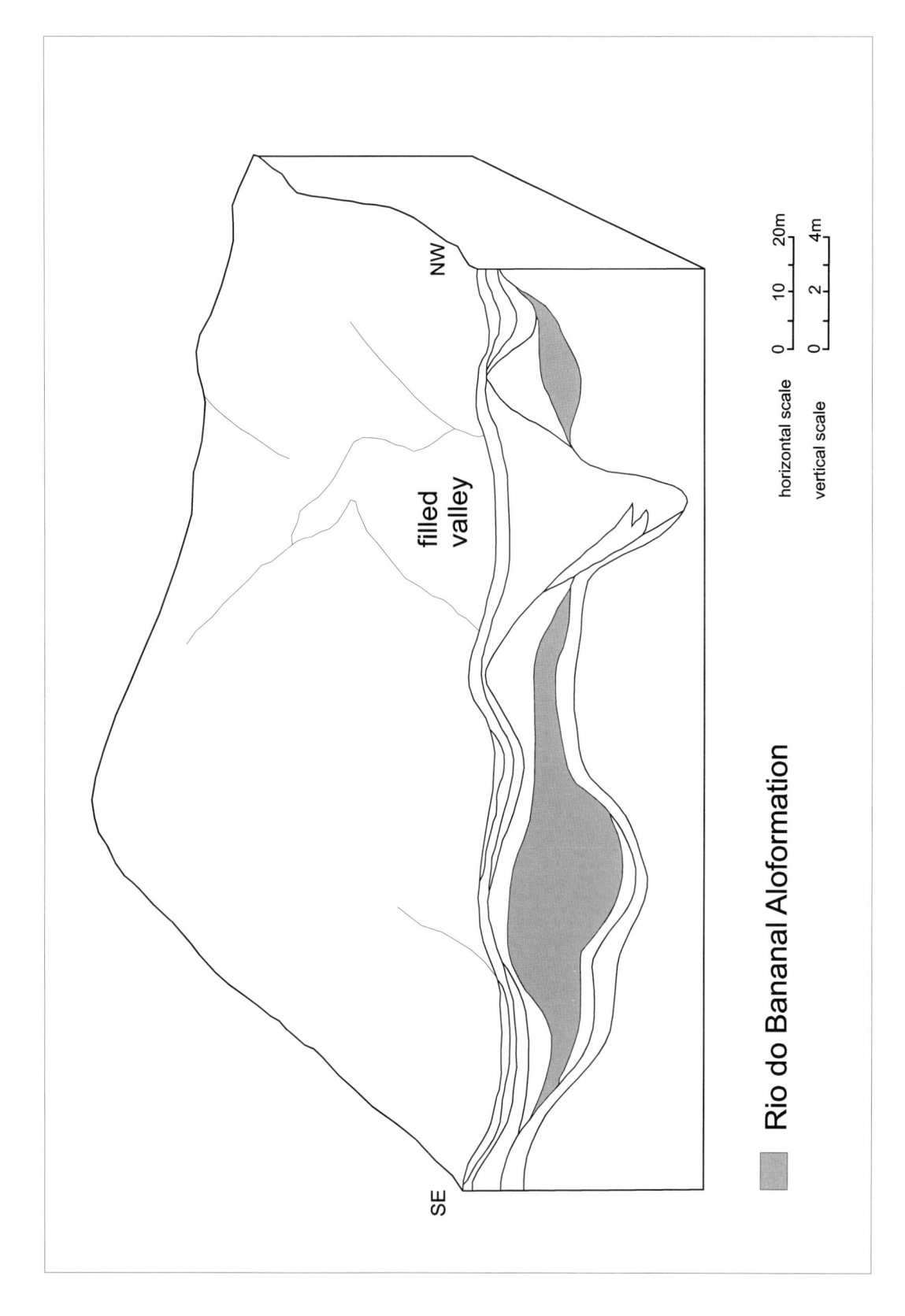

Figure 8: Example showing the complex spatial relationship between coluvial-aluvial deposits and paleosurfaces relative to the present surface. Note the stratigraphic positioning of Rio do Bananal aloformation, whose top age is 11,300 cal BP. Modified from Moura et al. (1991).

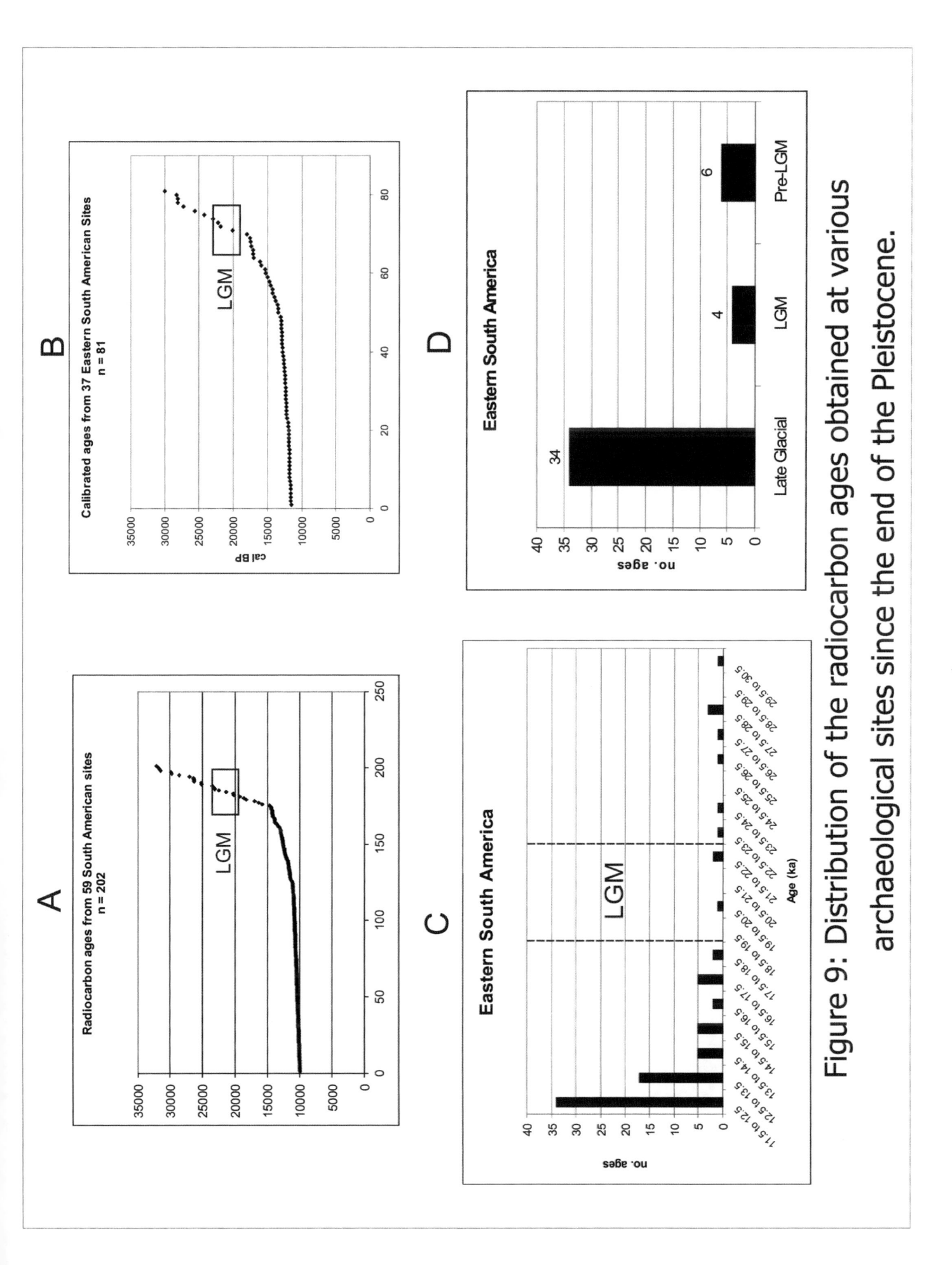

Figure 9: Distribution of the radiocarbon ages obtained at various archaeological sites since the end of the Pleistocene.

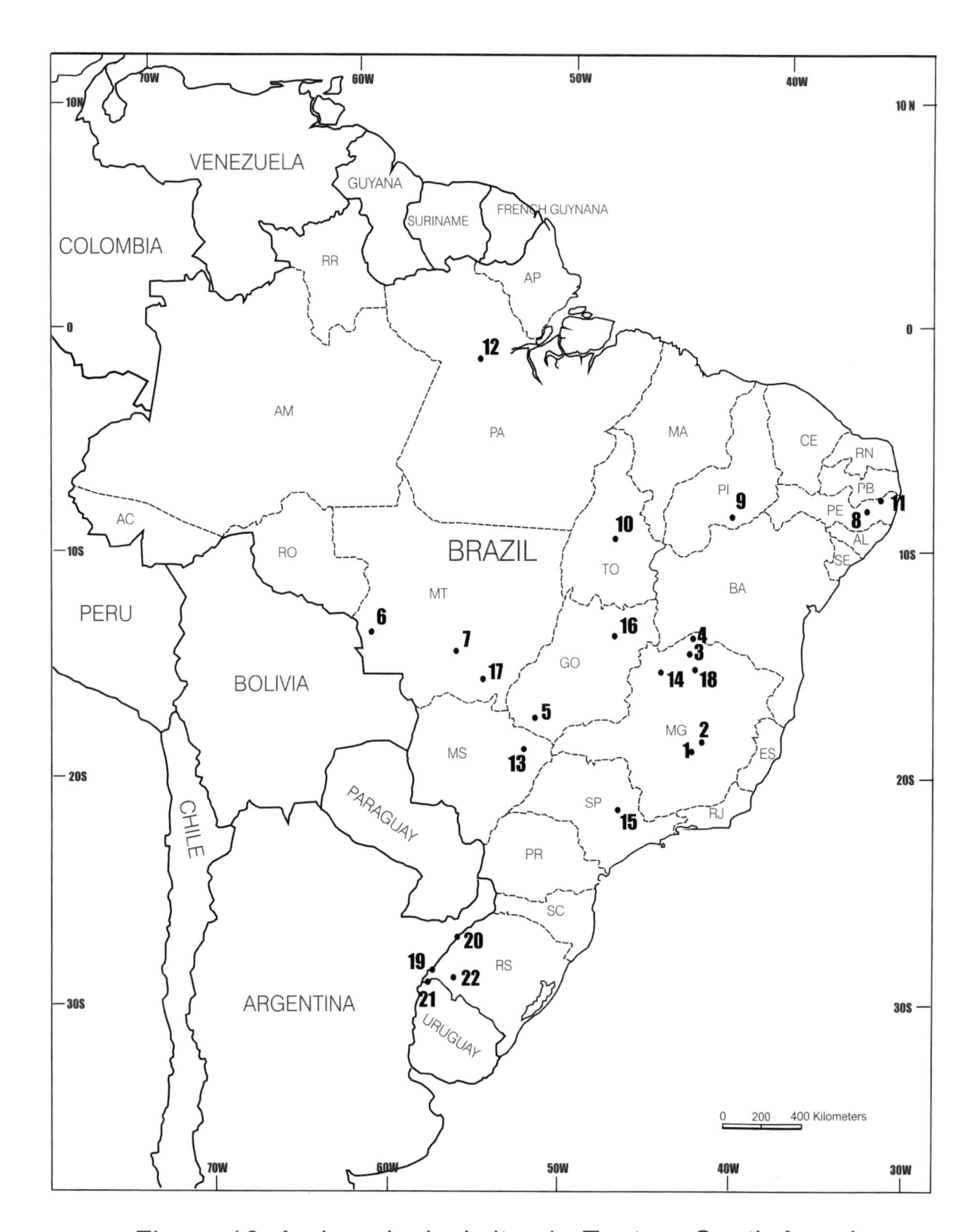

Figure 10. Archaeological sites in Eastern South America with ages older than 11,500 cal BP.

1= Lapa do Santo, Lapa Vermelha IV, Lapa Boleiras, Coqueirinho; 2= Santana do Riacho; 3= Lapa do Boquete; 4= Lapa do Dragão; 5= GO-JA-01, GO-JA-02, GO-JA-14; 6= Abrigo do Sol (MT-GU-01); 7= Santa Elina ; 8= Furna do Estrago and Brejo da Madre de Deus; 9= Pedra Furada, Tb2, and Sitio do Meio; 10 =Lajeado 18, Miracema 1, Jiboia and Capivara 5 ;11= Chã do Caboclo; 12= Pedra Pintada ; 13= MS-PA-02; 14= Lapa do Gentio; 15= Alice Boer; 16= Go-NI-08, Go-NI-49 and Go-NI-148; 17= Morro da Janela (MT-SL-31); 18= Lapa do Varal; 19= RS-I-66, RS-I-69; 20= RS-IJ-68; 21= RS-Q-2; 22= RS-I-50.

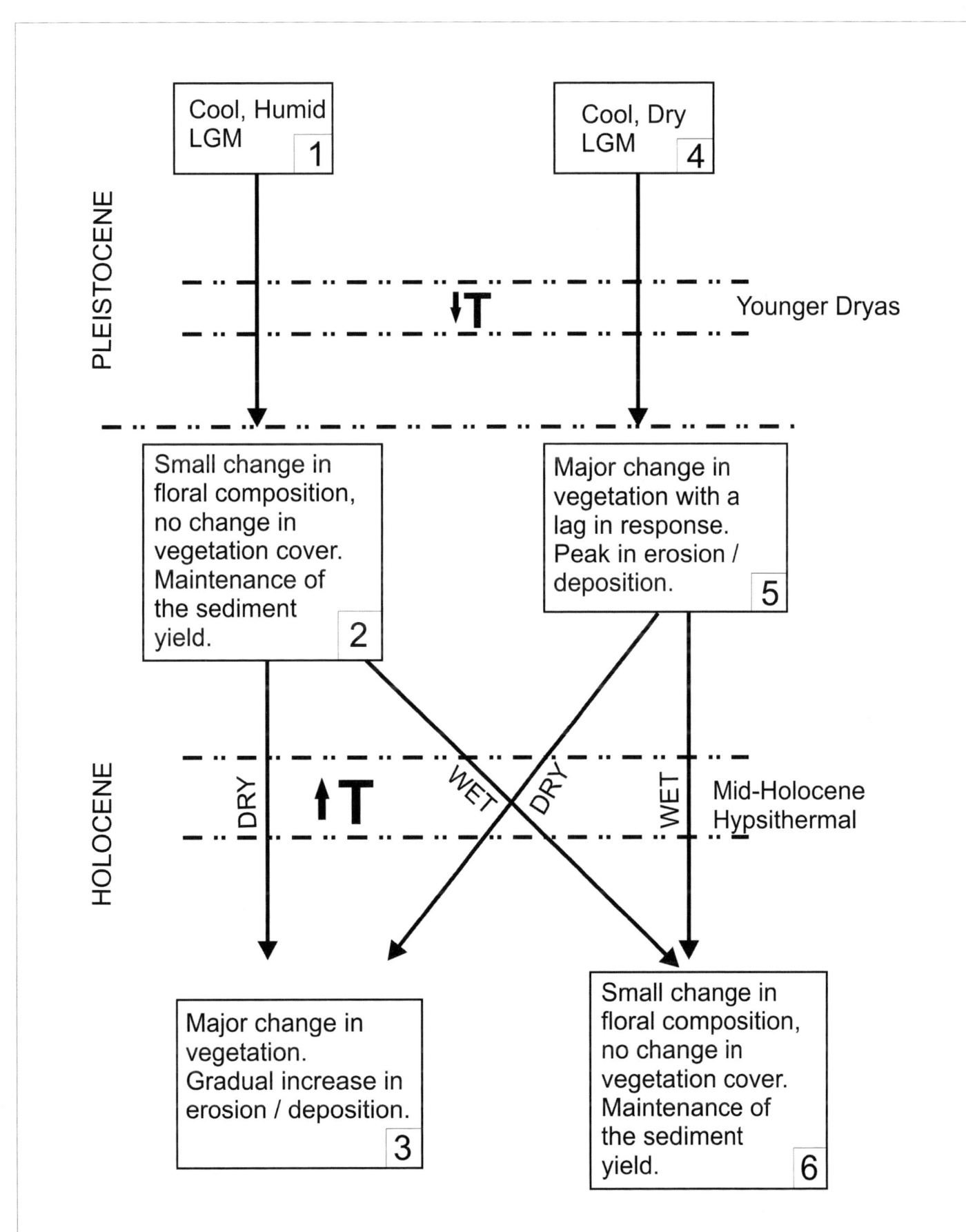

Figure 11: Hipothetical situation where two environmental systems undergo climate changes across time.

Submerged Lithic Tools Indicate Alternative Procurement Strategies
Alison T. Stenger, Institute for Archaeological Studies, Portland, Or. USA

ABSTRACT

The documentation of multiple lithic tool types from underwater locations contradicts the subsistence paradigm of later hunter-gatherer societies, in many regions. The variety of functions and cultural periods represented by these submerged materials support many researchers' suggestion of population replacement over time. As demonstrated by other archaeological evidence, changes are indicated through tool types, development of habitation areas, skeletal morphologies, and molecular (biological) indicators. Expansive water based food gathering strategies further add to these lines of investigation, advocating for population change over time. The content of this monograph, which was discussed after the papers presented at the Pre-Clovis in the Americas Conference, describes some of the differing patterns of subsistence that are represented by these underwater lithic assemblages.

At least five distinct time periods are represented by the lithic tools associated with underwater localities in western North America. It is probable that these materials represent distinct cultural phases, and that their presence underwater indicates differing food resource activities between ancient and recent populations. Similar observations occur for the eastern coast of the continent, with submerged findings extending from multiple Chesapeake Bay sites to areas offshore of Florida (Stanford and Bradley 2012; Hemmings and Adovasio 2012). While rising sea levels make the interpretation of some coastal sites difficult, water based tool usage is also evidenced at numerous interior sites, such as high mountain lakes. In these and some river sites that are far inland, it is less challenging to interpret the presence of tools, as ocean activity is not a factor.

Although it was initially assumed that the majority of the lithic materials were redeposited from terrestrial sites, there is increasing evidence that these artifact types were intentionally used in the water and at shoreline margins. The lithic assemblages discussed in this paper include projectile points, other biface tool types such as end scrapers and knives, and debitage. The materials range from basalt to CCS, with some minerals being from local resources and others classified as exotics.

Over a period of several years, there have been an increasing number of reports describing culturally modified materials from underwater environments. While organics are observed, the majority of the reports are of lithics (Adovasio 2012). When identifiable, many of the lithics are stylistically early. Cultural phases of the early types in the West include Windust, Haskett, and Lind Coulee. Older knife styles and crescents are also represented. Many early tool types are also well documented for the East coast (Stanford and Bradley 2012). While in the Northeast only large bipoints are currently documented from underwater, terrestrial assemblages include other early tool types such as scrapers made on blades (figure 1).[1] However, numerous lithic tool types are represented in underwater site assemblages to the South, as well as from inland waters in mid-western states and throughout much of the Great Basin (Smith, et al 2013; Bell 1980; Wisner 1997). The significantly greater time depth of the eastern sites, while not explored in this monograph, is considered elsewhere in this publication.

---

[1] This may be due to the method of collection of the early bipoints, as they were recovered with seines having large spacing. As suggested recently by Dennis Stanford, the probability of recovering even slightly smaller, or differently shaped, materials is therefore eliminated (personal communication 2012; 2013).

Figure 1a. Two early bipoints from underwater New England sites. Internet images shown, with permission for publication by the Smithsonian Institution and Dennis Stanford.

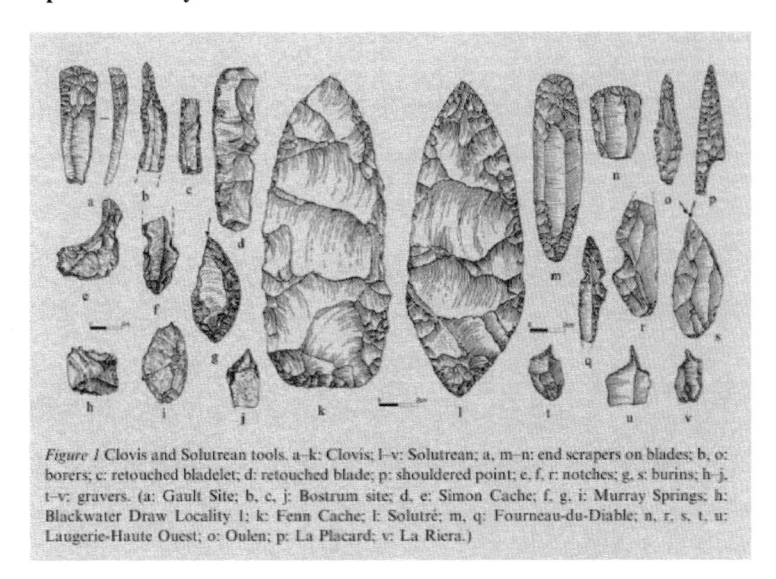

Figure 1 Clovis and Solutrean tools. a–k: Clovis; l–v: Solutrean; a, m–n: end scrapers on blades; b, o: borers; c: retouched bladelet; d: retouched blade; p: shouldered point; e, f, r: notches; g, s: burins; h–j, t–v: gravers. (a: Gault Site; b, c, j: Bostrum site; d, e: Simon Cache; f, g, i: Murray Springs; h: Blackwater Draw Locality 1; k: Fenn Cache; l: Solutré; m, q: Fourneau-du-Diable; n, r, s, t, u: Laugerie-Haute Ouest; o: Oulen; p: La Placard; v: La Riera.)

Figure 1b. East coast and other North American assemblages include large bi-points, gravers, and scrapers made on blades. Here, they are shown with selected early European material. Images courtesy of Dennis Stanford and Bruce Bradley. Not intended for duplication beyond this text.

The submerged early lithics are often isolated from more recent material, and the condition of the older specimens is consistently excellent (figure 2). Worn or battered objects represent a very small portion of the total number of early examples that are reported.

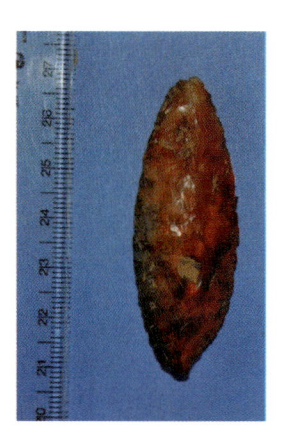

Figure 2. The styles and technologies demonstrated by these water-curated lithics are different, but the types are consistently early. The condition is excellent, with scars from transportation rarely observed. Identification courtesy of David Rice, specimen access courtesy of Mike Full.

Where the early bifaces differ is in their depositional environments. Many distinctive aquatic ecosystems, and vastly different elevations, are represented. Lithics have been reported from streams, rivers, estuaries, both low and high elevation lakes, and Pleistocene lakebeds. Site elevations extend from sea level to more than 9,500' msl (Stanford and Bradley 2012; Stenger 1993, 1988). Importantly, the lack of scarring from transportation is consistent among most of the lithics from every area. Inspection of the available material from all sites, and from all reported elevations, defines objects that have been well curated by their underwater environments. Primary deposition is represented by nearly every specimen.

More recent lithics have a higher frequency of damage. Stylistically, these materials predominately represent two different periods. These are ca. 3500-2500 yBP and 1000-500 yBP. Notably, the later specimens have only been reported from one lake and one river system in the Northwest. Data from other regions have not yet been collected.

Observations from the river identify 7 isolated bifaces of more recent types, made of several different materials. CCS predominates, but obsidian and other materials are also present. Represented are a knife, an exhausted core, a scraper, and four projectile points. Condition of the material varies, with some that are severely battered and/or rounded, but with a few specimens that are undamaged.

The assemblage from the lake had a very high frequency of lithics, from both early and late periods. As this recovery was the result of limited dredging, it is not possible to determine the temporal and spatial relationships of the excavated materials. However, several hundred specimens were contained in an initial sampling of an 8 m x 12 m area. When the area of investigation expanded, the total number of modified materials expanded to over 2,000. This location was approximately 150-250 m out into the water, offset from the low lying shoreline (Wessen 1983; 2012 personal communication).

Multiple tool types were identified within the dredge spoils from the lake. Most of these were processing tools rather than projectile points. Unifacially flaked small tools were documented, as were utilized flakes. Basalt was the dominant material among points, although CCS was represented. Debitage accounted for approximately 75% of the material, while 9% of the assemblage included small cutting and scraping tools. Cut bone was recorded in direct association with these lithics. This included both avifauna and terrestrial fauna. Fish bone was also observed. Organics such as fish weirs and canoe tie-ups have also been identified within this body of water (Wessen 2013; Stenger and Hibbs 1991).

At three lakes containing only older material, the cultural deposits are evidenced a similar distance from the shoreline. These water resources are at high elevation, and do not have steep slopes or hillsides in the areas from which the assemblages were documented. The lithics are not broadly distributed over the lake bottoms, and the concentrations are heaviest about 150m out into the water. Notably, there are broad areas where no cultural materials are observed (Stenger 1995).[2]

A database was established to synthesize the information associated with the older and the more recent lithics from underwater areas in the Northwest. Projectile forms were the first bifaces to be reported. Thus, it was initially thought that these lithics reflected terrestrial hunting strategies, with bifaces carried into the water by either game or misaimed weapons that did not find their targets. As data accumulated, however, a number of tool types were identified. The non-projectile forms fit well with hunting strategy assemblages, but as processing tools (figure 3). Although they were often considered stylistically early, they were not the tools of procurement. Further, when classified by form/function, at least 85% of the specimens reflected processing activities.

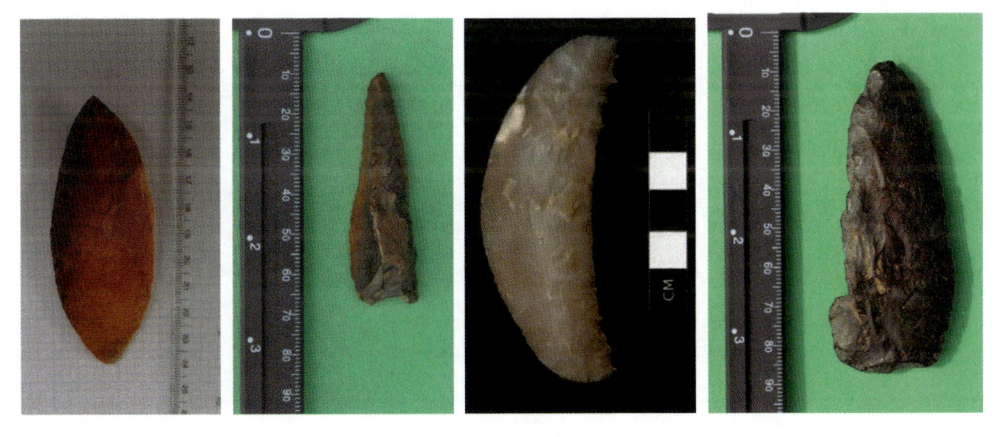

Figure 3. Processing tools include knives, gravers, and crescents. These may have been utilized as multiple-purpose tools.

The data demonstrate several things. The three most consistent observations are that (1) a broad range of geographic areas and time periods are represented, (2) different cultural styles and technologies are indicated, and (3) several different activities are represented by these tool types.

The majority of the material could not have eroded out of nearby landforms. This is primarily because most of the lithics, regardless of type, were located a significant distance from shore, with no sharply angled banks in the area. The distant shorelines were neither steep nor backed by significantly higher hills. Further, while riverbank slumping may have been the source of a few of the bifaces, it would have been necessary

---

[2] Significant information on high and low elevation sites was shared by David Rice and Jorie Clark, but no reports were referenced at that time (2012).

for that cultural material to immediately settle into a protected environment to avoid the scarring that accompanies movement within the associated river systems.

It is important to note that the deposits of lithics are discrete. Clusters of material, or isolates that lack redepositional scarring, have been documented in areas where an adjacent 80,000 m^2 lack any cultural indicators (Wessen 2013, Stenger 1994).[3] These archaeological localities reflect intentionally selected use areas. Further, the elevation above sea level of the waterway is not a factor. It is now clear that these use areas are not geographic anomalies, nor are they regional. Similar observations of underwater lithics are now being reported from bodies of water across the country.

When the localities of the submerged lithics within Oregon were placed on a map, it was immediately clear that all four quadrants of the state were represented (figure 4). The distribution of sites, and the many elevations reported, made it clear that neither a proximity to the ocean nor a specific environment were factors in the establishment of these sites.

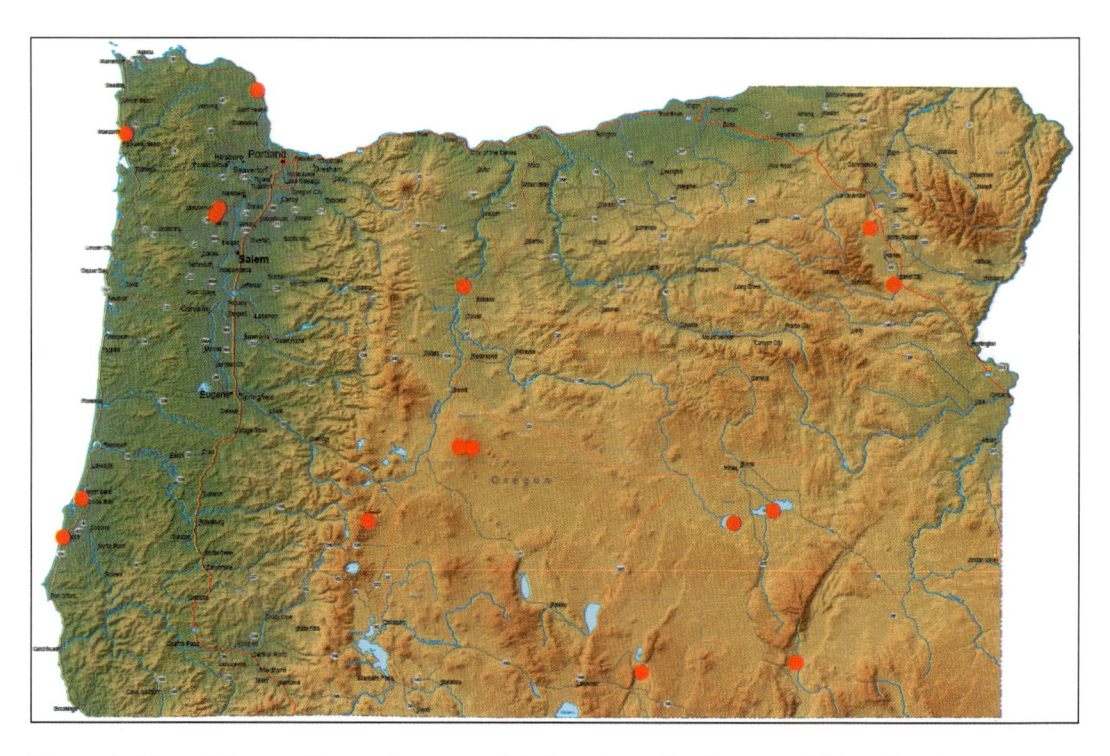

Figure 4. Map of Oregon, illustrating some of the locations of underwater lithics. Notably, every environment and elevation within the State is represented. Map courtesy of the Willamette Valley Pleistocene Project, and director, Mike full.

While late Northwest prehistory lacks proxies for the water based uses of most lithic tool types, the ethnographic and ethnohistoric records do demonstrate the use of projectiles over and into the water. There is little suggestion, however, of processing tools within that environment.[4] Yet when models from other areas are applied to the

---

[3] This discussion is also representative of high elevation lake beds from both Harney and Malheur counties, where early resources are known, but not yet listed in the database.

[4] Gathering activities continued to involve stone tools, although many of the forms changed over time.

submerged lithic assemblages from this region, the tool types represented can then be explained. In other cultures, the hunting of aquatic game is often accompanied by the processing of catches, *in situ*. The mending of catchment equipment also occurs. While this happens from boats and from pedestrian positions, both employ tools such as knives, drills, and scrapers.

One artifact type, specifically, may suggest alternative processing strategies. The crescent was probably a dual purpose tool. Phytoliths, retained on some edges, are residual to grasses and marsh plants, and were assumedly utilized in the collection and management of vegetation. However, the shape and often the edge wear of this tool type are also suggestive of use in the processing of water fowl.[5] The form of this tool lends itself to the cutting and scraping of ovate, or bird-form, carcasses. This type of function could occur both over water and on land, which helps explain the distribution of this tool type both in lakebeds and on inland environments. Thus, crescents exposed on relictual lakebeds at high elevation, and within terrestrial sites such as on the Channel Islands, may actually reflect the same functionality. Regardless of specific use, this is another example of a processing tool that has been documented from an aquatic environment, as well as one that is terrestrial.

Ignoring the division between terrestrial based hunting traditions and maritime subsistence methods allows the material to be considered without bias. This is especially important, when considering the preponderance of non-projectile lithics in assemblages from underwater. It is hoped that edge wear studies and blood protein analysis will provide further insight into this issue, and that researchers will continue to provide information on submerged lithic materials.

Acknowledgements: Thanks to Dennis Stanford, Bruce Bradley, James Adovasio, David Rice, Gary Wessen, Mike Full, James Chatters, Leslie Hickerson, Jorie Clark, and Paul Claeyssens for their input.

References

Adovasio, James M.
2012 Presentation and Questions: Plant Fiber Technologies and the Initial Colonization of the New World. Pre-Clovis in the Americas Conference, Smithsonian Institution, Washington, D.C.

Bell, Robert E. and George L. Cross
1980 Oklahoma Indian Artifacts. Contributions from the Stovall Museum, University of Oklahoma, #4, Norman, Ok

Hemmings, C. Andrew and J. M. Adovasio
2012 Inundated landscapes and the Colonization of the Northeastern Gulf of Mexico. Pre-Clovis in the Americas Conference, Smithsonian Institution, Washington, D.C.

---

[5] An article in the Journal of World Prehistory, published in 2013, discusses a use-relationship between waterfowl and crescents in, "Waterfowl and Lunate Crescents in Western North America: The Archaeology of the Pacific Flyway." Thanks to Jon Erlandson for forwarding this article.

Hibbs, Charles H. Jr. and Alison T. Stenger
1991   Description and Distribution of Storage Pits and Canoe Tie-ups in Lake River, Adjacent to 45 CL 12. Letter report to Washington SHPO office, describing bough lined pits, acorns, and lithics in the river bank. Mapping provided by Keith Garnet, Geographic Cartographer. (Description referenced here, as the letter report lacked a title.)

Moss, Madona L. and Jon Erlandson
2013   Waterfowl and Lunate Crescents in Western North America: The Archaeology of the Pacific Flyway. Journal of World Prehistory, September 2013, Volume 26, Issue 3, pp 173-211

Smith, Geoffrey M., Emily S. Middleton, Peter A. Carey
2013   Paleoindian technological provisioning strategies in the northwester Great Basin, Journal of Archaeological Science, 40:12. pp. 4182

Stanford, Dennis J. and Bruce Bradley
2012   Across Atlantic Ice. University of California Press, Berkley and Los Angeles, Ca. pp. 149-169

Stenger, Alison T.
1995   *Cultural Resource Survey of Underwater Areas at Elevation*, Report for U.S. Forest Service, Institute for Archaeological Studies, Portland, Or.
1994   *Pauline and East Lakes: Underwater Survey for Submerged Cultural Resources*, Report for U.S. Forest Service, Institute for Archaeological Studies, Portland, Or.
1993   *Paulina Lake: Preliminary Underwater Survey*, Report for U.S. Forest Service, Institute for Archaeological Studies, Portland, Or.
1988   *Nehalem Large Lithic Site.* Site Report for Oregon State Historic Preservation Office, Institute for Archaeological Studies, Portland, Or.

Wessen, Gary
2013   Email to author on details of fish weir from Vancouver Lake, Washington, 10/28/13, 4:02 P.M.

Wessen, Gary and Richard D. Daugherty
1983   Archaeological Investigations at Vancouver Lake, Washington, Western Heritage, Inc., Olympia, Washington pp. 76-98

Wisner, George
1997 Underwater Site Opens Window on Big environmental change. Mammoth Trumpet, v. 12, 2

Made in the USA
San Bernardino, CA
11 July 2014